Current Progress in Human Genetics

Current Progress in Human Genetics

Editor: Mark Williams

AMERICAN
MEDICAL PUBLISHERS
www.americanmedicalpublishers.com

AMERICAN
MEDICAL PUBLISHERS
www.americanmedicalpublishers.com

Cataloging-in-Publication Data

Current progress in human genetics / edited by Mark Williams.
    p. cm.
Includes bibliographical references and index.
ISBN 978-1-63927-245-7
1. Human genetics. 2. Genetics. 3. Heredity, Human. 4. Human biology. I. Williams, Mark.
QH431 .C87 2022
599.935--dc23

American Medical Publishers,
41 Flatbush Avenue,
1st Floor, New York,
NY 11217, USA

ISBN 978-1-63927-245-7 (Hardback)

Contents

Preface

This book has been an outcome of determined endeavour from a group of educationists in the field. The primary objective was to involve a broad spectrum of professionals from diverse cultural background involved in the field for developing new researches. The book not only targets students but also scholars pursuing higher research for further enhancement of the theoretical and practical applications of the subject.

The study of inheritance in human beings is referred to as human genetics. It is a multidisciplinary field that encompasses various fields such as molecular genetics, classical genetics, genomics, cytogenetics, biochemical genetics, population genetics, developmental genetics and clinical genetics. Genes are the common factors of the qualities of most human-inherited characteristics. A few major inheritance patterns are autosomal dominant inheritance, X-linked and Y-linked inheritance and autosomal recessive inheritance. Human genetics is an important field that helps to find the answers regarding human nature. It can also help in understanding various diseases and their effective treatment. This book outlines the processes and applications of human genetics in detail. From theories to research to practical applications, case studies related to all contemporary topics of relevance to this field have been included herein. Those in search of information to further their knowledge will be greatly assisted by this book.

It was an honour to edit such a profound book and also a challenging task to compile and examine all the relevant data for accuracy and originality. I wish to acknowledge the efforts of the contributors for submitting such brilliant and diverse chapters in the field and for endlessly working for the completion of the book. Last, but not the least; I thank my family for being a constant source of support in all my research endeavours.

Editor

GWIDD: a comprehensive resource for genome-wide structural modeling of protein-protein interactions

Petras J Kundrotas[1], Zhengwei Zhu[1,3] and Ilya A Vakser[1,2]*

Abstract

Protein-protein interactions are a key component of life processes. The knowledge of the three-dimensional structure of these interactions is important for understanding protein function. Genome-Wide Docking Database (http://gwidd.bioinformatics.ku.edu) offers an extensive source of data for structural studies of protein-protein complexes on genome scale. The current release of the database combines the available experimental data on the structure and characteristics of protein interactions with structural modeling of protein complexes for 771 organisms spanned over the entire universe of life from viruses to humans. The interactions are stored in a relational database with user-friendly interface that includes various search options. The search results can be interactively previewed; the structures, downloaded, along with the interaction characteristics.

Keywords: Protein-protein interactions, Structural modeling, Protein docking, Structural genomics, Interactome

Introduction

Proteins function by interacting with other biologically relevant molecules. Understanding the mechanisms of protein-protein interactions (PPI) is essential for studying life processes at the molecular level. Genome sequencing provided a vast amount of information on proteins at the sequence level. Currently, efforts focus on the function assignment of these proteins based on their three-dimensional (3D) structures and interactions. Interaction maps for specific organisms and biochemical pathways need to be complemented by the structural information. Experimental techniques are limited in their ability to produce the structures on the genome scale. Thus, computational methods are essential for this task [1].

Structural modeling of PPI has its origins in *ab initio* techniques based on shape and physicochemical complementarity. More recent approaches take advantage of statistical potentials and machine learning [2,3]. Despite progress in development of such template-free algorithms,

their accuracy in the high-throughput structure determination is limited.

Rapidly increasing amount of data on PPI makes possible application of the template-based methods. Such approaches are based on the observation that monomers with similar sequences and/or structures, generally, have similar binding modes. Several groups assessed the quality of PPI modeling based on sequence alignment to complexes with known structure [4-9]. Studies showed that the majority of such homology-docking models are of acceptable and medium quality, according to the established criteria [3]. An alternative template-based approach takes advantage of the structural similarity between the target and the template complexes [10-13].

The progress in 3D modeling of PPI is reflected in the Genome-Wide Docking Database (GWIDD) [14], which provides annotated collection of experimental and modeled PPI structures from the entire universe of life spanning from viruses to humans. The resource has user-friendly search interface, providing preview and download options for experimental and modeled PPI structures.

Database design

GWIDD imports PPI from external sources, including the last free release of BIND [15] and DIP [16,17].

* Correspondence: vakser@ku.edu
[1]Center for Bioinformatics, The University of Kansas, 2030 Becker Dr., Lawrence, KS 66047, USA
[2]Department of Molecular Biosciences, The University of Kansas, 2030 Becker Dr., Lawrence, KS 66047, USA
Full list of author information is available at the end of the article

Currently, we are working on interfacing GWIDD with MINT [18], BioGRID [19], and IntAct [20]. To provide the structures to PPI, the following scheme is utilized. If the complex is found in the Protein Data Bank (PDB), the X-ray structure is used, and no modeling is performed (10,924 GWIDD entries). Otherwise, a search for a pair of homologous sequences from complexes with known structure is performed, and the model is built by homology docking [6,7]. Statistical significance of the sequence alignments is assigned [7], with an additional requirement that both alignments contain at least 80% of the target sequences. This provides structures for 12,646 PPI. For the interactions not covered by these two steps, the interacting monomers are modeled independently by homology modeling, with subsequent docking of the models by structural alignment [12]. Incorporation of the structural alignment predictions (28,811 entries) into GWIDD is currently in progress (the structures are available from the authors by request). The graphical summary of the GWIDD coverage of genomes is in Figure 1.

User interface

The database (http://gwidd.bioinformatics.ku.edu) user interface (Figure 2) offers search by keywords, sequences (explicit input or upload in FASTA format), or structures (upload in PDB format), for one or both interacting proteins. The search by keywords can be performed using any word in the protein description (name of organism, cellular location, biological function etc.) or by selection from drop-down menus that are listing organisms currently in GWIDD. Repeated selection of the box 'Add another organism to the list' allows expansion of the search to several organisms. An option for search by standard taxonomy identification (ID) with link to taxonomy database http://www.uniprot.org is also provided. In case of input PDB file, the sequence is extracted from SEQRES tags or, if the SEQRES is not available, from ATOM tags of C^α atoms. The sequences from different sources can differ in length even for the same protein (e. g., due to unresolved fragments of the X-ray structure). Thus, advanced sequence search options are available. Figure 2 shows an example of search by organism.

The user can enable the second half of the search interface if information related to the interaction partner is available ('protein B,' Figure 2). The search results can be filtered by the structure availability (experimental, modeled, or no structures). Online help is provided in pop-up windows. The search result screen displays all interactions in the database satisfying the input search criteria in the form of an expandable list of GWIDD interaction IDs. For the homology-docking models, the alignments used to build the model are provided, and the model quality is assessed by the sequence identity criteria [5]. Links are provided to download the PDB-format files, along with the text file containing relevant information. Visualization screen is available to display the structures by different interactive representations. A link is provided to download the entire set of sequence-homology models in one gzipped archive.

Implementation

GWIDD unifies different external PPI data formats into a single data set, removing redundancy and retaining common data fields for all the sources. The

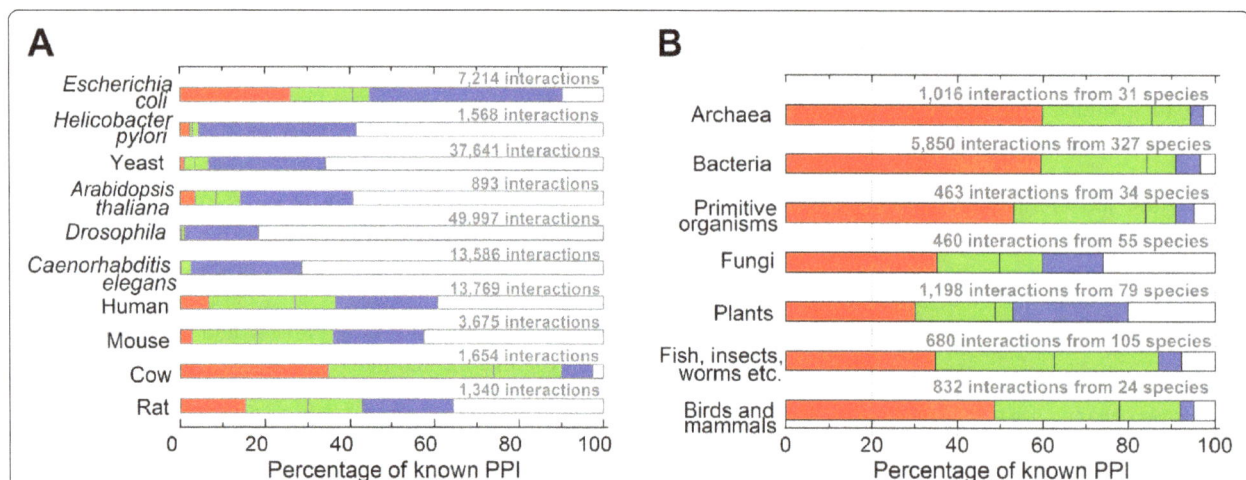

Figure 1 Structural coverage of different genomes in GWIDD. X-ray structures of complexes are in *red*, sequence-based models of complexes are in *green*, and interactions with structural models of the monomers are in *blue*. (**A**) Ten genomes with the largest number of known PPI. (**B**) The rest of the genomes (the data from A excluded).

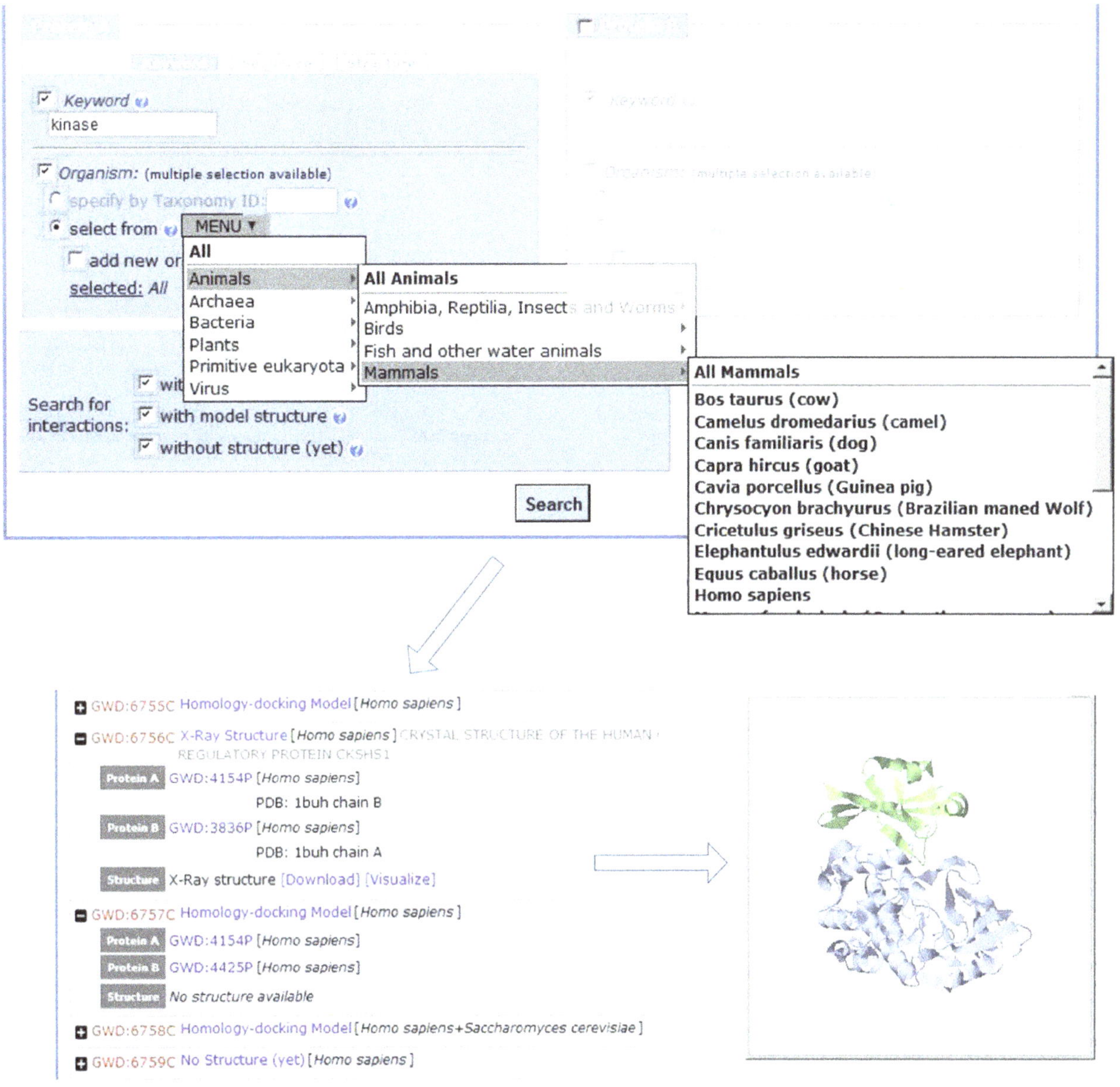

Figure 2 Example of a search.

interaction data are stored in a relational database, except for large files, such as structure coordinates, which are stored directly in the file system and are linked from the relational database. The web interface is implemented on the Linux-Apache-PostgreSQL -PHP software stack. Web user interface is built using hypertext preprocessor (PHP) and jQuery library, where PHP is for web presentation and logic as well as back-end database access; jQuery is responsible for AJAX and other JavaScript-based dynamic features. Visualization of protein structures is implemented in Jmol (www.jmol.org). Homology docking was performed by NEST [21], BLAST [22], and in-house

profile-to-profile alignment program. The procedures are joined by Python scripts.

Future directions
GWIDD development will incorporate other structural modeling techniques, such as multi-template/threading modeling of interacting proteins, partial structural alignment [12], and template-free docking by GRAMM [23-25]. A major expansion of GWIDD will be the incorporation of new PPI sources from other publicly available PPI databases. Large-scale systematic benchmarking of the high-through put modeling will be used to assign a confidence score to the modeled structures.

Competing interests

The authors declare that they have no competing interests.

Acknowledgments

Andrey Tovchigrechko and Tatiana Baronova made important contributions to the GWIDD project at the earlier stages of development. This work was supported by the National Institutes of Health grant R01 GM074255.

Author details

[1]Center for Bioinformatics, The University of Kansas, 2030 Becker Dr., Lawrence, KS 66047, USA. [2]Department of Molecular Biosciences, The University of Kansas, 2030 Becker Dr., Lawrence, KS 66047, USA. [3]Current address: Department of Genetics, Room 716B, Abramson Research Center, University of Pennsylvania, 3615 Civic Center Blvd., Philadelphia, PA 19104, USA.

Authors' contributions

ZZ performed calculations and implemented the web interface. PJK developed calculation pipelines, designed the web interface, analyzed data, and drafted the manuscript. IAV designed the research, analyzed data, and wrote the paper. All authors read and approved the final manuscript.

References

1. Russell RB, Alber F, Aloy P, Davis FP, Korkin D, Pichaud M, Topf M, Sali A: **A structural perspective on protein–protein interactions.** *Curr Opin Struct Biol* 2004, **14**:313–324.
2. Vakser IA, Kundrotas P: **Predicting 3D structures of protein-protein complexes.** *Curr Pharm Biotech* 2008, **9**:57–66.
3. Lensink MF, Wodak SJ: **Docking and scoring protein interactions: CAPRI 2009.** *Proteins* 2010, **78**:3073–3084.
4. Aloy P, Pichaud M, Russell RB: **Protein complexes: structure prediction challenges for the 21st century.** *Curr Opin Struct Biol* 2005, **15**:15–22.
5. Aloy P, Russell RB: **Interrogating protein interaction networks through structural biology.** *Proc Natl Acad Sci USA* 2002, **99**:5896–5901.
6. Kundrotas PJ, Alexov E: **Predicting 3D structures of transient protein-protein complexes by homology.** *Bioch Biophys Acta* 2006, **1764**:1498–1511.
7. Kundrotas PJ, Lensink MF, Alexov E: **Homology-based modeling of 3D structures of protein-protein complexes using alignments of modified sequence profiles.** *Int J Biol Macromol* 2008, **43**:198–208.
8. Lu L, Lu H, Skolnick J: **MULTIPROSPECTOR: an algorithm for the prediction of protein-protein interactions by multimeric threading.** *Proteins* 2002, **49**:350–364.
9. Mukherjee S, Zhang Y: **Protein-protein complex structure predictions by multimeric threading and template recombination.** *Structure* 2011, **13**:955–966.
10. Gunther S, May P, Hoppe A, Frommel C, Preissner R: **Docking without docking: ISEARCH - prediction of interactions using known interfaces.** *Proteins* 2007, **69**:839–844.
11. Keskin O, Nussinov R, Gursoy A: **PRISM: protein-protein interaction prediction by structural matching.** *Methods Mol Biol* 2008, **484**:505–521.
12. Sinha R, Kundrotas PJ, Vakser IA: **Docking by structural similarity at protein-protein interfaces.** *Proteins* 2010, **78**:3235–3241.
13. Korkin D, Davis FP, Alber F, Luong T, Shen M, Lucic V, Kennedy MB, Sali A: **Structural modeling of protein interactions by analogy: application to PSD-95.** *PLoS Comp Biol* 2006, **2**:1365–1376.
14. Kundrotas PJ, Zhu Z, Vakser IA: **GWIDD: genome-wide protein docking database.** *Nucl Acid Res* 2010, **38**:D513–D517.
15. Alfarano C, Andrade CE, Anthony K, Bahroos N, Bajec M, Bantoft K, Betel D, Bobechko B, Boutilier K, Burgess E, Buzadzija K, Cavero R, D'Abreo C, Donaldson I, Dorairajoo D, Dumontier MJ, Dumontier MR, Earles V, Farrall R, Feldman H, Garderman E, Gong Y, Gonzaga R, Grytsan V, Gryz E, Gu V, Haldorsen E, Halupa A, Haw R, Hrvojic A, *et al*: **The Biomolecular Interaction Network Database and related tools 2005 update.** *Nucl Acid Res* 2005, **33**:D418–D424.
16. Salwinski L, Miller CS, Smith AJ, Pettit FK, Bowie JU, Eisenberg D: **The Database of Interacting Proteins: 2004 update.** *Nucl Acid Res* 2004, **32**:D449–D451.
17. Xenarios I, Rice DW, Salwinski L, Baron NK, Marcotte EM, Eisenberg D: **DIP: the Database of Interacting Proteins.** *Nucleic Acids Res* 2000, **28**:289–291.
18. Ceol A, Aryamontri AC, Licata L, Peluso D, Briganti L, Perfetto L, Castagnoli L, Cesareni G: **MINT, the molecular interaction database: 2009 update.** *Nucl Acid Res* 2010, **38**:D532–D539.
19. Stark C, Breitkreutz BJ, Chatr-Aryamontri A, Boucher L, Oughtred R, Livstone MS, Nixon J, Van Auken K, Wang X, Shi X, Reguly T, Rust JM, Winter A, Dolinski K, Tyers M: **The BioGRID Interaction Database: 2011 update.** *Nucl Acid Res* 2011, **39**:D698–D704.
20. Aranda B, Achuthan P, Alam-Faruque Y, Armean I, Bridge A, Derow C, Feuermann M, Ghanbarian AT, Kerrien S, Khadake J, Kerssemakers J, Leroy C, Menden M, Michaut M, Montecchi-Palazzi L, Neuhauser SN, Orchard S, Perreau V, Roechert B, van Eijk K, Hermjakob H: **The IntAct molecular interaction database in 2010.** *Nucl Acid Res* 2010, **38**:D525–D531.
21. Petrey D, Xiang Z, Tang CL, Xie L, Gimpelev M, Mitros T, Soto CS, Goldsmith-Fischman S, Kernytsky A, Schlessinger A, Koh IY, Alexov E, Honig B: **Using multiple structure alignments, fast model building, and energetic analysis in fold recognition and homology modeling.** *Proteins* 2003, **53**:430–435.
22. Altschul SF, Madden TL, Schaffer AA, Zhang J, Zhang Z, Miller W, Lipman DJ: **Gapped BLAST and PSI-BLAST: a new generation of database programs.** *Nucleic Acids Res* 1997, **25**:3389–3402.
23. Katchalski-Katzir E, Shariv I, Eisenstein M, Friesem AA, Aflalo C, Vakser IA: **Molecular surface recognition: determination of geometric fit between proteins and their ligands by correlation techniques.** *Proc Natl Acad Sci USA* 1992, **89**:2195–2199.
24. Vakser IA, Matar OG, Lam CF: **A systematic study of low-resolution recognition in protein-protein complexes.** *Proc Natl Acad Sci USA* 1999, **96**:8477–8482.
25. Tovchigrechko A, Wells CA, Vakser IA: **Docking of protein models.** *Protein Sci* 2002, **11**:1888–1896.

2

Validation and assessment of variant calling pipelines for next-generation sequencing

Mehdi Pirooznia[1], Melissa Kramer[2], Jennifer Parla[2], Fernando S Goes[1], James B Potash[3], W Richard McCombie[2,4] and Peter P Zandi[1,5*]

Abstract

Background: The processing and analysis of the large scale data generated by next-generation sequencing (NGS) experiments is challenging and is a burgeoning area of new methods development. Several new bioinformatics tools have been developed for calling sequence variants from NGS data. Here, we validate the variant calling of these tools and compare their relative accuracy to determine which data processing pipeline is optimal.

Results: We developed a unified pipeline for processing NGS data that encompasses four modules: mapping, filtering, realignment and recalibration, and variant calling. We processed 130 subjects from an ongoing whole exome sequencing study through this pipeline. To evaluate the accuracy of each module, we conducted a series of comparisons between the single nucleotide variant (SNV) calls from the NGS data and either gold-standard Sanger sequencing on a total of 700 variants or array genotyping data on a total of 9,935 single-nucleotide polymorphisms. A head to head comparison showed that Genome Analysis Toolkit (GATK) provided more accurate calls than SAMtools (positive predictive value of 92.55% vs. 80.35%, respectively). Realignment of mapped reads and recalibration of base quality scores before SNV calling proved to be crucial to accurate variant calling. GATK HaplotypeCaller algorithm for variant calling outperformed the UnifiedGenotype algorithm. We also showed a relationship between mapping quality, read depth and allele balance, and SNV call accuracy. However, if best practices are used in data processing, then additional filtering based on these metrics provides little gains and accuracies of >99% are achievable.

Conclusions: Our findings will help to determine the best approach for processing NGS data to confidently call variants for downstream analyses. To enable others to implement and replicate our results, all of our codes are freely available at http://metamoodics.org/wes.

Keywords: Variant calling pipelines, Next-generation sequencing, Exome sequencing

Background

Advances in next-generation sequencing (NGS) technology are beginning to provide a cost-effective approach for identifying and cataloging the full spectrum of genetic variation across the genome at a scale not previously attainable by more traditional techniques such as Sanger sequencing or single-nucleotide polymorphism (SNP) arrays, thus creating a foundation for a profound understanding of human diseases [1-4]. The ability to comprehensively examine the genome in a high-throughput and unbiased manner has generated a great deal of interest in the use of NGS platforms to sequence entire exome or genome of large numbers of individuals to search variation in common disease, mutations underlying rare Mendelian disease [5,6], or spontaneously arising variation for which no gene-mapping shortcuts are available (e.g., somatic mutations in cancer [7,8] or *de novo* mutations in autism [9-13] and schizophrenia [14]).

Although NGS is a powerful approach, there are many technical challenges involved in obtaining a complete and accurate record of sequence variation from NGS data and in turning raw sequence reads into biologically meaningful information [15-17]. Given accurately mapped and calibrated reads, identifying simple SNPs, let alone more complex variation such as multiple base pair substitutions, insertions, deletions, inversions, and copy number

* Correspondence: pzandi@jhsph.edu
[1]Department of Psychiatry and Behavioral Sciences, Johns Hopkins University, Baltimore, MD 21205, USA
[5]Department of Mental Health, Johns Hopkins Bloomberg School of Public Health, Baltimore, MD 21205, USA
Full list of author information is available at the end of the article

variation, requires complex statistical models and sophisticated bioinformatics tools to implement these models on large amounts of data [16,18]. A number of such tools have recently been developed, including the short oligonucleotide alignment program (SOAP) [19,20], SAMtools [21], and the Genome Analysis Toolkit (GATK) [22]. However, many questions remain about how well these different tools work in identifying and accurately calling sequence variation and what are the best strategies for optimizing their use. Several recent studies have begun to evaluate and compare the performance of these tools [23-25].

We sought to add to these studies in order to determine best processes for identifying and calling sequence variants from NGS data. We carried out a comparative analysis of 130 whole exome subjects from an ongoing bipolar disorder exome sequencing project. We developed a

multi-stage pipeline for processing the exome data on these subjects and then examined the accuracy of calls derived from different implementations of the pipeline by validation with Sanger sequencing of a total of 700 variants using the ABI capillary sequencing platform and SNP genotyping on a total of 9,935 variants using the Affymetrix microarray platform. The goal was to critically evaluate and optimize processes for generating valid single nucleotide variant (SNV) calls from NGS data. Our results provide useful information and guidance for future studies analyzing data from next-generation sequencing experiments.

Results and discussion
Pipeline development
We developed a modular pipeline for processing NGS as shown in Figure 1 and described in Additional file 1 and

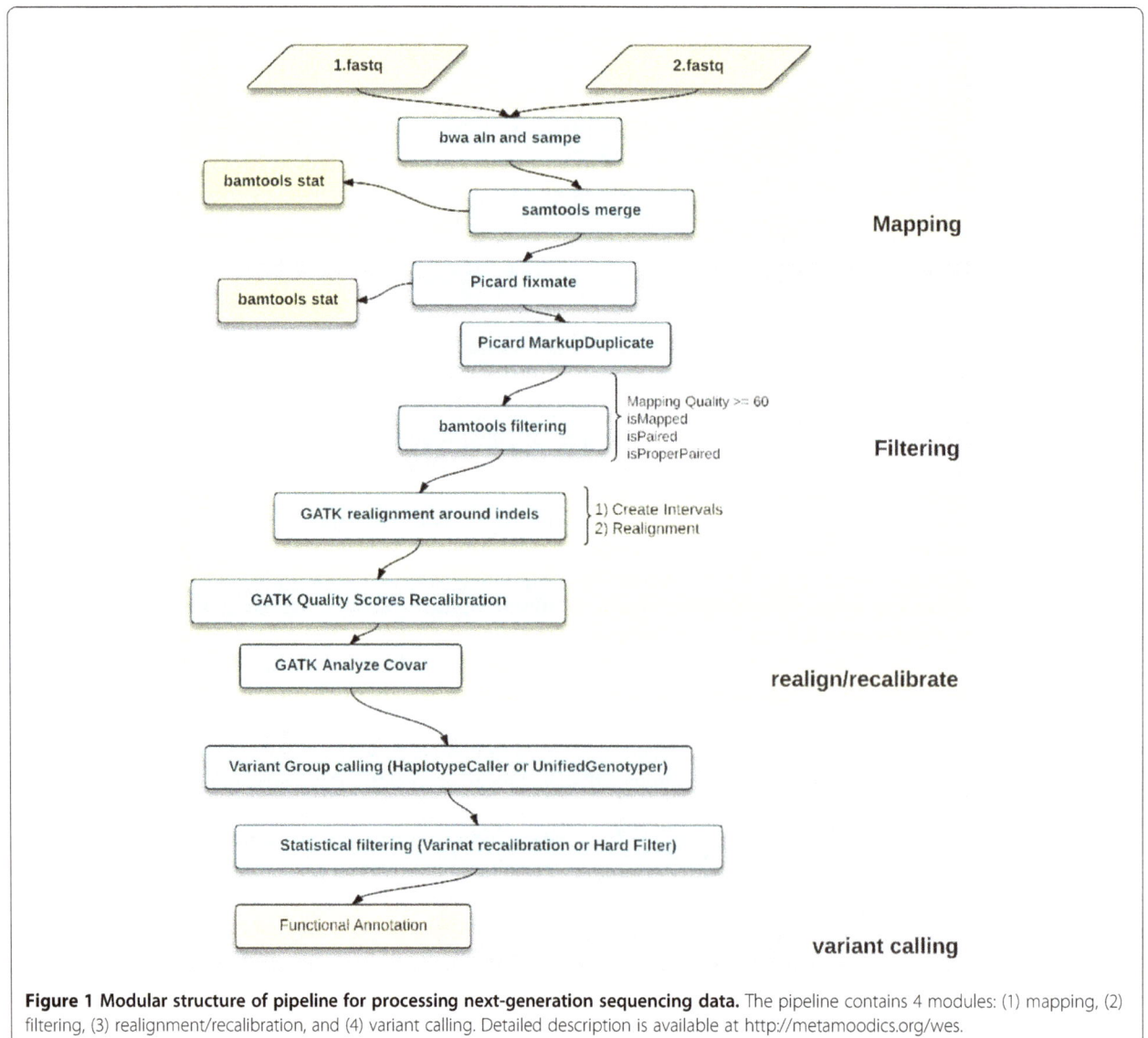

Figure 1 Modular structure of pipeline for processing next-generation sequencing data. The pipeline contains 4 modules: (1) mapping, (2) filtering, (3) realignment/recalibration, and (4) variant calling. Detailed description is available at http://metamoodics.org/wes.

in more detail at our Wiki site (http://metamoodics.org/wes). First, raw read data with well-calibrated base error estimates in fastq format are mapped to the reference genome. The BWA mapping (version 0.7.0) application [26] is used to map reads to the human genome reference, allowing for two mismatches in 30-base seeds, and generate a technology-independent SAM/BAM reference file format [21]. Next, duplicate fragments are marked and eliminated with Picard (version 1.8) (http://picard.sourceforge.net), mapping quality is assessed and low-quality mapped reads are filtered, and paired read information is evaluated to ensure that all mate-pair information is in sync between each read. We then refine the initial alignments by local realignment and identify suspicious regions. Using this information as a covariate along with other technical covariates and known sites of variation, the GATK base quality score recalibration (BQSR) is carried out. Lastly, SNV calling is performed using the recalibrated and realigned BAM files.

In this study, we evaluated different components of the pipeline that may influence the accuracy of the SNV calls in order to optimize the pipeline. We did this by comparing SNV call sets from the pipeline versus 'gold standard' calls either from targeted Sanger sequencing or previously available genome-wide association study (GWAS) data. In particular, we compared two of the most commonly used tools for variant calling (SAMtools versus GATK), different algorithms for variant calling implemented by GATK (UnifiedGenotyper versus HaplotypeCaller and hard filtering versus VariantRecalibration), and the influence of several sequence parameters (read depth, allele balance, and mapping quality).

GATK versus SAMtools

A number of tools have been developed for variant calling from aligned sequence reads, including GATK [22], SAMtools [21], MAQ [27], VarScan [28], SNVer [29], GNUMAP [30], and SOAPsnp [31]. We sought to compare GATK (version 2.6) and SAMtools (version 0.1.18), which are among the most widely used. Before making this comparison, we first evaluated the effect of realignment and recalibration of sequences on the accuracy of downstream variant calling. We did this by comparing SNV call sets from SAMtools with and without realignment/recalibration on a sample of 30 subjects with an average of 14,730 SNVs per subject. As shown in Figure 2, the majority of SNVs, approximately 96% of all SNVs called by either of the call sets, were called by both. Less than 1% of all SNVs were called only by the pipeline that did not use realignment/recalibration, while another 3% of all SNVs were called only by the pipeline with realignment/recalibration. We resequenced with Sanger methods a random selection of identified variants to evaluate the accuracy of these calls. A total of 341 individual SNV calls were available to evaluate the pipeline with realignment/recalibration, for which we observed a positive predictive value of 88.69% among variants that were called only after realignment/recalibration. By contrast, we found a positive predictive value of only 35.25% among individual SNV calls for the pipeline without realignment/recalibration only. Similar to others [23,32], we concluded based on these findings that realignment/recalibration improves the accuracy of calls and implemented these steps in our pipeline as standard practice moving forward.

We then compared SNV calls from GATK versus SAMtools using data from the same 30 subjects (Figure 3). For these comparisons, we used the UnifiedGenotyper algorithm in GATK and mpileup in SAMtools. We resequenced 336 individual calls from GATK and observed a true-positive rate of 95.00%. By contrast, from calls only made by SAMtools (1.23% of the total calls), we resequenced 341 individual calls and observed a much lower

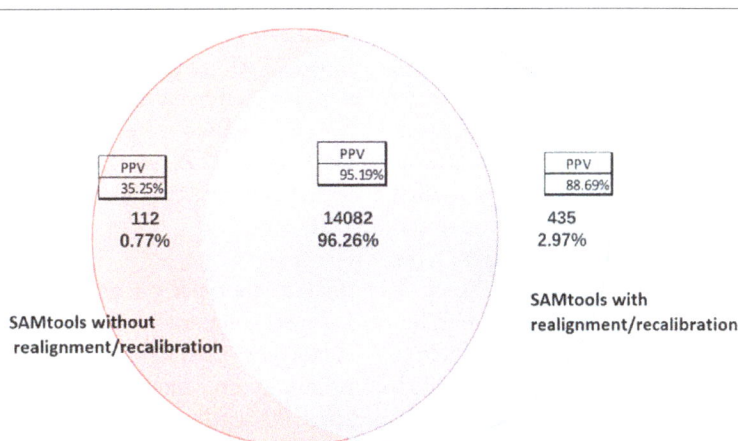

Figure 2 Comparison of SNV calling using SAMtools with and without realignment/recalibration on a sample of 30 subjects. Sanger sequencing was performed to evaluate the accuracy of these calls.

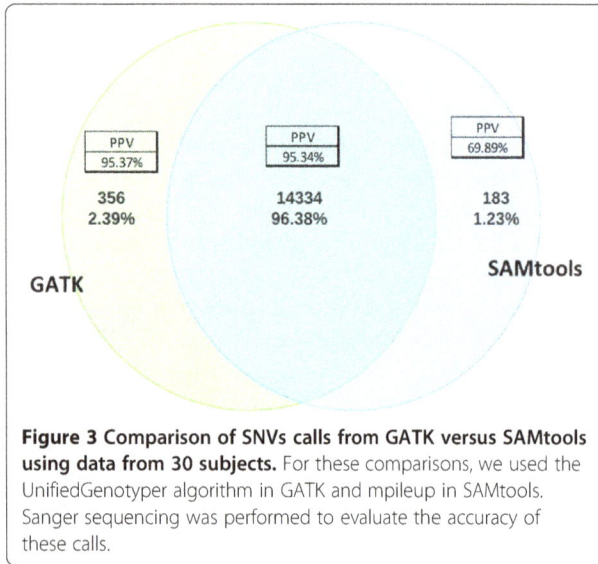

Figure 3 Comparison of SNVs calls from GATK versus SAMtools using data from 30 subjects. For these comparisons, we used the UnifiedGenotyper algorithm in GATK and mpileup in SAMtools. Sanger sequencing was performed to evaluate the accuracy of these calls.

true-positive rate of 69.89%. We considered whether it would be better to make calls using both tools and take the intersection as the final call set. Just over 96.38% of all SNVs called by either tool were called by both. We resequenced 165 individual calls of these SNVs and observed a positive predictive value of 95.34%. Another 2.39% of all SNVs were called only by GATK. Resequencing of 171 individual calls of these variants yielded a positive predictive value of 95.37%. As a result, we decided to go with GATK exclusively as our variant calling tool. Additional file 2: Table S1 provides a breakdown of the characteristics of the SNV calls that were concordant and discordant with the Sanger sequencing by the different calling methods.

Variant quality score recalibration versus hard filter
Moving forward with GATK, we examined the accuracy of calls when using hard filtering with recommended thresholds from GATK (variant confidence score ≥30, mapping quality ≥40, read depth ≥6, and strand bias FSfilter <60); a full description is provided in Additional file 1 versus using GATK's Variant Quality Score Recalibration (VQSR), which builds a Gaussian mixture model by looking at the annotation values over a high-quality subset of the input call set and then uses this model to evaluate all input variants. We compared calls using both strategies against GWAS SNP genotype data previously obtained from 100 subjects and 9,930 SNVs. We used the UnifiedGenotyper algorithm for these comparisons. A total of 181,304 out of 191,361 (94.74%) total SNVs were called in common between the hard filtering and VQSR strategies. Table 1 shows a breakdown of genotypes for these 181,304 SNVs. Over 99% of individual genotype calls at the SNVs were concordant between both strategies. As a result, the sensitivity and specificity of VQSR versus hard filtering using the GWAS SNP genotype as the gold standard were very

Table 1 UnifiedGenotyper Variant Quality Score Recalibration (UGVR) versus Hard Filter (UGHF)

		UGHF		
		AA	**AB**	**BB**
UGVR	AA	513,601	31 [5, 0, 0, 26]	49 [0, 0, 0, 49]
	AB	0	296,714	0
	BB	0	1,235 [0, 6, 1,222, 7]	170,818

similar, with sensitivity of 99.87% for both VQSR and hard filtering, and specificity of 99.79% and 99.56% for VQSR and hard filtering, respectively. In order to evaluate the differences more closely, we examined the small percentage of discordant genotype calls between VQSR and hard filtering. Here, the calls from VQSR were almost always in better agreement with the available GWAS SNP genotype data than were the calls from hard filtering (1,227 out of 1,233 calls in agreement for VQSR vs. 6 out of 1,233 for hard filtering). To evaluate the differences with respect to rarer SNVs with minor allele frequency (MAF) <10% that are not available in the GWAS data, we randomly selected 50 rarer SNVs from the subset that were discordantly called between VQSR and hard filtering and performed Sanger sequencing to validate the calls. Again, the VQSR calls were in better agreement (70%) than the hard filtering calls (61%) with the reference calls from Sanger sequencing. Overall, the comparisons against data from both GWAS and Sanger sequencing showed that VQSR provides better calling accuracy than simply using hard filtering. Thus, we used variant recalibration moving forward.

Shown is a comparison of genotype calls from the two approaches for the 181,304 variants that were called by both and for which we had GWAS SNP genotypes. *A* refers to the reference allele and *B* to the alternative allele. The four values in brackets [*w*, *x*, *y*, *z*] refer to the genotype calls from the GWAS data, where *w* refers to homozygous reference (*AA*) calls, *x* to heterozygous (*AB*) calls, *y* homozygous alternative (*BB*) calls, and *z* to missing. The GWAS genotype calls are only shown for those calls that are discrepant between UGVR and UGHF. A total of 191,361 variants were called by both UGVR and UGHF. Of these, 181,304 (94.74%) were in common, 3,655 (1.91%) were unique to UGVR, and 6,402 (3.35%) were unique to UGHF.

UnifiedGenotyper versus HaplotypeCaller
We next compared the accuracy of calls using the UnifiedGenotyper (UGVR) versus HaplotypeCaller (HCVR) algorithms as implemented in GATK version 2.5 (Table 2). Here, we used variant recalibration with both algorithms. Again, comparisons were made against GWAS genotype data from 100 subjects and 9,935 single nucleotide variations (SNVs). HaplotypeCaller calls variants via a local *de*

Table 2 UnifiedGenotyper Variant Quality Score Recalibration (UGVR) versus HaplotypeCaller Variant Quality Score Recalibration (HCVR)

		UGVR		
		AA	AB	BB
HCVR	AA	510,296	194 [176, 17, 0, 1]	[0, 0, 0, 0]
	AB	196 [60, 133, 2, 1]	294,595	210 [0, 5, 204, 1]
	BB	5 [0, 0, 5, 0]	230 [0, 10, 219, 1]	171,086

novo assembly of haplotypes in an active region, while UnifiedGenotyper simply looks for a coincident haplotype event in the reads. Both methods evaluate haplotypes using an affine gap penalty Pair Hidden Markov Model [33]. However, UnifiedGenotyper uses a Bayesian genotype likelihood model and estimates the most likely genotype calls while HaplotypeCaller chooses the best two haplotypes which explain the read data [34]. Of the 190,352 SNVs called by either algorithm, 90.29% (171,867) were called in common. Among those SNVs called in common, the genotype calls were also highly concordant between the two algorithms (99.91%). Overall, the sensitivity and specificity of the calls from UnifiedGenotype versus HaplotypeCaller were nearly similar: 99.78% versus 99.80%, respectively, for sensitivity, and 99.68% versus 99.70%, respectively, for specificity. Among the few discordant genotype calls, the HCVR algorithm provided slightly more accurate calls than UGVR, when compared against the GWAS data. Of the 835 discordant genotype calls, the HCVR was correct 63.83% of the time as compared to 34.85% for UGVR. Both algorithms did equally well in calling homozygous alternative calls, but UGVR made a few more mistakes in making heterozygous calls when the true genotype was homozygous reference. Again, to evaluate the accuracy with respect to rarer SNVs (MAF <10%), we randomly selected 50 rarer SNVs from the subset that was discordantly called between UGVR and HCVR and performed Sanger sequencing to validate the call. The results were very similar to what we observed with comparisons against GWAS data. HCVR was correct 61% of the time as compared to 39% of the time for UGVR.

Shown is a comparison of genotype calls from the two approaches for the 465,681 variants that were called by both and for which we had GWAS SNP genotypes. *A* refers to the reference allele and *B* to the alternative allele. The four values in brackets [*w, x, y, z*] refer to the genotype calls from the GWAS data, where *w* refers to homozygous reference (*AA*) calls, *x* to heterozygous (*AB*) calls, *y* homozygous alternative (*BB*) calls, and *z* to missing. The GWAS genotype calls are only shown when the calls are discrepant between UGVR and HCVR. A total of 190,352 variants were called by both UGVR and UGHF. Of these, 171,867 (90.29%) were in common, 15,839 (8.32%)

were unique to UGVR, and 2,646 (1.39%) were unique to HCVR.

Sequencing parameters

Finally, we evaluated the effects of varying certain sequencing parameters such as read depth, allele balance, and mapping quality (Figure 4). We compared the accuracy and missing data rates of the sequencing calls after systematically varying these parameters using data from 100 subjects with valid genotype data from GWAS on 7,370 SNPs. Overall, the accuracy of the sequence calls, which were made using the UnifiedGenotyper algorithm and VQSR, was very high when compared with the GWAS genotype calls. However, several trends emerged. The accuracy of calls increased with both increasing read depth and allele balance towards 50-50. The increase in accuracy was most notable after read depths greater than 10 times, while it plateaued after allele balances between 20 and 80. The missing data rate similarly increased with read depth and allele balance as calls that did not meet the more stringent read depth or allele balance requirements were filtered. Thus, as expected, there was a trade-off between increasing accuracy and increasing missing data. This was not found for mapping quality. As mapping quality increased, the missing data rate also increased while the accuracy actually decreased. This might be explained by the fact that as the mapping quality criteria are increased, the number of reads that align to the reference genome decreases, leading to lower overall read depths on which to base downstream SNV calls and, as a result, lower accuracy calls. It is important to note, however, that these trends were relatively subtle and the overall accuracy of these calls made using best practices was well over 99%, regardless of the read depth, allele balance, and mapping quality thresholds.

Conclusions

Advances in next-generation sequencing technologies have improved our ability to characterize genomic sequence variation at a scale and resolution not previously possible. This has opened up new avenues for studying how genetic variation contributes to human disease. A major challenge is how to process the copious data generated by the new technologies to yield high-quality data for downstream analyses. A variety of computational tools have been developed for this purpose. We have implemented a semi-automated pipeline using these tools to manage and analyze next-generation sequence data, and here we evaluated how key elements of the pipeline influence data quality.

After comparing SNV calls from GATK and SAMtools, we decided to adopt GATK [22] as our primary variant calling platform. In general, we found that GATK yields very high quality variant call data. Similar to others [23,31], we observed that realignment of mapped sequence reads

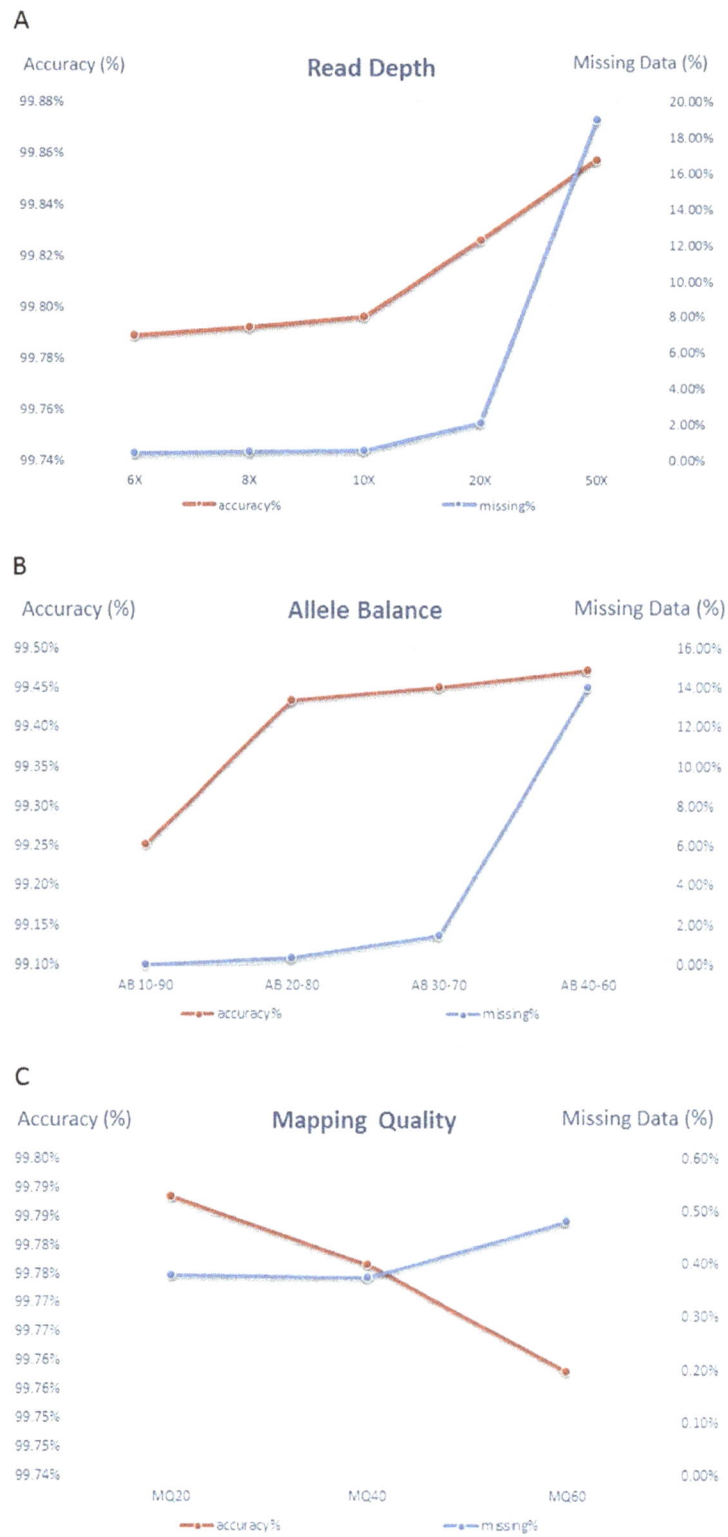

Figure 4 Evaluation of the effect of sequencing parameters. Read depth **(A)**, allele balance **(B)**, and mapping quality **(C)** on the calling accuracy. We compared the accuracy and missing data rates of the sequencing calls after systematically varying these parameters using data from 100 subjects with valid genotype data from GWAS on 7,370 SNPs.

around putative insertion/deletions (indels) and recalibration of base quality scores before variant calling are crucial to this performance. An example of the effects of realignment and recalibration on variant calling is illustrated in Additional file 3: Figure S1 and Additional file 4: Figure S2. For these comparisons between GATK and SAMtools with and without realignment/recalibration, we did not have SNP genotype data, and it was not practical to validate with Sanger sequencing non-calls by the different methods. As a result, we did not have information on false-negative and true-negative calls. Still, based on the available results from the validation of made calls, we felt confident in moving forward with GATK with realignment and recalibration.

GATK has developed several algorithms for variant calling from realigned and recalibrated sequence reads, including UnifiedGenotyper and HaplotypeCaller. Both performed well, but the HaplotypeCaller algorithm provided more accurate calls over all. Unlike UnifiedGenotype, HaplotypeCaller is capable of calling SNPs and insertion/deletion (indels) simultaneously. When the algorithm encounters a region that is highly variable, it discards the existing mapping information and reassembles the reads in the region *de novo*. The result is that HaplotypeCaller may be more accurate when calling regions that are traditionally difficult to call. This comes at a cost, however, as the HaplotypeCaller algorithm is currently computationally intensive, which limits the feasibility of using this approach with whole genome or larger exome sequencing studies. Improvements to the algorithm are needed to render it more efficient and practical to use with such studies. GATK has also implemented a Variant Quality Score Recalibration algorithm that uses machine learning methods for filtering variants that we demonstrated works better in terms of yielding a final set of accurate calls compared to hard filters based on pre-determined thresholds. Finally, we showed that there is a relationship between mapping quality, read depth and allele balance, and variant call accuracy, but if best practices are used throughout data processing, then additional filtering based on these metrics provides little gains.

Several previous studies have investigated factors that influence the accuracy of variant calling algorithms with sequence data [23-27]. One study sequenced 15 exomes from four families and processed the raw data using different alignment and variant-calling pipelines and found that there was a low concordance between approaches [25]. Another study used exome sequence data on 20 individuals and simulated whole genome sequence data to compare different algorithms for variant calling. Consistent with our results, this study found that GATK in particular outperformed SAMtools, especially for low coverage data, and yielded the most accurate data with multi-sample calling [27]. Still another study used whole genome sequence

data from monozygotic twins to determine optimal sequencing filters for achieving the greatest concordance in variant calling at the minimal costs of filtered data [29]. However, similar to our study, work by the group that developed GATK suggested that variant recalibration with their machine learning approach performed better than strategies using hard filtering [25].

Our study has several strengths including having been carried out with real rather than simulated sequence data and having utilized direct comparisons against calls from more traditional platforms such as Sanger sequencing and GWAS microarray data that were previously validated. One limitation is that the comparisons against GWAS data were only for more common variants. It is unclear if the observed accuracy rates would be different for rarer variants that are not well represented in GWAS data. However, we note that when the sensitivity and specificity of SNV calls for lower frequency variants among the GWAS data (<20%) were examined, the results were not materially different from the more common variants (results not shown). In addition, we did not evaluate the quality of indel calls which pose their own challenges. Overall, the results reported here provide reassurance that it is possible to generate highly accurate data from next-generation sequencing. Our findings will help inform researchers who are seeking to optimize their own pipelines for working with next-generation sequence data. As tools and methods for processing such data are constantly evolving, we will continue to evaluate them to determine which can yield the highest-quality sequencing data.

Methods
Samples
Samples for the validation experiments described herein came from an ongoing whole exome sequencing study of bipolar disorder. A total of 130 samples were selected from two collections of pedigrees with bipolar disorder from Johns Hopkins or from the National Institute of Mental Health (NIMH) Genetics Initiative Bipolar Disorder Collaborative Study.

Pre-capture library preparation
Genomic DNA samples were individually processed into Illumina paired-end or TruSeq DNA libraries using Illumina-compatible barcoded DNA adapters [17]. Purified genomic DNA, 1–3 µg, was initially fragmented using a Covaris S2 instrument (Covaris Inc, Woburn, MA, USA), followed by end-repair and ligation to paired-end adapters. As recommended by NimbleGen, pre-capture libraries were enriched with an additional 8 cycles of high-fidelity polymerase chain reaction (PCR). Pre-capture library quality and yield were assessed using the Bioanalyzer DNA 1000 Kit (Agilent Technologies, Santa Clara, CA, USA) and

the NanoDrop 1000 Spectrophotometer (Thermo Scientific, West Palm Beach, FL, USA).

Exome capture and sequencing

Due to ongoing changes in sequencing technology, sequencing was performed using two different exome capture kits and sequencing technologies. Our first set of analysis comparing SAMtools and GATK assessment was performed on sequencing data from 30 subjects captured with NimbleGen EZ exome v1.0 kit and sequenced with the Illumina Genome Analyzer (GA) II (Illumina Inc, San Diego, CA, USA). The NimbleGen EZ exome v1.0 kit was designed to capture approximately 33.8 Mb of hg18 genomic target, or approximately 180,000 coding exons from approximately 16,000 genes annotated in CCDS build 36.2 (April 2008 release). The remaining comparison analyses were carried out using 100 subjects that were captured with the NimbleGen EZ exome v2.0 kit and sequenced with the Illumina HiSeq 2000. The NimbleGen EZ exome v2.0 kit was designed to capture 36.0 Mb of hg19 genomic target, or approximately 300,000 coding exons from approximately 30,000 genes annotated across CCDS build 37.1 and RefSeq release 39. Sequencing generally produced enough coverage to obtain ≥80% of the target covered at ≥20X sequencing depth per sample. Samples that were just below this level (≥75% at 20X or more) were also included for further analyses. Variants were called using our pipeline as described in Additional file 1 and in more detail on our Wiki site (http://metamoodics. org/wes).

Validation sequencing and genotyping

Next-generation sequence variant calls were validated against either Sanger sequencing or microarray genotyping from a previous GWAS. Sanger sequencing was carried out on a random selection of variants identified through our sequencing pipeline in 30 subjects. Multiple SNPs were assayed across all individual samples. SNPs were validated by Sanger sequencing. Primer pairs flanking each SNP were designed using Primer3 software (http://primer3. sourceforge.net/). Template DNA, 25–50 ng, was then used for amplification with the NEB LongAmp PCR protocol. Following amplification, PCR products were visualized on 1% agarose gels, and products which showed a single clean band in the proper size range were selected for further processing. PCR products were then incubated with exonuclease I to remove excess primers and shrimp alkaline phosphatase to remove unincorporated nucleotides. Sequencing reactions were performed using ABI BigDye terminator chemistry (Life Technologies, Austin, TX, USA). Reactions were then precipitated with salt and washed with ethanol. Samples were sequenced with both forward and reverse primers on the ABI 3730 sequencer. SNPs were confirmed using the CONSED software [35] to align the Sanger reads to the reference sequence and visualize the alleles at the putative SNP position. In the first validation round, 400 total SNPs were assayed. 330 of those were confidently genotyped, 21 were potentially genotyped but suffered from slightly messy data, and 49 failed due to poor data quality. In the second validation round, 300 total SNPs were assayed. Our of 248 that were confidently genotyped, 11 were potentially genotyped but suffered from slightly messy data, 37 failed due to poor data quality, and 4 were reported as a possible indel rather than a SNP.

In addition, for comparisons, we used SNP genotype data from a previously conducted GWAS. Details of the GWAS have been described elsewhere [36]. Briefly, samples were genotyped using the Affymetrix Genome-Wide Human SNP Array 6.0 (Affymetrix, Santa Clara, CA, USA) [37]. Allele calling was performed using the BirdSeed algorithm [38]. Scans from the same production plate were clustered together. Rigorous quality control measures were carried out with the resulting genotype calls. Samples were not used in the analysis if they had low call rate (<98.5%), excessively high (>0.363) or low (<0.344) heterozygosity, or incompatibility between reported gender and genetically determined gender [36]. Samples were also checked for unexpected familial relationships using pairwise IBD (Identity by Descent) estimation in PLINK [39]. SNPs were not analyzed if the minor allele frequency (MAF) was <0.01, the call rate was <95%, the SNP violated Hardy-Weinberg equilibrium ($p < 1 \times 10^{-6}$) in control samples within an ancestry group, there were ≥3 Mendelian errors, or there was >1 discrepancy among duplicate samples. Each plate in the study was compared to all other plates with a Chi-square test to examine and remove any plate effects.

Additional files

Additional file 1: Whole Exome Sequencing Analysis Pipeline.

Additional file 2: Table S1. Characteristics of the true-positive (TP) and false-positive (FP) variant calls for the comparisons of SAMtools without realignment/recalibrations, SAMtools with realignment/recalibration calls, and GATK with realignment/recalibration. Characteristics include functional annotation (using NCBI RefSeq, release 63), average read depth, number of variants in putative indels, and number of variants in repeat regions defined by UCSC simple tandem repeats track (hg19).

Additional file 3: Figure S1. Illustration of SNVs at a specific locus using the integrated genomic viewer before (top) and after (bottom) applying realignment. Artefactual SNPs are recovered by realignment.

Additional file 4: Figure S2. Illustration of changes in the quality scores and the residual errors by machine cycle before (left top and bottom) and after (right top and bottom) applying quality score recalibration.

Abbreviations

BAM: binary alignment map; Indel: small insertion/deletion; NGS: next-generation sequencing; SAM: sequence alignment map; SNP: single-nucleotide polymorphism; SNV: single-nucleotide variant; VCF: variant call format; PLINK: population-based linkage analyses application.

Competing interests

The authors declare that they have no competing interests. WRM has participated in Illumina-sponsored meetings over the past 4 years and received travel reimbursement and an honorarium for presenting at these events. Illumina had no role in decisions relating to the study/work to be published, data collection and analysis of data and the decision to publish. WRM has participated in Pacific Biosciences-sponsored meetings over the past 3 years and received travel reimbursement for presenting at these events. WRM is a founder and shared holder of Orion Genomics, which focuses on plant genomics and cancer genetics.

Authors' contributions

PPZ, JBP, and WDM conceived, designed, and directed the project. MP and MK designed the pipeline, coded, and ran the analysis. JP and MK performed the sequencing. JBP, WDM, PPZ, and FSG coordinated to the project. MP drafted the manuscript. All authors read, contributed to, and approved the final manuscript.

Acknowledgements

This project is supported by the NIH funding from R01MH087979 (JBP), R01MH087992 (WRM), and K01MH093809 (MP).

Author details

[1]Department of Psychiatry and Behavioral Sciences, Johns Hopkins University, Baltimore, MD 21205, USA. [2]Stanley Institute for Cognitive Genomics, Cold Spring Harbor Laboratory, Woodbury, NY 11797, USA. [3]Department of Psychiatry, Carver College of Medicine, University of Iowa School of Medicine, Iowa City, IA 52242, USA. [4]Watson School of Biological Science, Cold Spring Harbor Laboratory, Cold Spring Harbor, NY 11724, USA. [5]Department of Mental Health, Johns Hopkins Bloomberg School of Public Health, Baltimore, MD 21205, USA.

References

1. Hodges E, Xuan Z, Balija V, Kramer M, Molla MN, Smith SW, Middle CM, Rodesch MJ, Albert TJ, Hannon GJ, McCombie WR: Genome-wide in situ exon capture for selective resequencing. Nat Genet 2007, 39(12):1522–1527.
2. Henson J, Tischler G, Ning Z: Next-generation sequencing and large genome assemblies. Pharmacogenomics 2012, 13(8):901–915.
3. Ku CS, Cooper DN, Polychronakos C, Naidoo N, Wu M, Soong R: Exome sequencing: dual role as a discovery and diagnostic tool. Ann Neurol 2012, 71(1):5–14.
4. Ross JS, Cronin M: Whole cancer genome sequencing by next-generation methods. Am J Clin Pathol 2011, 136(4):527–539.
5. Ku CS, Naidoo N, Pawitan Y: Revisiting Mendelian disorders through exome sequencing. Hum Genet 2011, 129(4):351–370.
6. Morris JA, Barrett JC: Olorin: combining gene flow with exome sequencing in large family studies of complex disease. Bioinformatics 2012, 28(24):3320–3321.
7. Vignot S, Frampton GM, Soria JC, Yelensky R, Commo F, Brambilla C, Palmer G, Moro-Sibilot D, Ross JS, Cronin MT, Andre F, Stephens PJ, Lazar V, Miller VA, Brambilla E: Next-generation sequencing reveals high concordance of recurrent somatic alterations between primary tumor and metastases from patients with non-small-cell lung cancer. J Clin Oncol Off J Am Soc Clin Oncol 2013, 31(17):2167–2172.
8. Wang Q, Jia P, Li F, Chen H, Ji H, Hucks D, Dahlman KB, Pao W, Zhao Z: Detecting somatic point mutations in cancer genome sequencing data: a comparison of mutation callers. Genome Medicine 2013, 5(10):91.
9. Iossifov I, Ronemus M, Levy D, Wang Z, Hakker I, Rosenbaum J, Yamrom B, Lee YH, Narzisi G, Leotta A, Kendall J, Grabowska E, Ma B, Marks S, Rodgers L, Stepansky A, Troge J, Andrews P, Bekritsky M, Pradhan K, Ghiban E, Kramer M, Parla J, Demeter R, Fulton LL, Fulton RS, Magrini VJ, Ye K, Darnell JC, Darnell RB, et al: De novo gene disruptions in children on the autistic spectrum. Neuron 2012, 74(2):285–299.
10. Bi C, Wu J, Jiang T, Liu Q, Cai W, Yu P, Cai T, Zhao M, Jiang YH, Sun ZS: Mutations of ANK3 identified by exome sequencing are associated with autism susceptibility. Hum Mutat 2012, 33(12):1635–1638.
11. O'Roak BJ, Deriziotis P, Lee C, Vives L, Schwartz JJ, Girirajan S, Karakoc E, Mackenzie AP, Ng SB, Baker C, Rieder MJ, Nickerson DA, Bernier R, Fisher SE, Shendure J, Eichler EE: Exome sequencing in sporadic autism spectrum disorders identifies severe de novo mutations. Nat Genet 2011, 43(6):585–589.
12. Neale BM, Kou Y, Liu L, Ma'ayan A, Samocha KE, Sabo A, Lin CF, Stevens C, Wang LS, Makarov V, Polak P, Yoon S, Maguire J, Crawford EL, Campbell NG, Geller ET, Valladares O, Schafer C, Liu H, Zhao T, Cai G, Lihm J, Dannenfelser R, Jabado O, Peralta Z, Nagaswamy U, Muzny D, Reid JG, Newsham I, Wu Y, et al: Patterns and rates of exonic de novo mutations in autism spectrum disorders. Nature 2012, 485(7397):242–245.
13. Sanders SJ, Murtha MT, Gupta AR, Murdoch JD, Raubeson MJ, Willsey AJ, Ercan-Sencicek AG, DiLullo NM, Parikshak NN, Stein JL, Walker MF, Ober GT, Teran NA, Song Y, El-Fishawy P, Murtha RC, Choi M, Overton JD, Bjornson RD, Carriero NJ, Meyer KA, Bilguvar K, Mane SM, Sestan N, Lifton RP, Gunel M, Roeder K, Geschwind DH, Devlin B, State MW: De novo mutations revealed by whole-exome sequencing are strongly associated with autism. Nature 2012, 485(7397):237–241.
14. Rees E, Kirov G, O'Donovan MC, Owen MJ: De novo mutation in schizophrenia. Schizophr Bull 2012, 38(3):377–381.
15. Johansen Taber KA, Dickinson BD, Wilson M: The promise and challenges of next-generation genome sequencing for clinical care. JAMA Intern Med 2014, 174(2):275–280.
16. Wang Z, Liu X, Yang BZ, Gelernter J: The role and challenges of exome sequencing in studies of human diseases. Front Genet 2013, 4:160.
17. Parla JS, Iossifov I, Grabill I, Spector MS, Kramer M, McCombie WR: A comparative analysis of exome capture. Genome Biol 2011, 12(9):R97.
18. Panoutsopoulou K, Tachmazidou I, Zeggini E: In search of low-frequency and rare variants affecting complex traits. Hum Mol Genet 2013, 22(R1):R16–21.
19. Li R, Li Y, Kristiansen K, Wang J: SOAP: short oligonucleotide alignment program. Bioinformatics 2008, 24(5):713–714.
20. Luo R, Liu B, Xie Y, Li Z, Huang W, Yuan J, He G, Chen Y, Pan Q, Liu Y, Tang J, Wu G, Zhang H, Shi Y, Liu Y, Yu C, Wang B, Lu Y, Han C, Cheung DW, Yiu SM, Peng S, Xiaoqian Z, Liu G, Liao X, Li Y, Yang H, Wang J, Lam TW, Wang J: SOAPdenovo2: an empirically improved memory-efficient short-read de novo assembler. GigaScience 2012, 1(1):18.
21. Li H, Handsaker B, Wysoker A, Fennell T, Ruan J, Homer N, Marth G, Abecasis G, Durbin R: Genome Project Data Processing S: The sequence alignment/map format and SAMtools. Bioinformatics 2009, 25(16):2078–2079.
22. McKenna A, Hanna M, Banks E, Sivachenko A, Cibulskis K, Kernytsky A, Garimella K, Altshuler D, Gabriel S, Daly M, DePristo MA: The Genome Analysis Toolkit: a MapReduce framework for analyzing next-generation DNA sequencing data. Genome Res 2010, 20(9):1297–1303.
23. DePristo MA, Banks E, Poplin R, Garimella KV, Maguire JR, Hartl C, Philippakis AA, del Angel G, Rivas MA, Hanna M, McKenna A, Fennell TJ, Kernytsky AM, Sivachenko AY, Cibulskis K, Gabriel SB, Altshuler D, Daly MJ: A framework for variation discovery and genotyping using next-generation DNA sequencing data. Nat Genet 2011, 43(5):491–498.
24. O'Rawe J, Jiang T, Sun G, Wu Y, Wang W, Hu J, Bodily P, Tian L, Hakonarson H, Johnson WE, Wei Z, Wang K, Lyon GJ: Low concordance of multiple variant-calling pipelines: practical implications for exome and genome sequencing. Genome Medicine 2013, 5(3):28.
25. Liu X, Han S, Wang Z, Gelernter J, Yang BZ: Variant callers for next-generation sequencing data: a comparison study. PLoS One 2013, 8(9):e75619.
26. Li H, Durbin R: Fast and accurate short read alignment with Burrows-Wheeler transform. Bioinformatics 2009, 25(14):1754–1760.
27. Li H, Ruan J, Durbin R: Mapping short DNA sequencing reads and calling variants using mapping quality scores. Genome Res 2008, 18(11):1851–1858.
28. Koboldt DC, Chen K, Wylie T, Larson DE, McLellan MD, Mardis ER, Weinstock GM, Wilson RK, Ding L: VarScan: variant detection in massively parallel sequencing of individual and pooled samples. Bioinformatics 2009, 25(17):2283–2285.
29. Wei Z, Wang W, Hu P, Lyon GJ, Hakonarson H: SNVer: a statistical tool for variant calling in analysis of pooled or individual next-generation sequencing data. Nucleic Acids Res 2011, 39(19):e132.
30. Clement NL, Snell Q, Clement MJ, Hollenhorst PC, Purwar J, Graves BJ, Cairns BR, Johnson WE: The GNUMAP algorithm: unbiased probabilistic

mapping of oligonucleotides from next-generation sequencing. *Bioinformatics* 2010, **26**(1):38–45.

31. Li R, Li Y, Fang X, Yang H, Wang J, Kristiansen K, Wang J: **SNP detection for massively parallel whole-genome resequencing.** *Genome Res* 2009, **19**(6):1124–1132.

32. Liu Q, Guo Y, Li J, Long J, Zhang B, Shyr Y: **Steps to ensure accuracy in genotype and SNP calling from Illumina sequencing data.** *BMC Genomics* 2012, **13 Suppl 8**:S8.

33. Krogh A, Brown M, Mian IS, Sjolander K, Haussler D: **Hidden Markov models in computational biology. Applications to protein modeling.** *J Mol Biol* 1994, **235**(5):1501–1531.

34. **GATK Documentation.** [http://www.broadinstitute.org/gatk/2013]

35. Gordon D, Abajian C, Green P: **Consed: a graphical tool for sequence finishing.** *Genome Res* 1998, **8**(3):195–202.

36. Smith EN, Bloss CS, Badner JA, Barrett T, Belmonte PL, Berrettini W, Byerley W, Coryell W, Craig D, Edenberg HJ, Eskin E, Foroud T, Gershon E, Greenwood TA, Hipolito M, Koller DL, Lawson WB, Liu C, Lohoff F, McInnis MG, McMahon FJ, Mirel DB, Murray SS, Nievergelt C, Nurnberger J, Nwulia EA, Paschall J, Potash JB, Rice J, Schulze TG, *et al*: **Genome-wide association study of bipolar disorder in European American and African American individuals.** *Mol Psychiatry* 2009, **14**(8):755–763.

37. Nishida N, Koike A, Tajima A, Ogasawara Y, Ishibashi Y, Uehara Y, Inoue I, Tokunaga K: **Evaluating the performance of Affymetrix SNP Array 6.0 platform with 400 Japanese individuals.** *BMC Genomics* 2008, **9**:431.

38. Korn JM, Kuruvilla FG, McCarroll SA, Wysoker A, Nemesh J, Cawley S, Hubbell E, Veitch J, Collins PJ, Darvishi K, Lee C, Nizzari MM, Gabriel SB, Purcell S, Daly MJ, Altshuler D: **Integrated genotype calling and association analysis of SNPs, common copy number polymorphisms and rare CNVs.** *Nat Genet* 2008, **40**(10):1253–1260.

39. Purcell S, Neale B, Todd-Brown K, Thomas L, Ferreira MA, Bender D, Maller J, Sklar P, de Bakker PI, Daly MJ, Sham PC: **PLINK: a tool set for whole-genome association and population-based linkage analyses.** *Am J Hum Genet* 2007, **81**(3):559–575.

3

Molecular signatures that correlate with induction of lens regeneration in newts: lessons from proteomic analysis

Konstantinos Sousounis[1], Rital Bhavsar[1], Mario Looso[2], Marcus Krüger[2], Jessica Beebe[1], Thomas Braun[2] and Panagiotis A Tsonis[1*]

Abstract

Background: Amphibians have the remarkable ability to regenerate missing body parts. After complete removal of the eye lens, the dorsal but not the ventral iris will transdifferentiate to regenerate an exact replica of the lost lens. We used reverse-phase nano-liquid chromatography followed by mass spectrometry to detect protein concentrations in dorsal and ventral iris 0, 4, and 8 days post-lentectomy. We performed gene expression comparisons between regeneration and intact timepoints as well as between dorsal and ventral iris.

Results: Our analysis revealed gene expression patterns associated with the ability of the dorsal iris for transdifferentiation and lens regeneration. Proteins regulating gene expression and various metabolic processes were enriched in regeneration timepoints. Proteins involved in extracellular matrix, gene expression, and DNA-associated functions like DNA repair formed a regeneration-related protein network and were all up-regulated in the dorsal iris. In addition, we investigated protein concentrations in cultured dorsal (transdifferentiation-competent) and ventral (transdifferentiation-incompetent) iris pigmented epithelial (IPE) cells. Our comparative analysis revealed that the ability of dorsal IPE cells to keep memory of their tissue of origin and transdifferentiation is associated with the expression of proteins that specify the dorso-ventral axis of the eye as well as with proteins found highly expressed in regeneration timepoints, especially 8 days post-lentectomy.

Conclusions: The study deepens our understanding in the mechanism of regeneration by providing protein networks and pathways that participate in the process.

Keywords: Regeneration, Lens, Newt, Proteomics, Gene expression, Regeneration program

Background

Several amphibian species own the ability to regenerate multiple different organs during adulthood making them excellent models to study the molecular mechanisms of tissue regeneration. Although regulation of regeneration might diverge among tetrapods, a deeper understanding of regenerative processes in amphibians will provide valuable clues for organ repair and regeneration in other organisms such as mammals [1,2].

Regeneration of the eye lens in newts provides a superb model to study regeneration. After surgical removal of the lens, the whole organ regenerates by transdifferentiation of dorsal iris pigmented epithelial (IPE) cells. Interestingly, the lens is never regenerated from the ventral iris, which provides a number of experimental options [3,4]. Most importantly, gene expression differences between the dorsal and the ventral part of the iris can be identified to unravel the molecular program enabling regeneration. Similarly, changes in the expression profile between the iris while the lens is intact and during lens regeneration allow characterization of regulatory pathways initiating regeneration. In addition, dorsal IPE cells retain their ability to form a lens after *in vitro* culturing, aggregation, and orthotopic transplantation or implantation into 3-D collagen lattices while ventral IPE aggregates fail to do so. Hence, gene expression profiles of cultured dorsal and ventral IPE cells might provide additional insights into lens regeneration [5].

* Correspondence: ptsonis1@udayton.edu
[1]Department of Biology and Center for Tissue Regeneration and Engineering at Dayton, University of Dayton, 300 College Park, Dayton, OH 45469, USA
Full list of author information is available at the end of the article

Recently, the first newt transcriptome was assembled, and RNA sequencing of newt iris has been used to study differences in gene expression between the dorsal and ventral iris. Analysis of gene expression identified genes exclusively expressed in either the dorsal or ventral iris as well as genes expressed in a gradient along the dorsoventral axis of the iris during lens regeneration [6,7]. In another study, custom newt microarrays were used to detect 467 genes that were differentially expressed during lens regeneration [8]. Although these studies provided essential information about the expression of genes during newt lens regeneration, protein data were missing so far. Mass-spectrometry-based protein analysis closes this gap providing information about changes in protein concentrations and potential post-transcriptional regulation during lens regeneration. Here, we computed the newt proteome from the assembled transcriptome and performed liquid chromatography followed by tandem mass spectrometry (LC-MS/MS) to identify proteins differentially expressed between dorsal and ventral iris as well as between regenerating and intact iris. We chose to use dorsal and ventral iris 4 and 8 days post-lentectomy (dpl) since during these timepoints iris cells re-enter the cell cycle and transdifferentiate. We then compared the expression data with the previously reported gene expression data at the mRNA level collected at the same timepoints. In addition, we performed LC-MS/MS with samples collected from *in-vitro*-cultured dorsal and ventral IPE cells. We compared the expression profiles between the *in vitro* and *in vivo* experiments. Lastly, we compared available high-throughput mRNA and protein expression data obtained from amphibian organ model systems undergoing regeneration, identifying a common regeneration program.

Results and Discussion
LC-MS/MS identifies novel newt proteins

Lenses were removed from newt eyes in order to initiate the regeneration process at the dorsal iris. Dorsal and ventral iris samples were collected at 4 and 8 dpl as well as from intact tissue (day 0). At 4 dpl, dorsal and ventral iris cells re-enter the cell cycle while at 8 dpl only dorsal iris cells initiate dedifferentiation. The iris samples were prepared, and LC-MS/MS was performed in order to investigate changes in protein concentrations (Figure 1A). The newly assembled newt transcriptome was used for peptide identification [7]. LC-MS/MS identified 8,167 different proteins. These proteins were uniquely annotated to 4,734 different human proteins (e-value < 1E-10). Direct comparisons with proteins found in previous proteomic studies in newts revealed that 701 of these annotated proteins have not been detected in newts before [7]. Our dataset also includes 143 proteins lacking annotations in other species, which raises the possibility that they are unique to

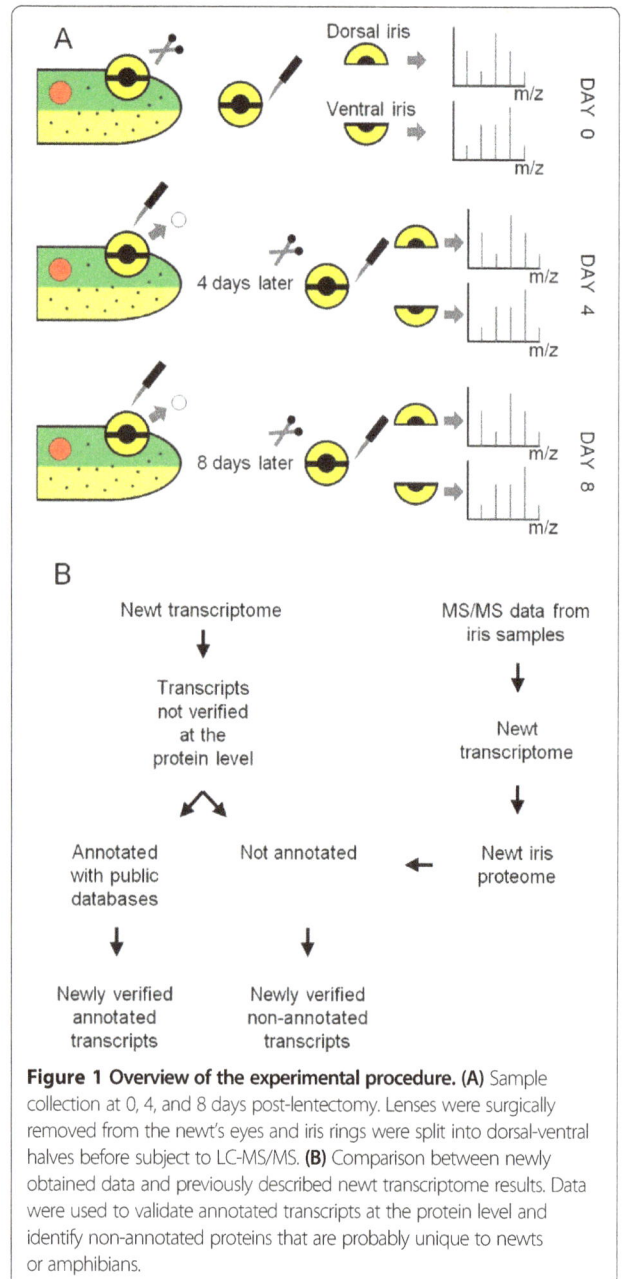

Figure 1 Overview of the experimental procedure. (A) Sample collection at 0, 4, and 8 days post-lentectomy. Lenses were surgically removed from the newt's eyes and iris rings were split into dorsal-ventral halves before subject to LC-MS/MS. **(B)** Comparison between newly obtained data and previously described newt transcriptome results. Data were used to validate annotated transcripts at the protein level and identify non-annotated proteins that are probably unique to newts or amphibians.

newts. These data are summarized as a whole and per sample in Table 1 (Workflow: Figure 1B).

Protein expression patterns during regeneration

The primary interest of our study was to identify proteins that might play a role during tissue regeneration. To achieve this goal, we compared proteins that were present at 4 and 8 dpl (both dorsal and ventral iris) and were expressed at least twofold higher than proteins in the intact iris at day 0 (regeneration group). Likewise, we examined proteins that were down-regulated (at least twofold) during regeneration (control group). To investigate the trends that our analysis yielded, we used the gene ontology (GO) terms

Table 1 Number of annotated and non-annotated proteins found by LC-MS/MS

	Total[a]	Day 0		Day 4		Day 8	
		Dorsal	Ventral	Dorsal	Ventral	Dorsal	Ventral
Proteins expressed	8,167	3,454	4,899	4,082	3,616	4,474	5,374
New verified proteins	1,479	248	604	361	285	436	682
Human proteins expressed	4,734	2,638	3,425	2,997	2,705	3,269	3,621
New verified human proteins	701	150	333	199	159	242	353
New verified newt proteins	143	13	37	29	30	28	54

Values are given per sample and for all experiments together.
[a]non-redundant.

of the annotated proteins and we performed enrichment analysis using Fisher's exact test corrected for multiple selections between the regeneration and control groups (false discovery rate, FDR < 0.05; Figure 2A). Metabolic process and gene expression were some of the GO terms enriched during regeneration in both dorsal and ventral iris samples while the GO term "cell periphery" was enriched in the control samples (Figure 2B and Additional file 1). Tables 2 and 3 (for dorsal iris) and Tables 4 and 5 (for ventral iris) list the genes reflecting the GO term enrichments for the regeneration group. These genes determine cellular functions in response to stress and reactive oxygen species. They are also involved in the regulation of translation, RNA maturation, and immune responses. Genes that were induced in the iris post-lentectomy can be grouped per function as follows:

a. Gene expression

Elongation factor 1-delta (EEF1D), elongation factor 1-gamma (EEF1G), valine-tRNA ligase, and ribosomal proteins RPL10, RPL13, RPL18A, RPL27A, RPL3, RPL4, RPL5, RPS23, and RPS8 are all known for their role in the translation of proteins. RNA-processing proteins include poly(U)-specific endoribonuclease (ENDOU), nuclease-sensitive element-binding protein 1 (YBX1), which is also implicated in cell proliferation [9], pre-mRNA-processing-splicing factor 8 (PRPF8), and probable ATP-dependent RNA helicase DDX5. In addition, two members of the ribonucleoprotein complex (ARC), polyadenylate-binding protein 1 (PABPC1) and cold shock domain-containing protein E1 (CSDE1), were detected. Consistent with these results, previous studies in mice have found PABPC1 to be up-regulated during liver regeneration [10]. Similarly, another stress-induced RNA processing protein, the heat shock cognate 71 kDa protein (HSPA8), was found to play a role during rat skeletal muscle regeneration and zebrafish caudal fin regeneration [11,12].

Basic leucine zipper and W2 domain-containing protein 1 (BZW1), protein arginine N-methyltransferase 5 (PRMT5), and SWI/SNF-related matrix-associated actin-dependent regulator of chromatin subfamily A

member 5 (SMARCA5) are known for regulation of gene expression. Interestingly, BZW1 has been previously found to induce histone H4 gene expression, which aids the progression of the G1/S phase of the cell cycle [13]. In addition, PRMT5 has been shown to play role in cell proliferation in planaria and be up-regulated post-injury in the kidneys [14,15]. Methionine aminopeptidase 1 (METAP1) and receptor-type tyrosine-protein phosphatase C (PTPRC), also known as CD45, have been shown to play a role in cell activation, proliferation, and cell cycle progression [16-18]. The protein expression data implicate regulation of gene expression as an important event for transdifferentiation.

b. ROS balance

Heme oxygenase 1 (HMOX1), redox-regulatory protein FAM213A, cis-aconitate decarboxylase (IRG1), and serpin B10 (SERPINB10) are known for their association with reactive oxygen species (ROS) and redox balance. IRG1 is activated by ROS to prevent infections [19]. SERPINB10 plays a role in protein processing and it is sensitive to redox stress [20]. HMOX1 is induced by ROS and has been linked to wound healing and regeneration in many regeneration model systems with the exception of mouse liver regeneration [21-24]. ROS stress and redox balance has gained a lot of attention since many studies have associated changes related to them with regenerative responses [25]. Our data indicate a potential role of ROS during newt lens regeneration as well.

c. Immune response

Argininosuccinate synthase (ASS1), complement factor B (CFB), acidic mammalian chitinase (CHIA), bis (5′-adenosyl)-triphosphatase ENPP4, eosinophil peroxidase (EPX), coagulation factor XIII A chain (F13A1), and myeloperoxidase (MPO) are up-regulated during lens regeneration and are known for their roles in preventing infections and generally to facilitate host defense and immune response.

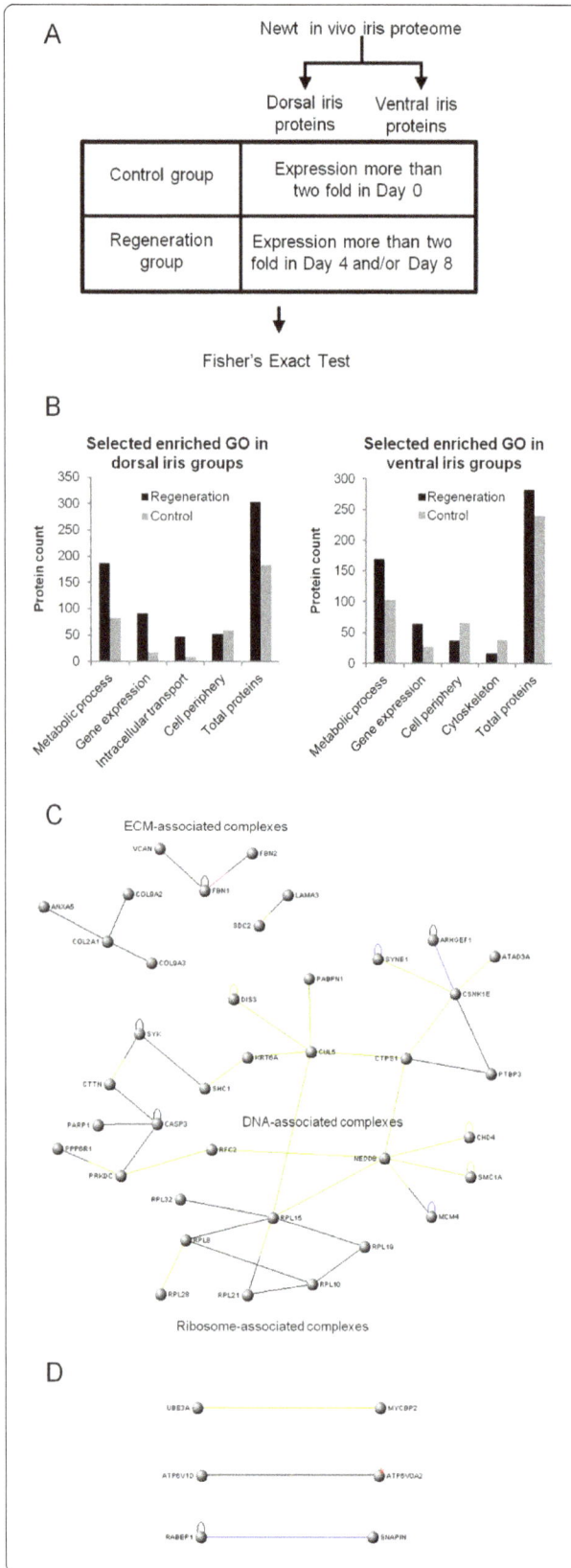

Figure 2 Comparisons between regenerating and control samples. (A) Workflow for selecting control and regeneration groups for Fisher's exact test. **(B)** Selected enriched gene ontology (GO) terms (FDR < 0.05) in regeneration or control groups for dorsal and ventral samples. *Bars* indicate the number of proteins in each group. **(C)** Network analysis of proteins expressed at higher levels in regenerating and dorsal samples. **(D)** Network analysis of proteins expressed at higher levels in regenerating and ventral samples. **(C, D)** Connections between nodes indicate protein-protein interaction. Only proteins that showed at least one interaction with another protein of the same group were displayed.

d. Metabolic processes

Sterol O-acyltransferase 1 (SOAT1), delta-1-pyrroline-5-carboxylate synthase (ALDH18A1), v-type proton ATPase subunit d 1 (ATP6V0D1), cytochrome b-245 heavy chain (CYBB), n-acylethanolamine-hydrolyzing acid amidase (NAAA), putative neutrophil cytosol factor 1C (NCF1C), neutrophil cytosol factor 2 (NCF2), and 6-phosphogluconate dehydrogenase, decarboxylating (PGD) participate in many metabolic processes and regulation of energy production. SOAT1 is also known to be involved during rat adrenal regeneration [26].

e. Other functions

Additional interesting proteins, potentially involved in the regulation of newt lens regeneration, are msx2-interacting protein (SPEN), which aids wound healing in *Drosophila* embryos [27]; cathepsin L1 (CTSL1), which mediates proteolysis; DNA topoisomerase 1 (TOP1), which is associated with DNA replication and transcription; epidermal retinol dehydrogenase 2 (SDR16C5), which is involved in making retinoic acid; and ATP-dependent DNA helicase Q1 (RECQL), which mediates DNA repair. Furthermore, we found an up-regulation of unconventional myosin-XVIIIa (MYO18A), which is involved in cell migration, as well as fibronectin (FN1) and integrin beta-2 (ITGB2), which mediate adhesion and cell migration and are essential for zebrafish heart regeneration [28].

The rapid up-regulation of all the aforementioned proteins during lens regeneration clearly underscores the importance of host defense, redox balance, and response to stress for the initiation of regeneration.

Proteins regulated in dorsal versus ventral iris

Since regeneration only proceeds from the dorsal iris, we wanted to find proteins that were specifically up-regulated in this tissue. To achieve this goal, we compared all proteins up-regulated during lens regeneration to proteins only up-regulated in the dorsal or the ventral iris. Proteins identified were imported in VisANT, a program that analyses protein-protein interactions [29]. Proteins that were expressed at higher levels in dorsal samples represented

Table 2 Genes with GO term related to gene expression that are up-regulated in the dorsal iris during regeneration versus control

GO:0010467; gene expression and parental GO							
Dorsal regeneration						Dorsal control	
AARS	ENDOU	KHSRP	RPL10A	RPLP2	SNRNP70	AHCYL2	GJA1
ANK3	ETF1	KIAA1967	RPL13	RPS2	SOAT1	BACE2	KANK2
BUD31	FUBP1	LUC7L3	RPL13A	RPS23	SPEN	CAV1	KRT17
BZW1	GARS	METAP1	RPL18	RPS24	SSB	CCDC88C	MPP6
C3	GCN1L1	MTA2	RPL18A	RPS28	SUGP2	CRK	NEO1
CSDE1	GLG1	MTOR	RPL19	RPS4X	SUPT16H	EEF1A1	PSMB9
DAD1	HBS1L	PABPC1	RPL21	RPS5	TARS	EIF4H	PTRF
DDX17	HCK	PES1	RPL26	RPS6	TMED2	FHL2	RBCK1
DPM1	HMOX1	POLR2A	RPL27A	RPS8	U2AF1L4	FKBP9	STAT5B
EEF1D	HNRNPM	POLR2E	RPL28	RPS9	UBE2I		
EEF1G	HNRNPU	PPP2R5C	RPL3	SERBP1	VARS		
EEF2	HSPA8	PRMT5	RPL35	SLTM	WDR77		
EIF3B	ILF3	PRPF8	RPL4	SMARCA5	WFS1		
EIF4G2	IPO9	PTBP3	RPL5	SMC1A	YARS		
EIF5A	KHDRBS1	RPL10	RPL8	SND1	YBX1		
EIF5B							

complexes associated with extracellular matrix, ribosomes, and DNA (Figure 2C) whereas proteins expressed at higher levels in ventral samples did not exhibit such a pattern (Figure 2D).

a. Extracellular matrix

Collagen proteins such as COL2A1, COL9A2, and COL9A3 are essential for structural support and provide the substrate for surrounding cells. In addition, COL9A2 and COL9A3 compose the vitreous area of the eye. Versican core protein (VCAN) is known for cell-extracellular matrix interactions allowing cell migration and growth. Laminin subunit alpha-3 (LAMA3) promotes migration. Fibrillin-1 (FBN1), fibrillin-2 (FBN2), and syndecan-2 (SDC2) are extracellular matrix proteins regulating the availability of growth factors to nearby cells. SDC2 also plays a role during rat periodontal wound healing [30]. Annexin A5 (ANXA5) is known for its anticoagulant properties and promotes wound healing in the cornea [31]. Previous studies in newt limb and lens regeneration also support these findings. A study in newt limb regeneration suggests that dynamic changes of the extracellular matrix provide a suitable microenvironment for regeneration [32], which is in-line with the up-regulation of several extracellular matrix genes during newt lens regeneration in the dorsal iris as determined by DNA microarray analysis [8]. These results suggest that remodeling an appropriate environment is a fundamental event during lens regeneration.

b. Cell activation

Rho guanine nucleotide exchange factor 1 (ARHGEF1), ATPase family AAA domain-containing protein 3A (ATAD3A), polypyrimidine tract-binding protein 3 (PTBP3, also known as ROD1), casein kinase I isoform epsilon (CSNK1E), src substrate cortactin (CTTN), replication factor C subunit 2 (RFC2), and NEDD8 have all been shown to have roles in cell proliferation, migration, and growth [33-36]. These cellular events are important since the lost tissue needs to be regenerated.

c. Gene expression

Cullin-5 (CUL5), polyadenylate-binding protein 2 (PABPN1), exosome complex exonuclease RRP44 (DIS3), SHC-transforming protein 1 (SHC1), tyrosine-protein kinase SYK, serine/threonine-protein phosphatase 6 regulatory subunit 1 (PPP6R1), chromodomain-helicase-DNA-binding protein 4 (CHD4), and ribosomal proteins RPL15, RPL32, RPL8, RPL28, RPL21, RPL10, and RPL19 are playing roles in regulating gene expression at various levels [37] and are all up-regulated in the dorsal iris. Interestingly, CHD4 has also been shown to be up-regulated during muscle regeneration in mice [38]. RNA sequencing during lens regeneration has previously revealed that genes associated with the regulation of gene expression are enriched in the dorsal iris, a pattern that we also found here at the protein level [6]. These results highlight the importance of rapid and impactful changes in the

Table 3 Genes with GO term related to metabolic process that are up-regulated in the dorsal iris during regeneration versus control

GO:0008152; metabolic process and parental GO

Dorsal regeneration				Dorsal control			
AASS	CYB5B	HSD17B13	PPID	ABAT	CP	ITPR2	PTPRA
ACTN1	CYBA	HSP90AB1	PPP3CC	ABCB7	CTSS	LANCL1	PTPRD
AFG3L2	CYBB	HSP90B1	PSMD1	ACAD10	CYP2A6	LTA4H	RAB27B
AKAP8L	DDX5	HSPA5	PTPRC	ACOT2	DCN	MAOA	RAB7A
ALDH16A1	DNAJA1	HSPA9	RAD23A	AGL	DCTN1	MINPP1	RECK
ALDH18A1	DNPEP	IRG1	RASA4	AK4	DHRS2	MRI1	SDHAF2
ALDOC	DPP3	ITGB2	RECQL	AKR1C4	ECHS1	MYH11	SDHD
ALOX5AP	DSP	ITPR2	RHO	ALDH3B1	EHD2	MYO5A	SH3GLB1
APMAP	DUSP11	LMNA	SCD5	ANXA1	ENTPD2	NEU3	SPTBN1
APOA1	ELOVL1	MAP3K15	SDHC	AOC3	FABP3	NPR3	SULT1B1
ASS1	ENPP4	MARCKS	SDR16C5	APPL2	GAPVD1	NRP1	SULT1C2
ATP2A2	ENTPD8	MPO	SERPINB10	ATP13A5	GGT5	PARG	TPM1
ATP6V0D1	EPX	MYO18A	SLC25A12	CDC42BPB	GRHPR	PGM5	TPP1
ATP6V1A	ERO1L	NAAA	SLC9A3R1	CECR1	GSTZ1	PIPOX	TUBB3
BIN1	F13A1	NAT8B	SQRDL	CLPX	HIBCH	PPIC	YWHAG
CANX	FAM213A	NCF1C	TIMM50				
CFB	FN1	NCF2	TNFAIP8				
CHIA	GNS	NUP93	TOP1				
COL3A1	GPD1L	PCCA	UBLCP1				
COX7A2L	GSTP1	PDPR	UQCRB				
CPT1A	H6PD	PFKP	USP5				
CTSL1	HK1	PGD	VRK1				
CUL3	HM13						

regulation of gene expression that will ultimately lead to transdifferentiation of iris cells to lens cells.

d. DNA repair

Poly [ADP-ribose] polymerase 1 (PARP1), DNA-dependent protein kinase catalytic subunit (PRKDC), and structural maintenance of chromosomes protein 1A (SMC1A) have known roles in DNA repair. DNA repair genes such as RAD1 have been previously found to be up-regulated in the dorsal iris using both microarrays and RNA sequencing during newt lens regeneration [6,8]. Such cellular machinery can play a role in maintaining the integrity of the genome an important aspect of regenerating an exact "clone" of the missing lens.

e. Other functions

Other proteins found to be up-regulated in the dorsal iris during lens regeneration were keratin, type II cytoskeletal 6A (KRT6A), and caspase-3 (CASP3) which

have been implicated in wound healing and regeneration [39,40].

All these proteins, grouped in the different functional categories, were up-regulated in the dorsal compared to the ventral iris during lens regeneration. Interestingly, several of these proteins formed interacting networks that can be linked to the regeneration process (Figure 2C).

Validation of changes in expression levels by qPCR

Since only few newt specific antibodies are available, we used quantitative real-time polymerase chain reaction (qPCR) to validate our data. Although qPCR measures mRNA and not protein concentrations, we reasoned that concomitant changes at the mRNA and protein levels might allow us to validate general patterns of gene activity during regeneration. We selected several proteins based on their putative function. Retinal dehydrogenase 1 (ALDH1A1), ephrin-B1 (EFNB1), and ephrin-B2 (EFNB2) were significantly up-regulated in the dorsal iris compared to the ventral irrespective of the timepoint ($p < 0.05$;

Table 4 Genes with GO term related to gene expression that are up-regulated in the ventral iris during regeneration versus control

GO:0010467; gene expression and parental GO

Ventral regeneration						Ventral control	
AARSD1	EIF3J	PFDN5	RPL27A	RQCD1	THBS1	CAT	ILK
ATP6AP2	ENDOU	POLR1C	RPL3	SEC11A	UBTF	CD44	KANK2
BZW1	GALNT12	PRKCI	RPL38	SMARCA5	VARS	CDH13	KRT17
CARS	HBS1L	PRMT5	RPL4	SNF8	VIPAS39	CHMP1A	PRKCA
CSDE1	HMOX1	PRPF8	RPL5	SNRPE	WARS	COG2	PRKDC
DDX39A	HSPA8	PSME3	RPS13	SOAT1	WDR61	COL4A6	PURA
DNAJC2	KDM1A	RCL1	RPS15	SPEN	WTAP	CTNNB1	RBCK1
EBNA1BP2	MARS	RPL10	RPS17	SRSF12	YBX1	DEK	SBDS
EEF1B2	METAP1	RPL13	RPS23	SSR4	YLPM1	DMD	SEC31A
EEF1D	NPC1	RPL17	RPS27	SUPT6H		EXOSC7	UGGT2
EEF1G	PABPC1	RPL18A	RPS8	TGFB1		GJA1	VPS36

Table 5 Genes with GO term related to metabolic process that are up-regulated in the ventral iris during regeneration versus control

GO:0008152; metabolic process and parental GO

Ventral regeneration				Ventral control			
ABCF2	DCN	LARP1	PRRC1	ABCB5	ECHDC2	NIT2	SLC22A2
ACSBG2	DDX5	LYN	PRSS16	ADH4	EHD2	OXCT1	SLC9A3R2
ACY1	DGAT1	LYZ	PRTN3	AGL	F8	PI4KA	SOD1
ADH1B	DNAJB11	MCM5	PTPN6	AK4	FAM162A	PLCG1	SULT1B1
ALDH18A1	DNM2	MCM6	PTPRC	ALG11	FBN1	PNP	SULT1C2
ALDH1A3	ECI2	MDN1	PZP	ATAD1	GALE	PNPO	SYTL2
ALOX5	ECM1	MOB1B	RCC1	ATG7	GLUL	PRKAB1	TGM1
ALPL	ENPP4	MPDU1	RECQL	C6orf130	GMPR2	PRKAR2B	TPP1
ARAP1	EPX	MPO	RFC5	CAPN5	GNA14	PTGR1	TTLL12
ASS1	F13A1	MYO18A	RNF213	CDIPT	GNAI1	PTK2	TUBB3
ATP2A1	FAM213A	MYO5A	RNLS	CECR1	GRHPR	PTPLAD1	VCP
ATP6V0D1	FASN	NAAA	SDR16C5	CLYBL	HAGH	PTPRD	XPNPEP1
ATP6V1F	FBP1	NADKD1	SERPINB10	COPS3	HIST1H2AG	PYGB	
C3	FEN1	NCF1C	SERPINB6	CP	HMOX2	RAB27B	
CFB	FKBP4	NCF2	SGPP1	CTSS	KLC4	RAB3D	
CHIA	FN1	NCKAP1L	SLC25A40	CUL5	LANCL1	RAB7A	
CKM	GBF1	NMT2	SMPD3	CYP2A13	MARCKS	RABGAP1	
CNEP1R1	GCAT	NRAS	SQSTM1	CYP2J2	MTCH1	RGN	
COX5B	GCLM	NUP155	SYK	DCTN1	MT-CYB	RTN4IP1	
CP	GLB1	NUP210	TBC1D9B	DLG1	MYH11	SEPHS1	
CTSA	GLUL	P2RX4	TOP1	DMGDH	NDUFB4	SERPINI1	
CTSL1	HECTD1	PAFAH1B2	UNC45A				
CUL2	HIST2H2AB	PGD	USP24				
CYBB	IDE	PLD3	VWA8				
CYP4F22	IRG1	PNP	WBSCR22				
CYP7B1	ITGB2						

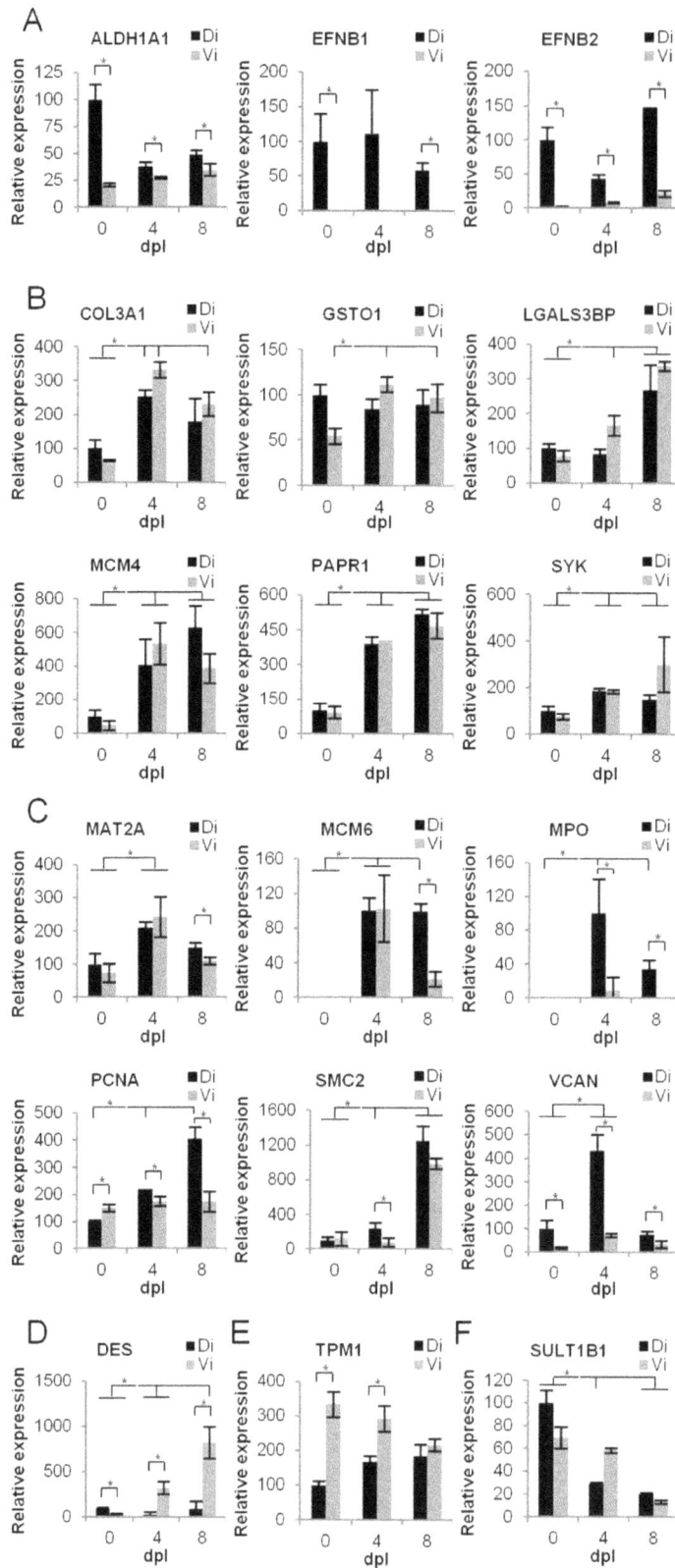

Figure 3 (See legend on next page.)

Figure 3 Validation of protein expression data by qPCR analysis. (A) Genes expressed at higher levels in dorsal iris. **(B)** Genes expressed at higher levels during regeneration. **(C)** Genes expressed at higher levels in the dorsal iris and during regeneration. **(D)** Gene expressed at higher levels in the ventral iris and during regeneration. **(E)** Gene expressed at higher levels in the ventral iris. **(F)** Gene expressed at higher levels in the intact iris. Student's *t*-test for independent samples was used for statistical significance. Homoscedasticity was assumed when Levene's test *p* value was greater than 0.05. Asterisk (*) indicates statistical significance (Student's *t*-test; *p* < 0.05). Each *bar* represents the average of triplicate values. *Error bars* represent standard deviation. *Lines on the top of the graph* compare samples during regeneration and control. *Lines on the top of bars* corresponding to a single day compare dorsal and ventral iris. For simplicity, only the statistics relevant for each group are presented on the graphs.

Figure 3A). Interestingly, these genes are also expressed during eye development in the dorsal optic cup [41,42] revealing a persisting pattern of gene expression from embryonic development to adulthood in the iris of newts. It is tempting to speculate that such genes may aid or repress regeneration hence providing the intrinsic regeneration potential of the dorsal iris. COL3A1, glutathione S-transferase omega-1 (GSTO1), galectin-3-binding protein (LGALS3BP), DNA replication licensing factor MCM4, PARP1, and SYK were up-regulated during regeneration both in the ventral and the dorsal parts of the iris (*p* < 0.05; Figure 3B). These genes are related to extracellular matrix, cell adhesion, redox balance, DNA maintenance, and DNA repair, processes required for regeneration and wound repair. S-adenosylmethionine synthase isoform type-2 (MAT2A), DNA replication licensing factor MCM6, MPO, proliferating cell nuclear antigen (PCNA), structural maintenance of chromosomes protein 2 (SMC2), and VCAN are also associated with the above-mentioned cellular processes but showed a higher expression in the dorsal versus the ventral iris and were expressed at higher levels during regeneration compared to undamaged controls (*p* < 0.05; Figure 3C) suggesting that the dorsal iris responds more robustly than the ventral iris to regenerative cues. Desmin (DES) is an intermediate filament found mostly in muscle tissue and has been associated with mitochondrial dysfunction and elevated ROS [43]. Desmin is only up-regulated during regeneration in the ventral iris (*p* < 0.05; Figure 3D). Tropomyosin alpha-1 chain (TPM1) is another cytoskeleton protein up-regulated in the ventral iris compared to the dorsal iris (*p* < 0.05; Figure 3E). Lastly, sulfotransferase family cytosolic 1B member 1 (SULT1B1), an enzyme catalyzing sulfonation, was expressed at higher levels in the intact iris (day 0), an expression pattern also found during liver regeneration in rats (Figure 3F) [44]. Overall, the qPCR data corroborated the expression changes found at the proteome level by mass spectrometry.

Protein expression patterns from the *in vitro* proteome
IPE cells retain their ability to transdifferentiate to lens cells *in vitro*, a process that can take up to 2 weeks. After re-aggregation and transplantation into a lentectomized eye, only the dorsal but not ventral iris aggregates

transdifferentiate to lens cells [45]. Similarly, only dorsal aggregates will transdifferentiate rapidly within 1–2 weeks to a structured lens when placed in 3-D collagen-based lattices like Matrigel [5]. Intriguingly, even IPE cells from higher animals, including humans, can be induced to transdifferentiate into lentoids (not structured lens) under certain culturing conditions [46]. We therefore examined protein expression in cultured IPE cells from the dorsal and ventral iris to monitor potential changes in the state of IPE cells (Figure 4A). In particular, we wanted to know whether culturing changed the protein profile of IPE cells and to identify markers that reflect transdifferentiation.

In total, we identified 2,269 annotated to known human proteins (e-value < 1E-10) in the cultured IPE cells. Proteins showing more than twofold higher expression either in the dorsal or ventral IPE cells are listed in Additional file 2. Numerous proteins were exclusively found either in the dorsal or ventral IPEs although GO terms analysis only revealed enrichment of cytoskeleton-associated terms in ventral versus dorsal IPE cells (Additional file 1 and Figure 4B).

Next, we compared the *in vitro* proteome with the *in vivo* proteome of 0, 4, and 8 dpl. Proteins with a similar expression pattern in respect to the dorsoventral axis under both *in vitro* and *in vivo* conditions are shown in Table 6. Some proteins have cell cycle, DNA replication, and splicing functions in the cell. EFNB1, DES, ALDH1A1, SMC2, and MCM4 proteins showed differences in expression levels between the dorsal and ventral IPE cells, an exact pattern as of their protein expression *in vivo*. As potential dorsoventral markers, we further validated these expression data by qPCR (Figure 4C). EFNB1, ALDH1A1, SMC2, and MCM4 were significantly up-regulated in dorsal IPE cells, while expression of DES was significantly up-regulated in ventral IPE cells (*p* < 0.05; Figure 4C). ALDH1A1, SMC2, and DES, which are similarly regulated in the iris during lens regeneration *in vivo* (Figure 3A,C), are involved in retinoic acid synthesis and DNA replication. Such cellular processes have been previously shown to be involved in lens regeneration from the dorsal iris [6,47]. Pearson correlation analysis of *in vivo* and *in vitro* datasets revealed that the R^2 correlation value increased from 0 to 4 dpl with the highest correlation at 8 dpl, indicating cells activated for tissue regeneration (Figure 4D). In contrast,

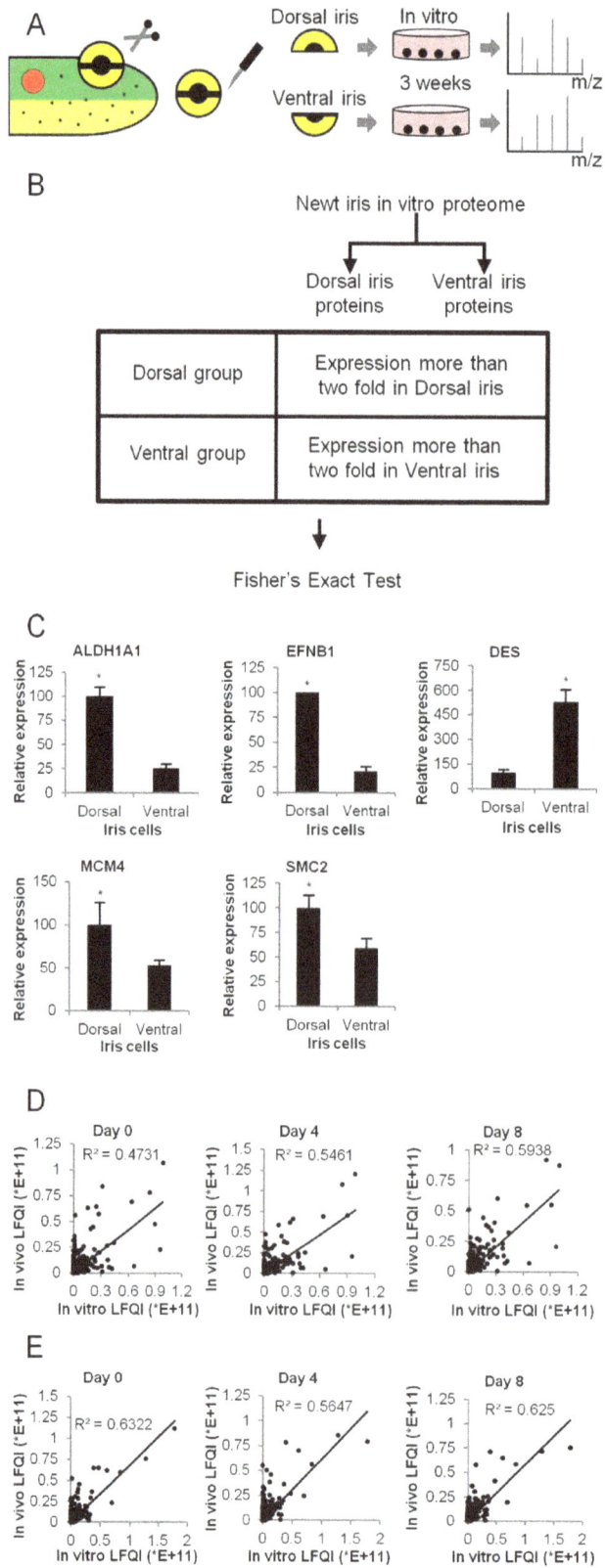

Figure 4 (See legend on next page.)

ventral IPE cells did not show such a correlation or trend (Figure 4E). The expression of dorsal markers ALDH1A1 and EFNB1 in dorsal IPE cells showed that they keep a "memory" of their origin, which consequently might be responsible for their ability to transdifferentiate to lens cells. The identified dorsal- or ventral-specific proteins might be used as markers in high-throughput screening using small molecules to identify agents inducing regeneration.

On the road for a common regeneration program
During the last two decades, several high-throughput methods including microarrays, next-generation RNA sequencing, and mass spectrometry have been developed to characterize gene expression profiles. We have used datasets from several different studies investigating organ regeneration in amphibians and extracted genes that were expressed at higher levels during tissue regeneration compared to intact controls. We focused on genes that were expressed more than twofold at any regeneration timepoint compared to intact tissue (for more information, see the "Methods" section). In addition, we annotated the genes based on human gene names serving as a common reference for the comparisons. Our search included seven microarray datasets from newt brain, spinal cord, hindlimb, forelimb, lens, heart and tail regeneration, one microarray and one RNA-seq dataset from axolotl limb regeneration [48,49], and two LC-MS/MS studies in newt heart regeneration and axolotl hindlimb regeneration [50-52]. We compared these datasets to proteins upregulated at least twofold in the dorsal iris during lens regeneration compared to intact iris (Figure 5 and Additional file 3). Surprisingly, the highest degree of similarity was found when RNA-seq data from limb regeneration were used (Table 7) [49]. Genes which were jointly activated in these, rather different, tissues during regeneration (Figure 5 and Table 7) most likely represent a part of

a canonical regeneration program. Hallmarks of the program include inflammation for host defense and cell activation, proliferation of new cells to replace lost tissue, migration for rearrangement of cells, generation of an appropriate extracellular matrix, regulation of ROS and DNA repair for tissue homeostasis, metabolic processes for cells to meet the needs of energy-consuming cellular processes, and changes in gene expression to shape the newly formed organ. We assume that these functional groups play a decisive role in the majority of all regeneration events.

Conclusions
In this study, we employed LC-MS/MS to identify proteins that are highly expressed during newt lens regeneration. Some of these proteins have similar functions and are arranged in protein networks associated with regulation of the extracellular matrix, DNA repair and maintenance, gene expression, and regulation of translation. Comparisons to other datasets collected during regeneration of a variety of different tissues from amphibians species revealed a putative canonical regeneration program, which seems to be required for regeneration to occur. Finally, we showed that cultured dorsal IPE cells *in vitro* maintain a molecular memory of their origin and show similar patterns as the 8-dpl *in vivo* lens regeneration dorsal iris. Taken together, our study provides information about proteins and protein groups that play an important role during tissue regeneration and deepens our understanding of the mechanism of regeneration.

Methods
Animal procedures
Animal handling and operations have been described previously [6,45]. Adult newts, *Notophthalmus viridescens*, were purchased from the Charles Sullivan Inc. Newt Farm. Anesthesia was performed with 0.1%(w/v) ethyl-3-aminobenzoate methanesulfonic acid (MS222; Sigma-Aldrich, St. Louis, MO) in phosphate buffered saline (PBS; 37 mM NaH_2PO_4 monohydrate, 58 mM Na_2HPO_4 anhydrous, pH 7.0). All procedures involving animals were approved by the University of Dayton Institutional Animal care and Use Committee.

Table 6 Genes with a similar expression pattern between iris during *in vivo* lens regeneration and *in vitro* cultured iris cells

Dorsal							Ventral
PHPT1	DDX46	DDX23	CHD4	CDK1	VCAN	SMC2	PLEC
MCM4	ALDH1A1	RANGAP1	APEH	PUS7	P4HA1	GSTO1	DES
PCNA	PARP1	MAT2A	SPTBN1				

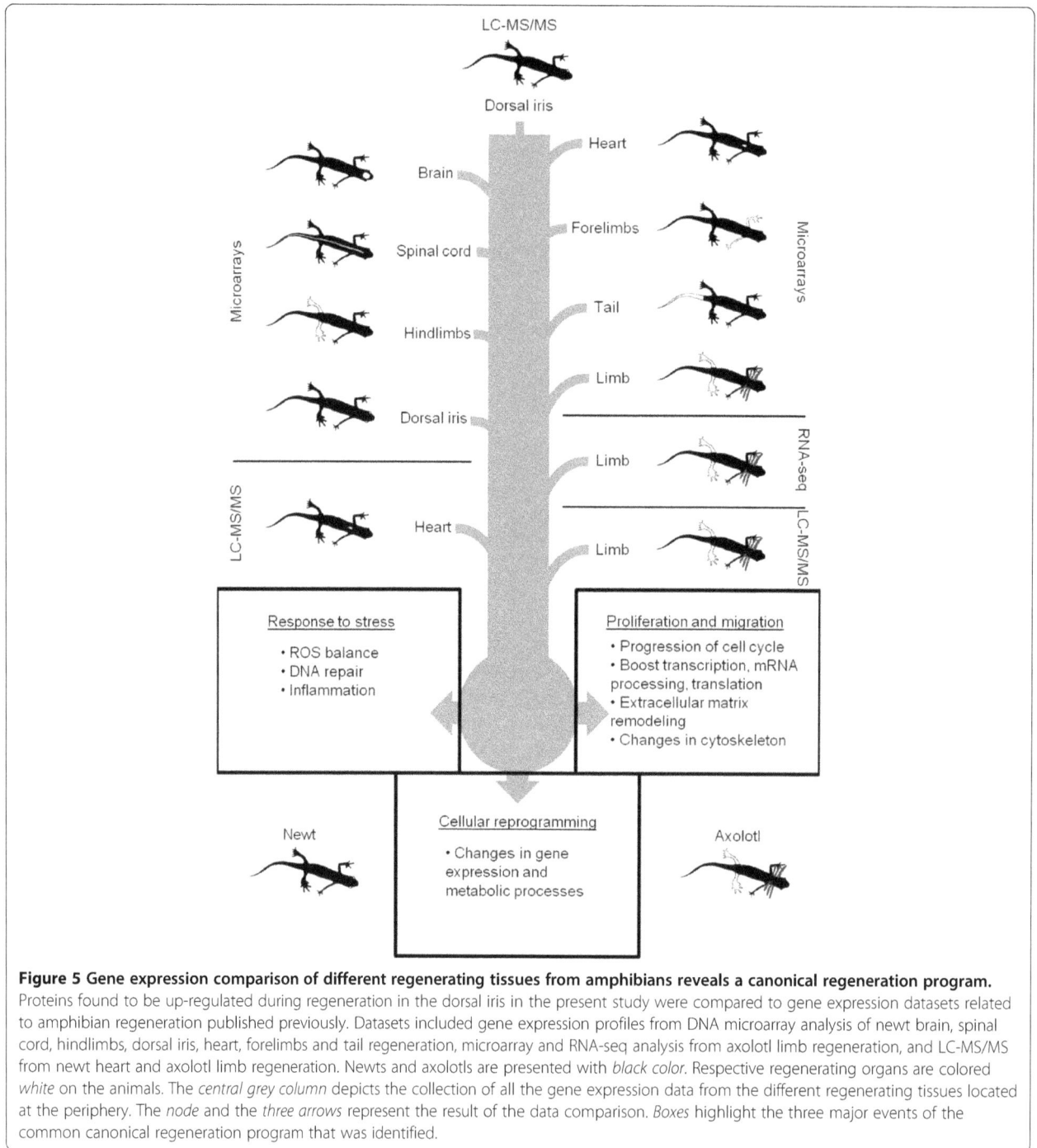

Figure 5 Gene expression comparison of different regenerating tissues from amphibians reveals a canonical regeneration program.
Proteins found to be up-regulated during regeneration in the dorsal iris in the present study were compared to gene expression datasets related to amphibian regeneration published previously. Datasets included gene expression profiles from DNA microarray analysis of newt brain, spinal cord, hindlimbs, dorsal iris, heart, forelimbs and tail regeneration, microarray and RNA-seq analysis from axolotl limb regeneration, and LC-MS/MS from newt heart and axolotl limb regeneration. Newts and axolotls are presented with *black color*. Respective regenerating organs are colored *white* on the animals. The *central grey column* depicts the collection of all the gene expression data from the different regenerating tissues located at the periphery. The *node* and the *three arrows* represent the result of the data comparison. *Boxes* highlight the three major events of the common canonical regeneration program that was identified.

Sample collection for LC-MS/MS

Newts were anesthetized and whole eye balls were removed and placed in calcium- and magnesium-free (CMF) Hank's solution. Using scissors, eye balls were dissected and iris rings were isolated. Using a scalpel, dorsal and ventral 135° sectors were extracted and kept frozen at –70°C until use.

Sample collection for qPCR, RNA extraction, reverse transcriptase reaction, qPCR reactions, and enrichment analysis

All procedures were performed as described previously [6]. For primers and quantitative real-time polymerase chain reaction (qPCR) settings see Additional file 4. Student's *t*-test

Table 7 Genes up-regulated during newt lens and axolotl limb regeneration

ABCA1[1]	C3[3]	EEF1E1[2]	HMOX1[3]	LSP1[3]	NCF2[1]	ROBO1[4]	TGFB1[2]
ABR[2]	CASP3[3]	EMILIN1[4]	HSPA5[3]	MARCO[3]	OPTC[4]	SACS[3]	TGM1[2]
ACSBG2[1]	CFB[3]	ERO1L[3]	HSPA9[3]	MCM4[5]	PAK2[5]	SAMD9L[5]	TGM2[2]
ACSL4[1]	CHIT1[3]	ETF1[2]	IDE[2]	MCM5[5]	PGD[1]	SERPINB10[2]	TGM6[2]
AIF1[3]	CLPTM1[3]	F13A1[3]	IFIH1[3]	MCM6[5]	PLEC[4]	SLC30A1[1]	TLR2[3]
ALOX5[1]	CPT1A[1]	FBLN1[4]	ITGAD[4]	MCM7[5]	PPID[3]	SOAT1[1]	TMEM43[6]
AQR[2]	CSE1L[5]	FBP1[1]	KIAA1967[5]	MOV10[2]	PSEN1[2]	TARS[2]	TNPO3[2]
ASTL[4]	CUL1[5]	FREM2[4]	LAMA3	MPO[3]	PTCD3[2]	TCN1[1]	VAMP8[2]
BCL2L1[5]	CYBA[1]	GIMAP7[2]	LGALS9[3]	MYO1F[6]	PTPRC[5]	TEX2[1]	VCAN[4]
BCS1L[1]	CYBB[1]	HK2[1]	LMNA[4]	NCCRP1[5]	RAI14[6]	TFRC[1]	VSIG4[3]
C1orf85[2]	DNAJC5[1]						

Gene function related to: metabolic process and transporters[1], gene expression and protein homeostasis[2], response to stress, host defense, immune response and reactive oxygen species[3], migration, adhesion and extracellular matrix[4], cell cycle, cell proliferation and DNA replication[5], cytoskeleton, cell shape, and organelle shape[6].

for independent samples was used to determine statistical significance for qPCR expression data. Equal variances were determined with Levene's test. Groups for enrichment analysis were selected as follows: For *in vivo* proteome analysis, protein expression was detected at 0 dpl and at least for one of the 4- and 8-dpl samples. Only differences more than twofold were used for further bioinformatical analysis. Annotation was assigned to newt proteins using BLAST2GO with e-value less than 1E-10. GO mapping and annotation was performed with default settings. GO enrichment analysis was calculated using Fisher's exact test corrected for multiple selections which is a built-in feature of BLAST2GO [53,54]. GO terms were considered enriched when FDR < 0.05.

Network analysis

Network analysis was performed using VisANT [29]. For construction of the dorsal regeneration network, only proteins with more than twofold change during regeneration (4 and/or 8 dpl) compared to the intact control and more than twofold change compared to the equivalent timepoints in the ventral iris samples were included. The selected proteins were used as input for the program. Human gene names and the human interactome were used for this analysis. Only proteins that had at least one interaction with a different protein of the group were displayed. The same procedure was used for the construction of the ventral regeneration network.

Newt IPE cell culture

Newt dorsal and ventral IPE cell culture was performed as described previously with minor modifications [45]. Dorsal and ventral IPE cells were plated separately in DMEM on collagen I coated plates (Becton Dickinson, Franklin Lakes, NJ). Cells were incubated at 27°C with 2% CO_2. The medium was changed every other day till day 21. On day 21, dispase (Gibco, Life technologies, Grand Island, NY) was added to the medium with a final concentration of 5% (v/v) and incubated overnight at 27°C with 2% CO_2. The collected cells were washed thrice with CMF Hank's solution and frozen with liquid nitrogen until use.

LC-MS/MS procedures

The iris tissue and cultured cells were isolated as described above. Proteins were isolated as described in [55] and processed for mass spectrometry (reverse-phase nano-LC-MS/MS, Thermo Velos and Q Exactive, Thermo Scientific, Waltham, MA) measurements. In brief, proteins were isolated by homogenizing tissue/cells in a buffer containing 1% Nonidet P-40, 0.1% sodium deoxycholate, 150 mm NaCl, 1 mm EDTA, and 50 mm Tris, pH 7.5 and protease inhibitor mixture (Roche, San Francisco, CA). Next, proteins were separated by 1D SDS-PAGE and stained by Coomassie Blue. The gel was cut into eight slices per lane (each timepoint *in vivo*, dorsal and ventral *in vitro*). Gel slices were washed by 100 μL 50 mM ammonium bicarbonate (ABC)/50% ethanol (EtOH) for 20 min at RT and dehydrated by incubating for 10 min in 100 μL absolute EtOH. Protein reduction was performed by incubating the gel pieces in 100 μL 10 mM DTT (in 50 mM ABC) for 45 min at 56°C. Alkylation was done by incubating the gel pieces in 100 μL 55 mM iodacetamide for 30 min at RT in darkness. After a final washing step, gel pieces were dried and proteins were in-gel digested using trypsin overnight. For desalting, peptides were loaded onto STAGE-tips and eluted with 80% acetonitril for mass spectrometry (MS) analysis [56,57]. Reversed-phase nano-LC-MS/MS was performed by using an Easy nanoflow HPLC system (Thermo Fisher Scientific, Odense Denmark; binary buffer system of A (0.1% (v/v) formic acid in H20) and B (0.1% (v/v) formic acid in 80% acetonitrile); 50-cm column (75-μm ID) packed in-house with 1.9-μm diameter C18 resin). The HPLC is coupled to Q Exactive mass spectrometer (Thermo Fisher Scientific, Bremen, Germany) with an electrospray ionization source (Thermo Fisher Scientific, Bremen, Germany). MS spectra were acquired at a resolution of 70,000 (200 m/z) in a mass range of 350–1,650 m/z and the top 10 most intense ions were selected for fragmentation. To identify mass-spectrometry-derived spectra, a *de-novo*-assembled transcriptome [7] was utilized by translating it into six reading frames generating an Andromeda search engine [58] compatible database. Only reading frames greater than 50 AA were used. Subsequent protein identification and label-free quantification was performed using MaxQuant software (Version 1.2.0.18) [59]. The maximum false discovery rate was set below 1% for peptide and protein identifications using the DECOY target database approach. Minimum peptide length was set to 7 AA and two peptides per protein group (at least one unique peptide). Carbamidomethyl at cysteine residues was set as a fixed modification. Oxidation at methionine and acetylation at the N-terminus were defined as variable modifications. Label-free quantification was based on at least two ratio counts. *In vivo* and *in vitro* LC-MS/MS data can be found in Additional file 2.

Comparisons with other datasets

Microarray, RNA sequencing, and LC-MS/MS data were extracted from the following papers: newt heart [50], forelimb, hindlimb, spinal cord, tail, brain, heart, tail [51], lens (dorsal iris) [8], and axolotl limb regeneration [48,49,52]. Genes were selected based on two expression criteria: expressed more than twofold in any of the regeneration timepoints compared to the control and not expressed more than twofold in the control compared to any of the regeneration timepoints. Human gene names were assigned to the extracted genes from all the datasets based on the annotation provided in the corresponding papers. Ambiguous annotations were discarded. The extracted genes can be found in Additional file 3. Comparisons, annotation assignments, and data mining were performed using custom Perl scripts.

Comparison between *in vitro* LC-MS/MS and *in vivo* LC-MS/MS data

In vitro LC-MS/MS data and *in vivo* LC-MS/MS data were normalized together for better correlation of expression levels. Pearson correlation was performed between dorsal IPE cell protein expression and the different timepoints of *in vivo* dorsal iris. Similar comparisons were performed with ventral samples. Tests were performed using Microsoft Office Excel spreadsheets.

Additional files

Additional file 1: Gene ontology enrichment analysis. Enrichment analysis using Fisher's exact test for genes up-regulated during *in vivo* regeneration versus control and *in vitro* ventral versus dorsal IPE cells.

Additional file 2: Protein expression data. Protein expression data from LC-MS/MS for 0, 4, and 8 dpl dorsal and ventral iris, dorsal and ventral IPE *in vitro* cultured cells, and genes up-regulated at least twofold in dorsal or ventral IPE cells.

Additional file 3: Gene pools for common regeneration program analysis. Genes from previous high-throughput amphibian studies, which were up-regulated during different timepoints of regenerating versus control.

Additional file 4: Primer sequences and qPCR/PCR settings. All primers and qPCR/PCR settings that were used for this study.

Competing interests

The authors declare that they have no competing interests

Authors' contributions

PAT and TB conceived the idea, designed the experiments, and wrote the manuscript. KS, RB, and ML performed experiments, analyzed data, and wrote the manuscript. MK and JB performed experiments. All authors read and approved the final manuscript.

Acknowledgements

This study is supported by NIH grant EY10450 to PAT. KS is supported in part by the University of Dayton Office of Graduate Academic Affairs through the Graduate Student Summer Fellowship Program.

Author details

[1]Department of Biology and Center for Tissue Regeneration and Engineering at Dayton, University of Dayton, 300 College Park, Dayton, OH 45469, USA. [2]Department of Cardiac Development and Remodeling, Max-Planck-Institute for Heart and Lung Research, Ludwigstrasse 43, 61231 Bad Nauheim, Germany.

References

1. Baddour JA, Sousounis K, Tsonis PA: **Organ repair and regeneration: an overview.** *Birth Defects Res C Embryo Today* 2012, **96**:1–29.
2. Barbosa-Sabanero K, Hoffmann A, Judge C, Lightcap N, Tsonis PA, Del Rio-Tsonis K: **Lens and retina regeneration: new perspectives from model organisms.** *Biochem J* 2012, **447**:321–334.
3. Eguchi G, Shingai R: **Cellular analysis on localization of lens forming potency in the newt iris epithelium.** *Dev Growth Differ* 1971, **13**:337–349.
4. Wolff G: **Entwickelungsphysiologische studien. I. Die regeneration der urodelenlinse.** *Arch EntwMech Org* 1895, **1**:380–390.
5. Hoffmann A, Nakamura K, Tsonis PA: **Intrinsic lens forming potential of mouse lens epithelial versus newt iris pigment epithelial cells in three-dimensional culture.** *Tissue Eng Part C Methods* 2014, **20**:91–103.
6. Sousounis K, Looso M, Maki N, Ivester CJ, Braun T, Tsonis PA: **Transcriptome analysis of newt lens regeneration reveals distinct gradients in gene expression patterns.** *PLoS One* 2013, **8**:e61445.
7. Looso M, Preussner J, Sousounis K, Bruckskotten M, Michel CS, Lignelli E, Reinhardt R, Hoeffner S, Krueger M, Tsonis PA, Borchardt T, Braun T: **A de novo assembly of the newt transcriptome combined with proteomic validation identifies new protein families expressed during tissue regeneration.** *Genome Biol* 2013, **14**:R16.
8. Sousounis K, Michel CS, Bruckskotten M, Maki N, Borchardt T, Braun T, Looso M, Tsonis PA: **A microarray analysis of gene expression patterns during early phases of newt lens regeneration.** *Mol Vis* 2013, **19**:135–145.
9. Frye BC, Halfter S, Djudjaj S, Muehlenberg P, Weber S, Raffetseder U, En-Nia A, Knott H, Baron JM, Dooley S, Bernhagen J, Mertens PR: **Y-box protein-1 is actively secreted through a non-classical pathway and acts as an extracellular mitogen.** *EMBO Rep* 2009, **10**:783–789.

10. Hsieh HC, Chen YT, Li JM, Chou TY, Chang MF, Huang SC, Tseng TL, Liu CC, Chen SF: Protein profilings in mouse liver regeneration after partial hepatectomy using iTRAQ technology. J Proteome Res 2009, 8:1004–1013.

11. Duguez S, Bihan MC, Gouttefangeas D, Feasson L, Freyssenet D: Myogenic and nonmyogenic cells differentially express proteinases, Hsc/Hsp70, and BAG-1 during skeletal muscle regeneration. Am J Physiol Endocrinol Metab 2003, 285:E206–E215.

12. Tawk M, Joulie C, Vriz S: Zebrafish Hsp40 and Hsc70 genes are both induced during caudal fin regeneration. Mech Dev 2000, 99:183–186.

13. Mitra P, Vaughan PS, Stein JL, Stein GS, van Wijnen AJ: Purification and functional analysis of a novel leucine-zipper/nucleotide-fold protein, BZAP45, stimulating cell cycle regulated histone H4 gene transcription. Biochemistry 2001, 40:10693–10699.

14. Braun MC, Kelly CN, Prada AE, Mishra J, Chand D, Devarajan P, Zahedi K: Human PRMT5 expression is enhanced during in vitro tubule formation and after in vivo ischemic injury in renal epithelial cells. Am J Nephrol 2004, 24:250–257.

15. Rouhana L, Vieira AP, Roberts-Galbraith RH, Newmark PA: PRMT5 and the role of symmetrical dimethylarginine in chromatoid bodies of planarian stem cells. Development 2012, 139:1083–1094.

16. Morikawa K, Oseko F, Morikawa S: The role of CD45 in the activation, proliferation and differentiation of human B lymphocytes. Int J Hematol 1991, 54:495–504.

17. Shivtiel S, Lapid K, Kalchenko V, Avigdor A, Goichberg P, Kalinkovich A, Nagler A, Kollet O, Lapidot T: CD45 regulates homing and engraftment of immature normal and leukemic human cells in transplanted immunodeficient mice. Exp Hematol 2011, 39:1161–1170. e1161.

18. Hu X, Addlagatta A, Lu J, Matthews BW, Liu JO: Elucidation of the function of type 1 human methionine aminopeptidase during cell cycle progression. Proc Natl Acad Sci U S A 2006, 103:18148–18153.

19. Li Y, Zhang P, Wang C, Han C, Meng J, Liu X, Xu S, Li N, Wang Q, Shi X, Cao X: Immune responsive gene 1 (IRG1) promotes endotoxin tolerance by increasing A20 expression in macrophages through reactive oxygen species. J Biol Chem 2013, 288:16225–16234.

20. Przygodzka P, Ramstedt B, Tengel T, Larsson G, Wilczynska M: Bomapin is a redox-sensitive nuclear serpin that affects responsiveness of myeloid progenitor cells to growth environment. BMC Cell Biol 2010, 11:30.

21. Ueno S, Campbell J, Fausto N: Reactive oxygen species derived from NADPH oxidase system is not essential for liver regeneration after partial hepatectomy. J Surg Res 2006, 136:260–265.

22. Lyoumi S, Puy H, Tamion F, Scotte M, Daveau M, Nordmann Y, Lebreton JP, Deybach JC: Nitric oxide synthase inhibition and the induction of cytochrome P-450 affect heme oxygenase-1 messenger RNA expression after partial hepatectomy and acute inflammation in rats. Crit Care Med 1998, 26:1683–1689.

23. Nada SE, Tulsulkar J, Shah ZA: Heme oxygenase 1-mediated neurogenesis is enhanced by Ginkgo biloba (EGb 761(R)) after permanent ischemic stroke in mice. Mol Neurobiol 2013, 49:945–956.

24. Wagener FA, van Beurden HE, von den Hoff JW, Adema GJ, Figdor CG: The heme-heme oxygenase system: a molecular switch in wound healing. Blood 2003, 102:521–528.

25. Love NR, Chen Y, Ishibashi S, Kritsiligkou P, Lea R, Koh Y, Gallop JL, Dorey K, Amaya E: Amputation-induced reactive oxygen species are required for successful Xenopus tadpole tail regeneration. Nat Cell Biol 2013, 15:222–228.

26. Tyczewska M, Rucinski M, Ziolkowska A, Trejter M, Szyszka M, Malendowicz LK: Expression of selected genes involved in steroidogenesis in the course of enucleation-induced rat adrenal regeneration. Int J Mol Med 2014, 33:613–623.

27. Mace KA, Pearson JC, McGinnis W: An epidermal barrier wound repair pathway in Drosophila is mediated by grainy head. Science 2005, 308:381–385.

28. Wang J, Karra R, Dickson AL, Poss KD: Fibronectin is deposited by injury-activated epicardial cells and is necessary for zebrafish heart regeneration. Dev Biol 2013, 382:427–435.

29. Hu Z, Snitkin ES, DeLisi C: VisANT: an integrative framework for networks in systems biology. Brief Bioinform 2008, 9:317–325.

30. Worapamorn W, Xiao Y, Li H, Young WG, Bartold PM: Differential expression and distribution of syndecan-1 and −2 in periodontal wound healing of the rat. J Periodontal Res 2002, 37:293–299.

31. Watanabe M, Kondo S, Mizuno K, Yano W, Nakao H, Hattori Y, Kimura K, Nishida T: Promotion of corneal epithelial wound healing in vitro and in vivo by annexin A5. Invest Ophthalmol Vis Sci 2006, 47:1862–1868.

32. Calve S, Odelberg SJ, Simon HG: A transitional extracellular matrix instructs cell behavior during muscle regeneration. Dev Biol 2010, 344:259–271.

33. Gilquin B, Cannon BR, Hubstenberger A, Moulouel B, Falk E, Merle N, Assard N, Kieffer S, Rousseau D, Wilder PT, Weber DJ, Baudier J: The calcium-dependent interaction between S100B and the mitochondrial AAA ATPase ATAD3A and the role of this complex in the cytoplasmic processing of ATAD3A. Mol Cell Biol 2010, 30:2724–2736.

34. Sadvakassova G, Dobocan MC, Difalco MR, Congote LF: Regulator of differentiation 1 (ROD1) binds to the amphipathic C-terminal peptide of thrombospondin-4 and is involved in its mitogenic activity. J Cell Physiol 2009, 220:672–679.

35. Tano K, Mizuno R, Okada T, Rakwal R, Shibato J, Masuo Y, Ijiri K, Akimitsu N: MALAT-1 enhances cell motility of lung adenocarcinoma cells by influencing the expression of motility-related genes. FEBS Lett 2010, 584:4575–4580.

36. Daub H, Olsen JV, Bairlein M, Gnad F, Oppermann FS, Korner R, Greff Z, Keri G, Stemmann O, Mann M: Kinase-selective enrichment enables quantitative phosphoproteomics of the kinome across the cell cycle. Mol Cell 2008, 31:438–448.

37. Querido E, Blanchette P, Yan Q, Kamura T, Morrison M, Boivin D, Kaelin WG, Conaway RC, Conaway JW, Branton PE: Degradation of p53 by adenovirus E4orf6 and E1B55K proteins occurs via a novel mechanism involving a Cullin-containing complex. Genes Dev 2001, 15:3104–3117.

38. Mammen AL, Casciola-Rosen LA, Hall JC, Christopher-Stine L, Corse AM, Rosen A: Expression of the dermatomyositis autoantigen Mi-2 in regenerating muscle. Arthritis Rheum 2009, 60:3784–3793.

39. Boland K, Flanagan L, Prehn JH: Paracrine control of tissue regeneration and cell proliferation by Caspase-3. Cell Death Dis 2013, 4:e725.

40. Pechter PM, Gil J, Valdes J, Tomic-Canic M, Pastar I, Stojadinovic O, Kirsner RS, Davis SC: Keratin dressings speed epithelialization of deep partial-thickness wounds. Wound Repair Regen 2012, 20:236–242.

41. Fan X, Molotkov A, Manabe S, Donmoyer CM, Deltour L, Foglio MH, Cuenca AE, Blaner WS, Lipton SA, Duester G: Targeted disruption of Aldh1a1 (Raldh1) provides evidence for a complex mechanism of retinoic acid synthesis in the developing retina. Mol Cell Biol 2003, 23:4637–4648.

42. Triplett JW, Feldheim DA: Eph and ephrin signaling in the formation of topographic maps. Semin Cell Dev Biol 2012, 23:7–15.

43. Maloyan A, Osinska H, Lammerding J, Lee RT, Cingolani OH, Kass DA, Lorenz JN, Robbins J: Biochemical and mechanical dysfunction in a mouse model of desmin-related myopathy. Circ Res 2009, 104:1021–1028.

44. Dunn RT 2nd, Kolaja KL, Klaassen CD: Effect of partial hepatectomy on the expression of seven rat sulphotransferase mRNAs. Xenobiotica 1999, 29:583–593.

45. Bhavsar RB, Nakamura K, Tsonis PA: A system for culturing iris pigment epithelial cells to study lens regeneration in newt. J Vis Exp 2011, 52:e2713.

46. Tsonis PA, Jang W, Del Rio-Tsonis K, Eguchi G: A unique aged human retinal pigmented epithelial cell line useful for studying lens differentiation in vitro. Int J Dev Biol 2001, 45:753–758.

47. Tsonis PA, Trombley MT, Rowland T, Chandraratna RA, del Rio-Tsonis K: Role of retinoic acid in lens regeneration. Dev Dyn 2000, 219:588–593.

48. Campbell LJ, Suarez-Castillo EC, Ortiz-Zuazaga H, Knapp D, Tanaka EM, Crews CM: Gene expression profile of the regeneration epithelium during axolotl limb regeneration. Dev Dyn 2011, 240:1826–1840.

49. Stewart R, Rascon CA, Tian S, Nie J, Barry C, Chu LF, Ardalani H, Wagner RJ, Probasco MD, Bolin JM, Leng N, Sengupta S, Volkmer M, Habermann B, Tanaka EM, Thomson JA, Dewey CN: Comparative RNA-seq analysis in the unsequenced axolotl: the oncogene burst highlights early gene expression in the blastema. PLoS Comput Biol 2013, 9:e1002936.

50. Looso M, Michel CS, Konzer A, Bruckskotten M, Borchardt T, Kruger M, Braun T: Spiked-in pulsed in vivo labeling identifies a new member of the CCN family in regenerating newt hearts. J Proteome Res 2012, 11:4693–4704.

51. Mercer SE, Cheng CH, Atkinson DL, Krcmery J, Guzman CE, Kent DT, Zukor K, Marx KA, Odelberg SJ, Simon HG: Multi-tissue microarray analysis identifies a molecular signature of regeneration. PLoS One 2012, 7:e52375.

52. Rao N, Jhamb D, Milner DJ, Li B, Song F, Wang M, Voss SR, Palakal M, King MW, Saranjami B, Nye HL, Cameron JA, Stocum DL: Proteomic analysis of blastema formation in regenerating axolotl limbs. BMC Biol 2009, 7:83.

53. Conesa A, Gotz S, Garcia-Gomez JM, Terol J, Talon M, Robles M: Blast2GO: a universal tool for annotation, visualization and analysis in functional genomics research. Bioinformatics 2005, 21:3674–3676.

54. Myhre S, Tveit H, Mollestad T, Laegreid A: **Additional gene ontology structure for improved biological reasoning.** *Bioinformatics* 2006, **22:**2020–2027.

55. Looso M, Borchardt T, Kruger M, Braun T: **Advanced identification of proteins in uncharacterized proteomes by pulsed *in vivo* stable isotope labeling-based mass spectrometry.** *Mol Cell Proteomics* 2010, **9:**1157–1166.

56. Shevchenko A, Wilm M, Vorm O, Mann M: **Mass spectrometric sequencing of proteins silver-stained polyacrylamide gels.** *Anal Chem* 1996, **68:**850–858.

57. Andersen JS, Lam YW, Leung AK, Ong SE, Lyon CE, Lamond AI, Mann M: **Nucleolar proteome dynamics.** *Nature* 2005, **433:**77–83.

58. Cox J, Neuhauser N, Michalski A, Scheltema RA, Olsen JV, Mann M: **Andromeda: a peptide search engine integrated into the MaxQuant environment.** *J Proteome Res* 2011, **10:**1794–1805.

59. Cox J, Mann M: **MaxQuant enables high peptide identification rates, individualized p.p.b.-range mass accuracies and proteome-wide protein quantification.** *Nat Biotechnol* 2008, **26:**1367–1372.

4

Whole-exome sequencing identifies de novo mutation in the *COL1A1* gene to underlie the severe osteogenesis imperfecta

Katre Maasalu[1,2*], Tiit Nikopensius[3,4], Sulev Kõks[5], Margit Nõukas[3,4], Mart Kals[3], Ele Prans[4], Lidiia Zhytnik[1], Andres Metspalu[3,4,6] and Aare Märtson[1,2]

Abstract

Background: Osteogenesis imperfecta (OI) comprises a clinically and genetically heterogeneous group of connective tissue disorders, characterized by low bone mass, increased bone fragility, and blue-gray eye sclera. OI often results from missense mutations in one of the conserved glycine residues present in the Gly-X-Y sequence repeats of the triple helical region of the collagen type I α chain, which is encoded by the *COL1A1* gene. The aim of the present study is to describe the phenotype of OI II patient and a novel mutation, causing current phenotype.

Results: We report an undescribed de novo *COL1A1* mutation in a patient affected by severe OI. After performing the whole-exome sequencing in a case parent–child trio, we identified a novel heterozygous c.2317G > T missense mutation in the *COL1A1* gene, which leads to p.Gly773Cys transversion in the triple helical domain of the collagen type I α chain. The presence of the missense mutation was confirmed with the Sanger sequencing.

Conclusions: Hereby, we report a novel mutation in the *COL1A1* gene causing severe, life threatening OI and indicate the role of de novo mutation in the pathogenesis of rare familial diseases. Our study underlines the importance of exome sequencing in disease gene discovery for families where conventional genetic testing does not give conclusive evidence.

Keywords: Osteogenesis imperfecta, Type I collagen, OI genotype–phenotype, *COL1A1*, *De novo* mutation

Introduction

Osteogenesis Imperfecta (OI), or "brittle bone" disease, is a heritable disorder of collagen type I metabolism with a generalized involvement of connective tissues. Collagen is the most abundant protein in mammals, constituting a quarter of the total protein weight [1]. Collagens are grouped into families based on their structural and functional features. Type I collagen is the major protein in bone, skin, tendon, ligament, sclera and cornea tissues, blood vessels, and hollow organs ENREF 2 [2]. OI is mostly caused by quantitative or qualitative collagen type I defects. The condition is characterized by low bone mass, bone fragility, and often short stature.

Extraskeletal manifestations may include blue-gray eye sclera and dental abnormalities. The clinical severity varies widely from nearly asymptomatic forms with a mild predisposition to fractures, normal stature, and normal lifespan to profoundly disabling and even lethal [3, 4].

The pathogenetic approach to OI is changed with the recent identification of non-collagenous genes, mutations in which may cause OI. In general, a clear genotype-phenotype correlation does not exist. General rules for genotype-phenotype correlations have been published only in *COL1A1/2*-related OI [5]. Approximately 90 % of individuals affected with OI are heterozygous for a causative variant in one of the two genes, *COL1A1* or *COL1A2*, which encode the pro-1(I) and pro-2(I) chains of type I procollagen, respectively [6].

The proportion of cases caused by a de novo *COL1A1* or *COL1A2* mutation varies according to the severity of the disease. Approximately 60 % of cases of classic non-

* Correspondence: Katre.maasalu@kliinikum.ee
[1]Clinic of Traumatology and Orthopaedics, University of Tartu, Puusepa 8, 51014 Tartu, Estonia
[2]Clinic of Traumatology and Orthopaedics, Tartu University Hospital, Puusepa 8, 51014 Tartu, Estonia

deforming OI with blue sclerae or common variable OI with normal sclerae, virtually 100 % of perinatally lethal OI, and close to 100 % of progressively deforming OI are caused by de novo mutations [7, 8].

In 1979, a classification of OI was introduced by David Sillence and the disease was divided into four types with a wide spectrum of clinical features, where OI type II is the most severe and prenatally lethal form of the disorder [9]. Initial classification, based on clinical and histological manifestations, was extended into five distinct types of OI [10]. By now, genetic studies described various OI phenotypes, and genetic OI classification is broadened up to 15 different OI types, according to the affected gene. In addition to collagen genes, OI is caused by mutations in the *CRTAP* (OI VII), *LEPRE 1* (OI VIII), *BMP1* (OI XIII), *TMEM38B* (OI XIV), *IFITM5* (OI V), *SERPINH1* (OI X), *WNT1* (OI XV), *SP7* (OI XII), *PPIB* (OI IX), *SERPINF1* (OI VI), *FKBP10* (OI XI) genes [11–14].

In addition to big genotype diversity, the phenotype manifestations may vary not only among representatives of the same OI type but also among carriers of the same mutation and even affected members of the same family. Some of the OI forms represent transitional phenotypes between different forms. Therefore, the genotype-phenotype relationships and clinical manifestations of OI are still difficult to explain.

The aim of present study is to describe the phenotype of OI II patient, with severe life threatening bone fragility, and report a novel mutation, causing current phenotype. We strongly believe that new additional information on current transitional OI form will deepen the understanding of the pathogenesis of the brittle bone disease.

Materials and methods

The Ethics Review Committee on Human Research of the University of Tartu approved the study and the participants, and their legal representatives gave prior consent to participate in the study and publish the results. Blood samples were obtained from the proband (716), her brother (715), and parents (710 and 711) from a family without previous history of OI (Fig. 1). Genomic DNA was extracted from EDTA-preserved blood according to standard high-salt extraction methods and stored at −80 °C.

The proband (716) was the first child (first pregnancy, first delivery) of non-consanguineous Estonian parents; age of the mother and father were 23 and 27 years, respectively. The parents were healthy without history of chronic or clinically significant diseases.

The girl was born with totally soft skull, the head was large (diameter 43 cm), and bones in occipital region were not palpable (skinhead). Her head was flat, being collapsed from front to back direction. She has exophthalmia and blue sclerae. The newborn had disproportional growth retardation (limbs > body): short, bowed arms and legs, and both hips were hyperflexed and turned outward.

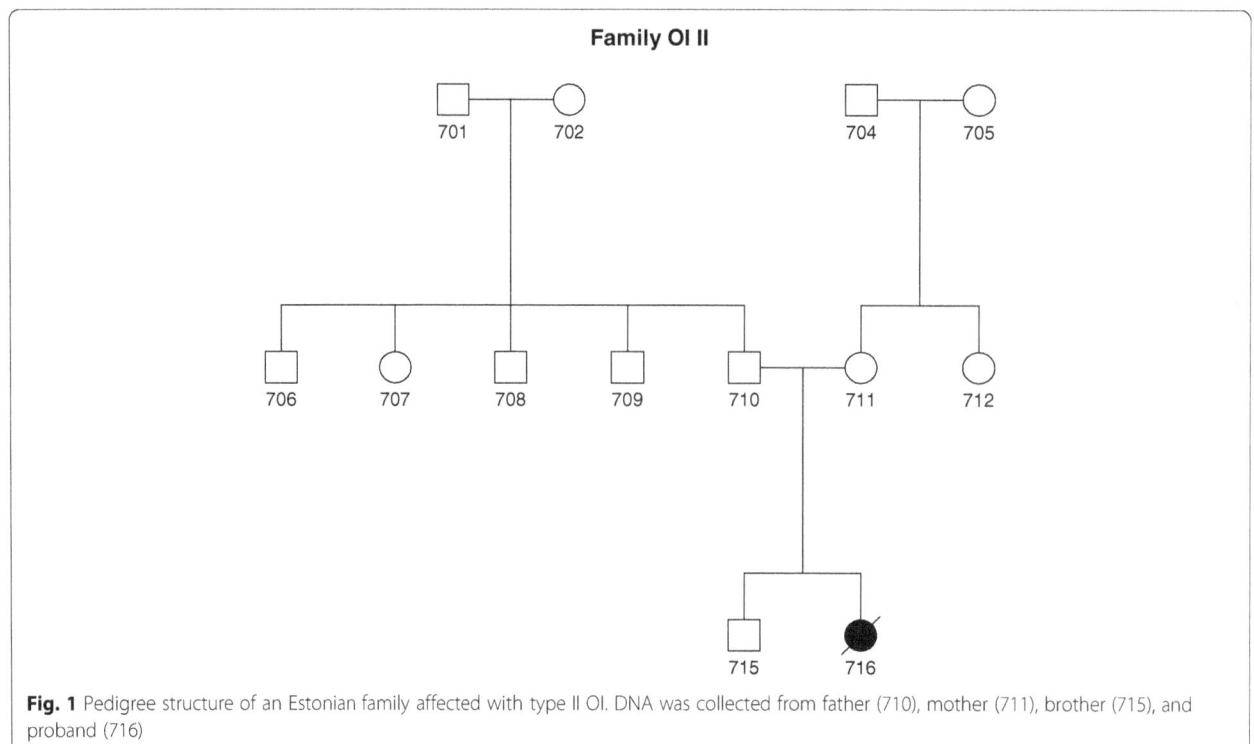

Fig. 1 Pedigree structure of an Estonian family affected with type II OI. DNA was collected from father (710), mother (711), brother (715), and proband (716)

Total skinhead and only partially developed upper part of parietal and mandibular bones were detected in Baby-gram X-ray investigation on date of birth. All long bones were extremely osteopenic; she had accordion-type ribs, and several fractures in different healing stages were detected in every bone. In addition, fresh right humeral fracture and left tibial fracture were confirmed.

In genetic counseling, at the age of 3 days, the baby was diagnosed with osteogenesis imperfecta type II based on clinical signs and X-ray data. As family history was negative for the OI, de novo autosomal dominant mutation was suspected.

Exome sequencing

Whole-exome sequencing was performed on an affected child and both unaffected parents at the NGS core facility of the Estonian Genome Center, University of Tartu. Exome capture was performed using the TruSeq Exome Enrichment kit (Illumina) following the manufacturer's protocol. The captured libraries were sequenced with Illumina HiSeq2000 with 100-bp paired-end reads. Over 10 Gb of sequence was generated from each individual, resulting in a coverage depth of 84× for both parents and 87× for an affected child. Sequence reads were aligned to the human reference genome (hg19, GRCh37) with the Burrows-Wheeler Aligner (BWA, version 0.6.1) [15]. Single-nucleotide substitutions and small indel variants were called with SAM tools (version 0.1.18), Picard tools (version 1.60), and a Genome Analysis Toolkit (GATK, version 1.5.21) [16, 17]. Genotypes were called at all positions with high-quality sequence bases and filtered to retain SNPs and insertion-deletions with Phred-like quality scores of at least 20. We focused on non-synonymous and canonical splice-site variants being absent from public datasets (including dbSNP135 and the 1000 Genomes Project) and in-house exome and full-genome data. We used the PolyPhen-2, SIFT, and Condel software tools to predict the functional effects of mutations [18–20]. Mutation analysis was performed with Sanger sequencing on an affected child (716), on both parents (710, 711), and an unaffected brother (715).

Results

We studied an Estonian family (Fig. 1) of a patient with severe OI in order to identify causative mutation for the disease. Using exome sequencing, we found a novel heterozygous G to T transversion in exon 33/34 at the position 2317 of the COL1A1 gene. This mutation leads to the substitution of glycine to cysteine at residue 773 in the triple helical domain of the alpha-1 chain of type I collagen. In the other previously known 15 AR OI genes, no potential pathogenic variants were found. The identified mutation was absent in both parents, demonstrating that this variant arose as de novo in the proband. Validation

by Sanger sequencing confirmed that the c.2317G > T (p.Gly773Cys) mutation was present in a heterozygous state in the index patient only (Fig. 2). This missense mutation affects a highly conserved amino acid (phyloP score 5.418), and in silico analysis predicted this variant to be deleterious. This mutation was not present neither in the Exome Variant Server of the NHLBI-ESP database or in the 1000 Genomes database. It was also not detected in 221 Estonian control exomes and in full genomes from 87 Estonians.

Discussion

Here, we report a heterozygous p.Gly773Cys mutation in COL1A1, affecting a highly conserved residue in the triple helical domain of collagen type I α chain, as a novel causative variant for severe OI with the lethal outcome. Therefore, given the large number of different

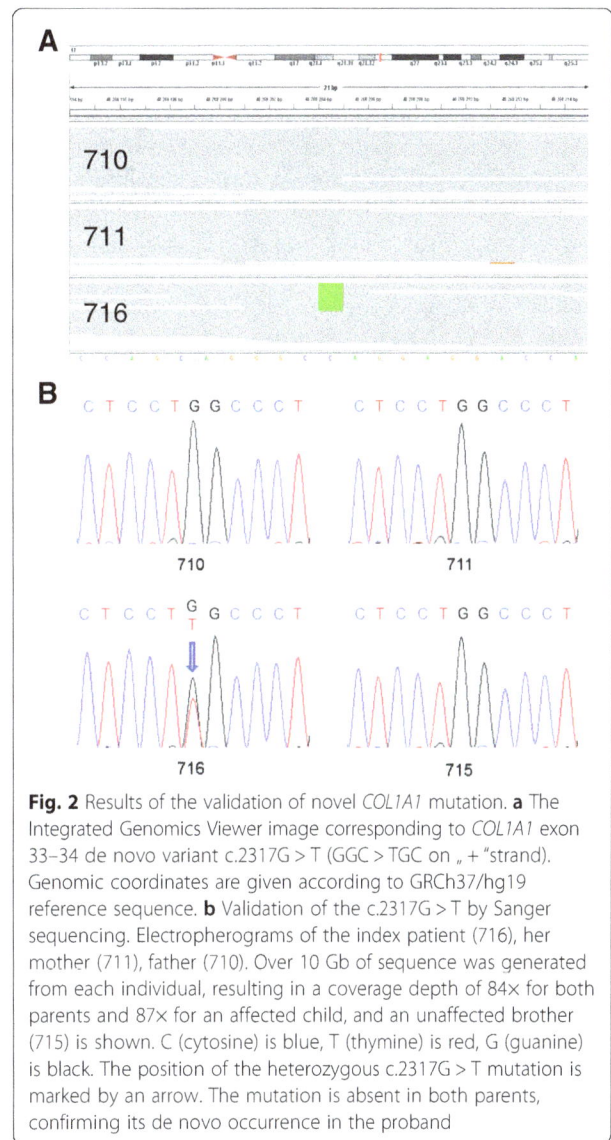

Fig. 2 Results of the validation of novel COL1A1 mutation. a The Integrated Genomics Viewer image corresponding to COL1A1 exon 33–34 de novo variant c.2317G > T (GGC > TGC on „ + "strand). Genomic coordinates are given according to GRCh37/hg19 reference sequence. b Validation of the c.2317G > T by Sanger sequencing. Electropherograms of the index patient (716), her mother (711), father (710). Over 10 Gb of sequence was generated from each individual, resulting in a coverage depth of 84× for both parents and 87× for an affected child, and an unaffected brother (715) is shown. C (cytosine) is blue, T (thymine) is red, G (guanine) is black. The position of the heterozygous c.2317G > T mutation is marked by an arrow. The mutation is absent in both parents, confirming its de novo occurrence in the proband

genes responsible for OI forms, the genotype-phenotype relationships and clinical manifestations are difficult to explain and present report gives additional information of pathophysiology of OI.

The severity of clinical signs (blue sclerae, totally soft, large and collapsed head; disproportional growth retardation, short and bowed arms and legs, severe breathing failure) and clinical findings (extremely osteopenic bones, several fractures in different healing stages in all bones, deformities, bone development retention) of our patient resembles those of the patients described with the lethal type of OI.

Our study found a novel heterozygous G to T transversion in exon at the position 2317 of the *COL1A1* gene (Fig. 3). There is another mutation described in dbSNP database—rs72651659—at the exact same position, with G to A substitution that leads to the substitution of glycine to serine at residue 773. Currently, altogether 756 unique DNA variants (substitutions, insertions, deletions, insertions/deletions, and duplications) in OI patients have been identified in the *COL1A1* gene, and the total number of reported variants is 1352 [11]. The overall pattern of severity of OI phenotypes that result from glycine substitutions in the triple helical domain of the α1(I) chain of type I procollagen is not uniform along the chain. Mutation can be non-lethal, when its position is smaller or equal to 688. If substitution occurs C-terminally to 688, its effect is lethal [21]. Approximately one-third (35.6 %) of all independent glycine substitutions in *COL1A1* are reported to result in lethal type of OI. However, whereas glycine substitutions to valine and to the charged amino acids (aspartic acid, glutamic acid, and arginine) are predominantly lethal (reaching up to 73 % of occurrences for valine), substitutions by the polar residues serine and cysteine have interspersed lethal and non-lethal outcomes with a frequency of lethal cases in ~30 % of occurrences for cysteine [8].

The p.Gly773Cys is located within a long stretch of helical residues 691–823, in which essentially mostly lethal substitutions are identified and also corresponds to Bodian's "COL1A1 decision tree" about the lethality of mutation [8, 21, 22]. This region correlates with the major ligand binding region 2 (MLBR 2) that extends from helical positions 682–830 in the type I collagen fibril and includes sites that are crucial for collagen self-assembly [8]. This region is important also for interactions of collagen monomers or fibrils with α1β1/α2β1 integrins, matrix metalloproteinases (MMPs), fibronectin, and cartilage oligomeric matrix protein [23]. The Gly773 residue lies within overlapping MMP interaction domain and cell interaction domain with human type I collagen fibril and within ligand binding sites for secreted protein, acidic, and rich in cysteine (SPARC), fibronectin, and MMP1, 12, and 13 [24]. Therefore, mutations in this region interfere severely with the function of protein.

Substitutions of the same glycine residue in *COL1A1* often have independent lethal and non-lethal outcomes. The described p.Gly773Cys mutation was lethal, excluding possibility that this family manifests OI type III. Moreover, neighboring Gly to Cys substitutions are reported to have diverse outcomes. P.Gly770Cys was reported to cause OI type II with a lethal outcome, whereas p.Gly788Cys represents a non-lethal variant causing OI type III/IV [8].

Conclusions

The present study describes a de novo p.Gly773Cys mutation in *COL1A1* related to the severe OI that gives additional information of pathophysiology of OI. This shows that phenotyping together with genotyping is important to identify special patients and relevant for counseling of parents.

Our study underlines the importance of exome sequencing in disease gene discovery for families where conventional genetic testing does not give conclusive evidence.

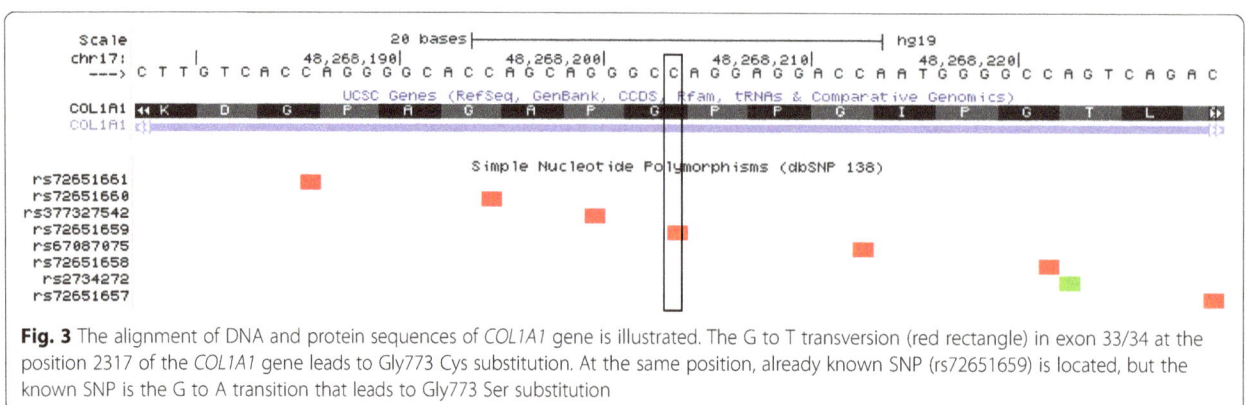

Fig. 3 The alignment of DNA and protein sequences of *COL1A1* gene is illustrated. The G to T transversion (red rectangle) in exon 33/34 at the position 2317 of the *COL1A1* gene leads to Gly773 Cys substitution. At the same position, already known SNP (rs72651659) is located, but the known SNP is the G to A transition that leads to Gly773 Ser substitution

Competing interests
The authors declare that they have no competing interests.

Authors' contributions
KM conceived of the study, participated in its design, interacted with patients, coordinated blood sample collection, and drafted the manuscript. TN, MN, MK, and AMe carried out the genetic studies and performed the statistical analysis. EP and LZ participated in the design of the study, participated in blood sample collection and DNA extraction, and helped to draft the manuscript. SK and AMä participated in the design of the study, coordinated data interpretation and statistical analysis, and helped to draft the manuscript. All authors read and approved the final manuscript.

Acknowledgments
This study was supported by the Estonian Science Agency project IUT20-46 (TARBS14046I), the European Regional Development Fund and Archimedes Foundation to the Centre of Excellence on Translational Medicine, the Targeted Financing from the Estonian Ministry of Education and Research (grant SF0180142s08), the HypOrth Project funded by the European Union 7th Framework Programme grant agreement n° 602398, the EU FP7 grant BBMRI-LPC (#313010), the Development Fund of the University of Tartu (grant SP1GVARENG), and the grant 3.2.0304.11-0312 funded by the European Regional Development Fund to the Centre of Excellence in Genomics (EXCEGEN).

Author details
[1]Clinic of Traumatology and Orthopaedics, University of Tartu, Puusepa 8, 51014 Tartu, Estonia. [2]Clinic of Traumatology and Orthopaedics, Tartu University Hospital, Puusepa 8, 51014 Tartu, Estonia. [3]Estonian Genome Centre, University of Tartu, Riia 23b, Tartu 51010, Estonia. [4]Institute of Molecular and Cell Biology, University of Tartu, Riia 23b, Tartu 51010, Estonia. [5]Department of Pathophysiology, University of Tartu, Ravila 19, Tartu 50411, Estonia. [6]Estonian Biocentre, Riia 23b, 51010 Tartu, Estonia.

References
1. Kielty CM, Grant ME. Connective Tissue and Its Heritable Disorders. Hoboken, NJ, USA: John Wiley & Sons, Inc; 2002. p. 159–221.
2. Chan T-F, Poon A, Basu A, Addleman NR, Chen J, Phong A, et al. Natural variation in four human collagen genes across an ethnically diverse population. Genomics. 2008;91:307–14.
3. Ben Amor IM, Glorieux FH, Rauch F. Genotype-phenotype correlations in autosomal dominant osteogenesis imperfecta. J Osteoporos. 2011;2011:540178.
4. Amor IMB, Rauch F, Gruenwald K, Weis M, Eyre DR, Roughley P, et al. Severe osteogenesis imperfecta caused by a small in-frame deletion in CRTAP. Am J Med Genet A. 2011;155A:2865–70.
5. Ben Amor M, Rauch F, Monti E, Antoniazzi F. Osteogenesis imperfecta. Pediatr Endocrinol Rev. 2013;10(2):397–405.
6. Van Dijk FS, Cobben JM, Kariminejad A, Maugeri A, Nikkels PGJ, van Rijn RR, et al. Osteogenesis imperfecta: a review with clinical examples. Mol Syndromol. 2011;2:1–20.
7. Byers PH, Steiner RD. Osteogenesis imperfecta. Annu Rev Med. 1992;43:269–82.
8. Marini JC, Forlino A, Cabral WA, Barnes AM, San Antonio JD, Milgrom S, et al. Consortium for osteogenesis imperfecta mutations in the helical domain of type I collagen: regions rich in lethal mutations align with collagen binding sites for integrins and proteoglycans. Hum Mutat. 2007;28:209–21.
9. Sillence DO, Senn A, Danks DM. Genetic heterogeneity in osteogenesis imperfecta. J Med Genet. 1979;16:101–16.
10. Rauch F, Glorieux F. Osteogenesis imperfecta. Lancet. 2004;363(9418):1377–85.
11. Switch gene. Osteogenesis Imperfecta Variant Database. Leiden Open Variation Database [https://oi.gene.le.ac.uk/home.php]
12. Laine CM, Wessman M, Toiviainen-Salo S, Kaunisto MA, Mäyränpää MK, Laine T, et al. A novel splice-mutation in PLS3 causes X-Linked early-onset low-turnover Osteoporosis. J Bone Miner Res. 2015;30(3):510–8.
13. Keupp K, Beleggia F, Kayserili H, Barnes AM, Steiner M, Semler O, et al. Mutations in WNT1 cause different forms of bone fragility. Am J Hum Genet. 2013;92:565–74.
14. Fahiminiya S, Majewski J, Mort J, Moffatt P, Glorieux FH, Rauch F. Mutations in WNT1 are a cause of osteogenesis imperfecta. J Med Genet. 2013;50:345–8.
15. Li H, Durbin R. Fast and accurate short read alignment with Burrows-Wheeler transform. Bioinformatics. 2009;25:1754–60.
16. McKenna A, Hanna M, Banks E, Sivachenko A, Cibulskis K, Kernytsky A, et al. The Genome Analysis Toolkit: a MapReduce framework for analyzing next-generation DNA sequencing data. Genome Res. 2010;20:1297–303.
17. Li H, Handsaker B, Wysoker A, Fennell T, Ruan J, Homer N, et al. The Sequence Alignment/Map format and SAMtools. Bioinformatics. 2009;25:2078–9.
18. Kumar P, Henikoff S, Ng PC. Predicting the effects of coding non-synonymous variants on protein function using the SIFT algorithm. Nat Protoc. 2009;4:1073–81.
19. Adzhubei IA, Schmidt S, Peshkin L, Ramensky VE, Gerasimova A, Bork P, et al. A method and server for predicting damaging missense mutations. Nat Methods. 2010;7:248–9.
20. González-Pérez A, López-Bigas N. Improving the assessment of the outcome of nonsynonymous SNVs with a consensus deleteriousness score. Condel Am J Hum Genet. 2011;88:440–9.
21. Bodian DL, Madhan B, Brodsky B, Klein TE. Predicting the clinical lethality of osteogenesis imperfecta from collagen glycine mutations. Biochemistry. 2008;47:5424–32.
22. Bodian DL, Chan T-F, Poon A, Schwarze U, Yang K, Byers PH, et al. Mutation and polymorphism spectrum in osteogenesis imperfecta type II: implications for genotype-phenotype relationships. Hum Mol Genet. 2009;18:463–71.
23. Di Lullo GA, Sweeney SM, Korkko J, Ala-Kokko L, San Antonio JD. Mapping the ligand-binding sites and disease-associated mutations on the most abundant protein in the human, type I collagen. J Biol Chem. 2002;277:4223–31.
24. Sweeney SM, Orgel JP, Fertala A, McAuliffe JD, Turner KR, Di Lullo GA, et al. Candidate cell and matrix interaction domains on the collagen fibril, the predominant protein of vertebrates. J Biol Chem. 2008;283:21187–97.

5

SIRT1 affects DNA methylation of polycomb group protein target genes, a hotspot of the epigenetic shift observed in ageing

Luisa A Wakeling[1], Laura J Ions[1], Suzanne M Escolme[1], Simon J Cockell[2], Tianhong Su[1], Madhurima Dey[1], Emily V Hampton[1], Gail Jenkins[4], Linda J Wainwright[4], Jill A McKay[3] and Dianne Ford[1*]

Abstract

Background: SIRT1 is likely to play a role in the extension in healthspan induced by dietary restriction. Actions of SIRT1 are pleiotropic, and effects on healthspan may include effects on DNA methylation. Polycomb group protein target genes (PCGTs) are suppressed by epigenetic mechanisms in stem cells, partly through the actions of the polycomb repressive complexes (PRCs), and have been shown previously to correspond with loci particularly susceptible to age-related changes in DNA methylation. We hypothesised that SIRT1 would affect DNA methylation particularly at PCGTs. To map the sites in the genome where SIRT1 affects DNA methylation, we altered SIRT1 expression in human intestinal (Caco-2) and vascular endothelial (HuVEC) cells by transient transfection with an expression construct or with siRNA. DNA was enriched for the methylated fraction then sequenced (HuVEC) or hybridised to a human promoter microarray (Caco-2).

Results: The profile of genes where SIRT1 manipulation affected DNA methylation was enriched for PCGTs in both cell lines, thus supporting our hypothesis. SIRT1 knockdown affected the mRNA for none of seven PRC components nor for DNMT1 or DNMT3b. We thus find no evidence that SIRT1 affects DNA methylation at PCGTs by affecting the expression of these gene transcripts. EZH2, a component of PRC2 that can affect DNA methylation through association with DNA methyltransferases (DNMTs), did not co-immunoprecipitate with SIRT1, and SIRT1 knockdown did not affect the expression of EZH2 protein. Thus, it is unlikely that the effects of SIRT1 on DNA methylation at PCGTs are mediated through direct intermolecular association with EZH2 or through effects in its expression.

Conclusions: SIRT1 affects DNA methylation across the genome, but particularly at PCGTs. Although the mechanism through which SIRT1 has these effects is yet to be uncovered, this action is likely to contribute to extended healthspan, for example under conditions of dietary restriction.

Keywords: Dietary restriction, Polycomb group proteins, Polycomb repressive complexes, Stem cells

Background

The DNA methylation profile of the vertebrate genome changes over time, reflected as a change in total methyl-cytosine content [1, 2]. These changes have been mapped to specific sites in species including mice [3, 4] and humans [5–8], revealing a drift in DNA methylation across most of the genome with components that are both tissue specific and tissue independent [4, 9].

A notable feature of the age-related drift in DNA methylation is that hypermethylation clusters at the gene targets of polycomb group proteins (PCGTs), as observed in human whole blood from postmenopausal women [7], mouse intestine [4] and mouse haematopoietic stem cells [3]. Several arguments and observations support the premise that epigenetic changes, such as changes in DNA methylation, contribute to the ageing process. For example, the fundamental role of epigenetic reprogramming in the process of gamete formation, which must reverse the ageing clock to prevent progressively shortened lifespan in each successive generation, provides a compelling

* Correspondence: dianne.ford@ncl.ac.uk
[1]Institute for Cell and Molecular Biosciences, Human Nutrition Research Centre, Newcastle University Medical School, Newcastle upon Tyne NE2 4HH, UK

argument to support this view; likewise the role of epigenetic reprogramming to restore pluripotency in the success of somatic cell nuclear transfer [10]. Also consistent with the premise that epigenetic changes contribute to ageing is that extended lifespan can be inherited trans-generationally in *Caenorhabditis elegans* via genes that are components of a major epigenetic modifier—the histone H3 lysine 4 trimethylation (H3K4me3) complex [11]. The polycomb group proteins bind to PCGTs as polycomb repressive complexes (PRCs). PCGTs are repressed by mechanisms involving chromatin modification in stem cells and must be expressed to achieve cell differentiation [12]. PCGTs also tend to be hypermethylated in cancer [13–15].

We showed recently that manipulating the expression of the histone deacetylase SIRT1 in human cells affected promoter DNA methylation of a small panel of genes that we tested, selected on the basis that they have been reported to show an age-related change in DNA methylation and to be expressed differentially in response to dietary restriction (DR), an intervention shown robustly in multiple species to increase lifespan and/or healthspan [16]. The view that SIRT1 contributes to increased healthspan and/or lifespan, including under conditions of DR, is controversial. The supporting literature is extensive and is covered by recent reviews (e.g. [17, 18]). Notable recent developments include the observation that male and female transgenic mice that overexpress Sirt1 specifically in the brain had extended lifespan and enhanced neural activity in the dorsomedial and lateral hypothalamic nuclei [19]. It appears, however, that some earlier work in model organisms proposed to demonstrate that the gene homologues of SIRT1 confer extended lifespan requires re-evaluation. For example, extended lifespan in strains of *C. elegans* transgenic for *Sir2* tracked with loci other than the transgene [20]. Also, confounding effects of genetic manipulation used to create *Sir2* transgenic *Drosophila*, rather than the *Sir2* transgene per se, appear to be responsible for the long-lived phenotype [20]. However, the debate has been re-opened by reports including that lifespan was extended in *Drosophila* when *Sir2* expression was manipulated using an inducible system that eliminated genetic background as a confounding factor [21]. Also, a body of other recent data show consistently effects on mammalian physiology commensurate with sirtuins having actions that protect against features of ageing (reviewed in [22]). Intermediates in pleiotropic cellular pathways and several key transcription factors with likely effects on healthspan are substrates for deacetylation by SIRT1. These substrates include PGC1α, which controls mitochondrial biogenesis, p53 [23] and many others [24]. Our discovery that SIRT1 affects DNA methylation with a bias towards genes that also show altered expression in response to dietary restriction [16] uncovers a novel and fundamental function of SIRT1 with likely

particular relevance to its effects on healthspan. Recent reviews provide a fuller exposition of evidence supporting the view that SIRT1 has a role in healthspan (e.g. [25]).

Here we hypothesised that altering the level of SIRT1 expression would affect DNA methylation on a genome-wide basis and target preferentially genes, including PCGTs, where DNA methylation is affected by increasing age. Supporting our hypothesis, we made the fundamentally important observation that effects of SIRT1 on DNA methylation do indeed cluster particularly at PCGTs.

Results

Manipulating SIRT1 expression affects DNA methylation across the genome

We increased SIRT1 expression by transient transfection with a plasmid construct or reduced expression using siRNA (as in our previous work [16]) to measure the effect on DNA methylation across the genome in two different human cell line models—HuVECs (vascular endothelial) and (as used in our previous work) Caco-2 (intestinal) cells. Efficacy of overexpression or knockdown for HuVECs was confirmed by RT-qPCR and Western blotting (Fig. 1) and has been shown previously for Caco-2 cells [16]. DNA was enriched for the methylated component and compared to the input sample. For HuVECs, a recombinant H6-GST-MBD protein was bound to fragmented DNA, and then the methylated fraction was captured on magnetic beads coated with GSH. Input and enriched samples were then sequenced, and reads were mapped to the human genome then filtered to those within 2 kb of a transcription start site or within genes (between the TSS and stop codon). The data are deposited under GEO accession number GSE54072 [26]. Differentially methylated genes were identified using the package MEDIPS (Bioconductor) then classified as hypomethylated or hypermethylated when SIRT1 expression was increased or reduced. For Caco-2 cells, DNA was enriched for the methylated fraction by MeDIP using an antibody recognising 5-methylcytidine (5mC), and efficacy was confirmed by measuring enrichment by qPCR of a lambda phage DNA added as a spike to all samples in both a demethylated and in vitro methylated form and of two loci known to be hypermethylated (*H19* and *L1.2*) relative to a hypomethylated locus (*UBE2B*) [27]. Comparing input and immunoprecipitated samples, the lambda phage spike was enriched 1000–12,000-fold and the endogenous hypermethylated versus hypomethylated loci were enriched 40–270-fold (see Additional file 1), thus confirming efficacy. Input and enriched samples were co-hybridised to the human 3x720K CpG Island Plus RefSeq Promoter Array (NimbleGen). The data are deposited under GEO accession number GSE53569 [28]. Genes were scored as methylated differently under conditions of SIRT1 knockdown

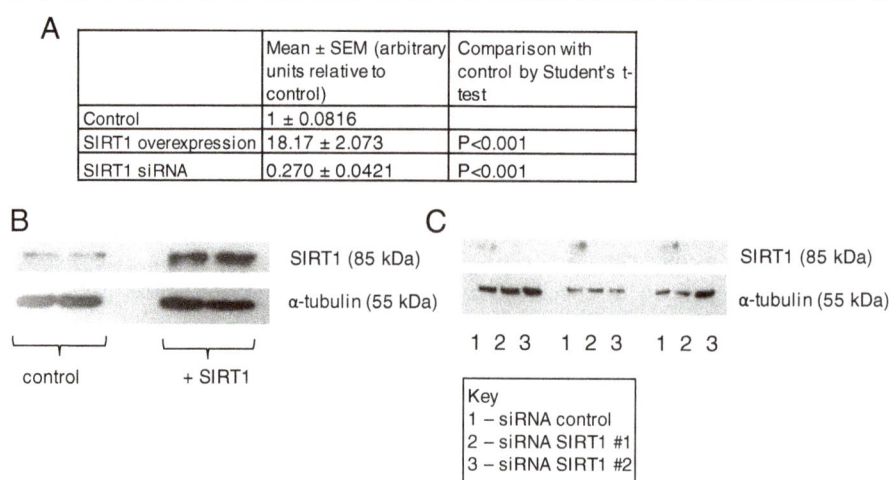

		Mean ± SEM (arbitrary units relative to control)	Comparison with control by Student's t-test
	Control	1 ± 0.0816	
	SIRT1 overexpression	18.17 ± 2.073	P<0.001
	SIRT1 siRNA	0.270 ± 0.0421	P<0.001

Fig. 1 Confirmation of SIRT1 overexpression and knockdown in HuVECs. **a** Measurement of SIRT1 mRNA by RT-qPCR. Data are for n = 4–8. Measurement of SIRT1 by Western blotting following transient transfection with plasmid pCMV6-ENTRY-SIRT1 (Origene) or with vector control (**b**) or with siRNA targeted to SIRT1 or with control siRNA (**c**). Approximately 10 μg of protein was loaded in each lane. Three biological replicates are presented for each condition. Approximate molecular weights are indicated. "+SIRT1" indicates cells transfected with pCMV6-ENTRY-SIRT1; "control" indicates cells tranfected with vector control; "siRNA" indicates cells transfected with one of two siRNAs (#1 or #2) targeted to SIRT1 or with a control siRNA. Data are representative of multiple independent repeats of the procedure

compared with control where they appeared only in one or other list of enriched genes.

For ease of reference, we refer to genes that lost DNA methylation with SIRT1 knockdown and/or gained DNA methylation with SIRT1 overexpression as having a positive DNA methylation response to SIRT1. Conversely, we classify genes that responded to SIRT1 in an opposite direction as having a negative response to SIRT1. A total of 1554 genes in HuVECs [29] and 1845 genes in Caco-2 cells [29] showed a positive DNA methylation response to SIRT1[29], of which 139 (a larger number than expected by chance; $P < 0.001$) were common to both cell lines (Fig. 2). Similarly, the two different cell lines shared a subset of genes that showed a negative DNA methylation response to SIRT1 that was greater than expected by chance ($P = 0.005$), comprising 49 genes from a total of 1475 in HuVECs [29] and 873 in Caco-2 cells [29] (Fig. 2).

PCGTs are over-represented among genes for which DNA methylation is affected by SIRT1 manipulation

We determined if PCGTs were over-represented in our list of genes that responded to SIRT1 using lists derived by performing genome-wide location analysis of DNA immunoprecipitated by antibodies against core components of polycomb repressive complex 1 (PRC1) (Phc1 and Rnf2) and PRC2 (Suz12 and Eed) [30]. This analysis is summarised in Table 1. Gene targets of each individual component of PRC1 or PRC2, or targets of at least one component, were all enriched 1.3 to 1.8-fold in genes we found to show a positive DNA methylation

response to SIRT1 in both cell lines (with the exception of targets of Rnf2 in HuVECs). Similarly, genes identified as targets of ALL components of PRC1 and PRC2 were enriched (1.6 to 2.2-fold) in the genes showing a positive DNA methylation response to SIRT1. We found a similar relationship between genes that showed a negative DNA methylation response to SIRT1 and PCGTs. Gene targets of each individual component of PRC1 or PRC2 as well as gene targets of at least one component were all enriched 1.3 to 1.7-fold in these gene lists derived

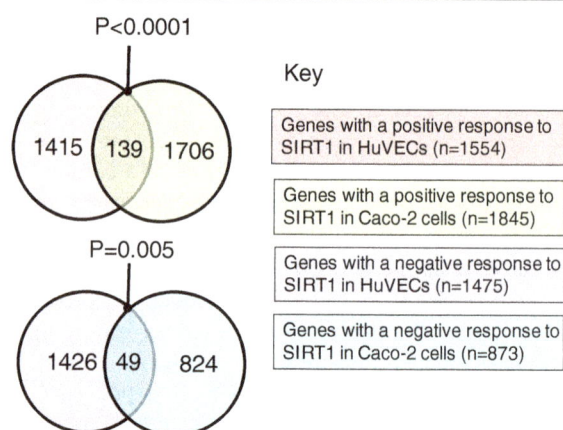

Fig. 2 Intersections between lists of genes that showed positive or negative DNA methylation responses to SIRT1. Data are shown for HuVECs and Caco-2 cells, as defined in the key. P values were derived using chi-square analysis applying Yates' correction

Table 1 Analysis of the size of intersections between polycomb group protein target genes (PCGTs) and genes with higher levels of DNA methylation

	HuVEC				Caco-2			
	Positive response to SIRT1		Negative response to SIRT1		Positive response to SIRT1		Negative response to SIRT1	
	RF	P	RF	P	RF	P	RF	P
Suz12 targets	1.5	<0.0001	1.5	=0.0008	1.6	<0.0001	1.5	=0.0088
Eed targets	1.4	=0.0102	1.1	=0.4573	1.8	<0.0001	1.5	=0.0347
Phc1 targets	1.4	=0.0069	1.3	<0.0001	1.6	<0.0001	1.7	=0.0004
Rnf2 targets	1.2	=0.0853	1.3	=0.0341	1.7	<0.0001	1.7	=0.0012
Targets of all polycomb group proteins	1.6	<0.0001	1.1	=0.6135	2.2	<0.0001	1.4	=0.1355
Targets of at least one polycomb group protein	1.3	=0.0090	1.3	=0.0046	1.5	<0.0001	1.6	<0.0001

Analysis of the size of intersections between polycomb group protein target genes (PCGTs) and genes with higher levels of DNA methylation under control conditions and/or reduced levels of DNA methylation under conditions of SIRT1 knockdown (positive response to SIRT1) or genes with reduced levels of DNA methylation under control conditions and/or higher levels of DNA methylation under conditions of SIRT1 knockdown (negative response to SIRT1). PCGT lists were compiled from published data (Boyer et al. 2006). RF (representation factor) values show the ratio of observed to expected number of genes in the intersection. P values were derived using chi-square analysis applying Yates' correction. Italicized cells indicate where data did not reach statistical significance

from both cell lines (with the exception of Eed in HuVECs). Genes identified as targets of ALL components of PRC1 and PRC2 [30], however, were not enriched in the lists of genes that responded negatively to SIRT1.

Polycomb group protein mRNA levels are not affected by SIRT1 manipulation

The chromatin modifications that repress PCGTs in stem cells result partly from actions of the polycomb group proteins themselves to effect epigenetic modification [31], including DNA methylation [32]. Thus, we proposed that SIRT1 may affect DNA methylation at PCGTs by changing the level of expression of polycomb group proteins. We investigated this hypothesis by determining the effect of SIRT1 knockdown in HuVECs and Caco-2 cells on the relative level of mRNA for individual polycomb group proteins (components of PRC1 and PRC2). We also determined if SIRT1 knockdown affected the mRNA for the histone demethylase KDM2B, which has been shown to recruit PRC1 to CpG islands [33]. None of the mRNAs measured was changed consistently when SIRT1 expression was reduced using both siRNAs (separately) (Fig. 3). Increases in SUZ12, EZH2, BMI1 and PHC1 mRNAs in HuVECs and in RNF2 mRNA in Caco-2 cells were observed using only one of the two siRNAs in each instance. Given that the second siRNA was equally effective in reducing SIRT1 expression, then these responses cannot be attributed to SIRT1 knockdown. Off-target effects of the siRNA on genes that influence the expression of these polycomb group protein mRNAs is a possible explanation for these observations. We thus found no evidence to support the idea that SIRT1 affects DNA methylation at PCGTs by effects on the expression of PRC components.

SIRT1 does not affect the quantity of EZH2 protein in the cell nor associate with EZH2

It has been shown that the histone methyltransferase EZH2 (a component of PRC2) associates with DNA methyltransferases (DNMTs) and is necessary to recruit DNMTs to EZH2-repressed genes [32]. Also, an intermolecular association between recombinant SIRT1 and EZH2 was observed in HeLa cells. [34]. We thus reasoned that effects on EZH2 were a likely point of action through which SIRT1 affects DNA methylation at PCGTs. To explore further if SIRT1 affects the expression of EZH2, we determined by Western blotting if SIRT1 knockdown affected EZH2 protein expression in Caco-2 cells and saw no effect (Fig. 4). We also investigated if EZH2 co-immunoprecipitated with SIRT1. We achieved successful immunoprecipitation of both SIRT1 and EZH2 from Caco-2 cells and HuVECs but detected no EZH2 in the protein fraction enriched using anti-SIRT1 antibody and no SIRT1 in the protein fraction enriched using anti-EZH2 antibody (Fig. 5). We thus found no evidence that SIRT1 and EZH2 form an intermolecular complex in these cell lines.

Discussion

Our findings make an important contribution to key developments in understanding how age-related changes in DNA methylation contribute to the process of ageing, a field where the importance of PCGTs is just beginning to emerge. Salient points are that (1) the DNA methylation signatures of a mixed cell population from human blood during ageing and mouse intestinal cells mimicked features common to both stem cells and cancer with respect to PCGTs [4, 7]; (2) changes in DNA methylation during the ageing of haematopoietic stem cells clustered

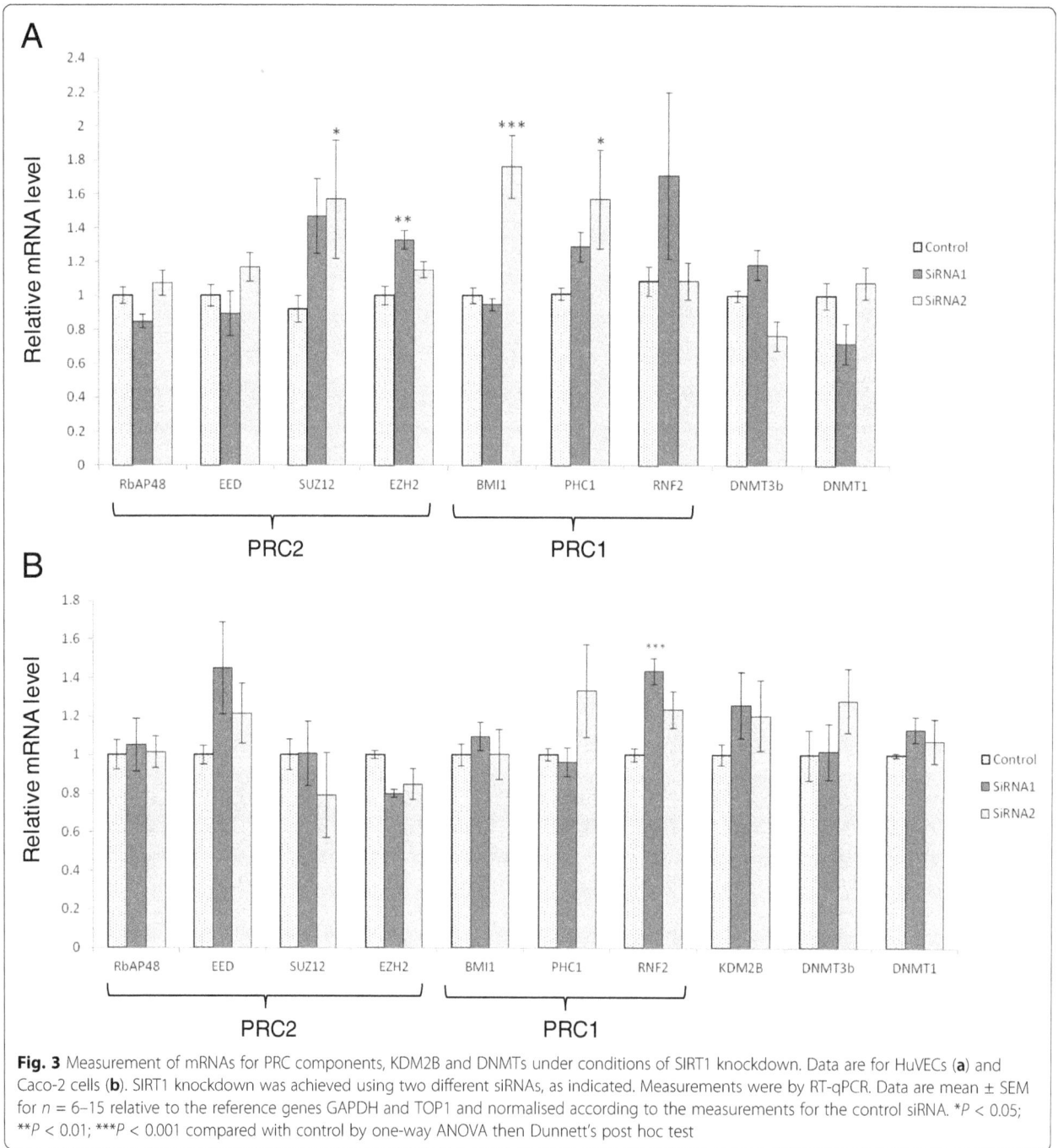

Fig. 3 Measurement of mRNAs for PRC components, KDM2B and DNMTs under conditions of SIRT1 knockdown. Data are for HuVECs (**a**) and Caco-2 cells (**b**). SIRT1 knockdown was achieved using two different siRNAs, as indicated. Measurements were by RT-qPCR. Data are mean ± SEM for $n = 6$–15 relative to the reference genes GAPDH and TOP1 and normalised according to the measurements for the control siRNA. *$P < 0.05$; **$P < 0.01$; ***$P < 0.001$ compared with control by one-way ANOVA then Dunnett's post hoc test

particularly at PCGTs [3]; (3) reversal of age-driven accumulation of DNA methylation changes in stem cells ("epigenetic rejuvenation") may reverse the ageing process [10]. An effect of SIRT1 on DNA methylation at PCGTs is a highly novel and important finding and is pertinent to evaluating how the impact of SIRT1 on pleiotropic cellular processes may affect healthspan.

We made this fundamentally important discovery concerning actions of SIRT1 in two different cell lines using approaches based on different principles both to enrich

DNA for the methylated fraction and for downstream detection of the enriched sequences. The discovery is thus highly robust, and we make no attempt to confirm effects by direct measurement of DNA methylation at specific PCGTs. In contrast to techniques we could apply for such measurements, neither approach we used reports on DNA methylation at specific CpG sites. Both approaches sample the total level of DNA methylation across a fragment of DNA, whereas approaches to targeted measurements (such as pyrosequencing, which we

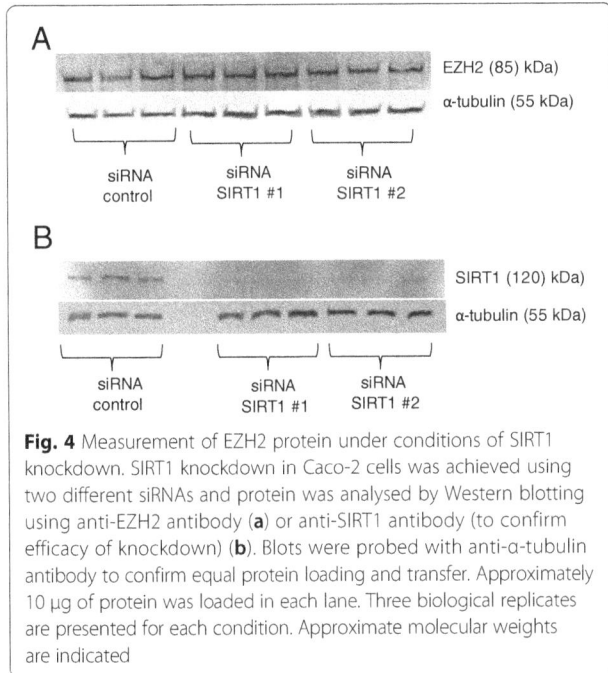

Fig. 4 Measurement of EZH2 protein under conditions of SIRT1 knockdown. SIRT knockdown in Caco-2 cells was achieved using two different siRNAs and protein was analysed by Western blotting using anti-EZH2 antibody (**a**) or anti-SIRT1 antibody (to confirm efficacy of knockdown) (**b**). Blots were probed with anti-α-tubulin antibody to confirm equal protein loading and transfer. Approximately 10 μg of protein was loaded in each lane. Three biological replicates are presented for each condition. Approximate molecular weights are indicated

use routinely [16]), sample only a very limited number of CpG sites. Thus, failure of targeted approaches to validate findings based on genome-wide analysis may reflect sampling of unaffected CpG sites in the vicinity of affected sites. Despite these caveats, we did observe that 4 of a panel of 10 genes (*IRX3*, *PTPRG*, *STK10* and *KLF3*) we found previously to show differential DNA methylation when the expression of SIRT1 was manipulated in

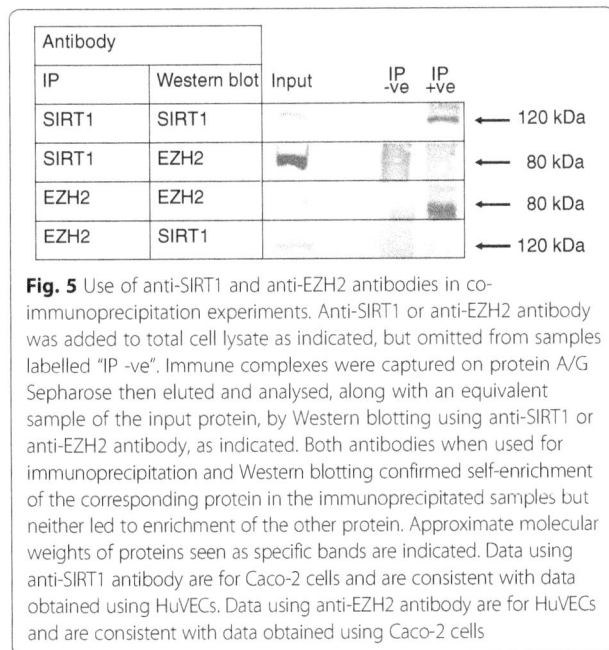

Fig. 5 Use of anti-SIRT1 and anti-EZH2 antibodies in co-immunoprecipitation experiments. Anti-SIRT1 or anti-EZH2 antibody was added to total cell lysate as indicated, but omitted from samples labelled "IP -ve". Immune complexes were captured on protein A/G Sepharose then eluted and analysed, along with an equivalent sample of the input protein, by Western blotting using anti-SIRT1 or anti-EZH2 antibody, as indicated. Both antibodies when used for immunoprecipitation and Western blotting confirmed self-enrichment of the corresponding protein in the immunoprecipitated samples but neither led to enrichment of the other protein. Approximate molecular weights of proteins seen as specific bands are indicated. Data using anti-SIRT1 antibody are for Caco-2 cells and are consistent with data obtained using HuVECs. Data using anti-EZH2 antibody are for HuVECs and are consistent with data obtained using Caco-2 cells

Caco-2 cells [16] were also detected as differentially methylated in the current analysis.

We reported previously the effect of SIRT1 knockdown in Caco-2 cells on the transcriptome [16]. Comparison of the list of genes that underwent DNA hypermethylation or hypomethylation in response to SIRT1 knockdown with the list of genes for which we detected a parallel change in expression revealed no significant correlation. This finding is consistent with a wider body of published data that reveals at best a weak correlation between effects on DNA methylation and gene expression. For example, correlation between age-related changes in genome-wide DNA methylation in haematopoietic stem cells, which clustered at genes regulated by PRC2, and changes in gene expression was low [3], suggesting that effects are manifest at the level of the transcriptome only when passed on to downstream progeny or indirectly. Furthermore, studies in diverse cell types have revealed that there is generally little correlation between changes in genome-wide DNA methylation and gene expression [35–37]. Reported weak correlations between DNA methylation and gene expression were more pronounced for cell lineage-specific genes where DNA methylation changes were in regulatory elements [3, 35]. The resolution of our current data does not allow the identification of DNA methylation changes that are specifically within regulatory elements. Thus, attempting to validate our data on SIRT1-driven effects on DNA methylation at PCGTs by measurement of the response at the RNA level of these genes would thus be of limited value.

A future priority should be to uncover in detail the mechanism through which SIRT1 affects DNA methylation at PCGTs. Action mediated through the polycomb group proteins is a highly plausible suggestion, given that the PRCs affect the epigenetic status, including DNA methylation, of PCGTs [31, 32]. We found no evidence that the level of SIRT1 in the cell affects expression at the mRNA level of any of the components of PRC1 or PRC2 or of KDM2B, which targets PRC1 to CpG islands [33]. Our data do not exclude the possibility that SIRT1 affects polycomb group protein expression downstream of mRNA, and thus, measurement of the effect of SIRT1 knockdown on polycomb group protein level (e.g. by Western blotting) should be a future priority. Of the multiple components of the PRCs, EZH2 is arguably the most likely candidate as the point at which SIRT1 interacts to modify actions of the PRCs on DNA methylation at their gene targets because EZH2 has been shown to control DNA methylation through association with DNMTs [32]. Also, direct intermolecular association between recombinant, epitope-tagged SIRT1 and EZH2 was observed in HeLa cells [34]. Moreover, trimethylation of H3K27 by EZH2 is an early event in the sequence of epigenetic modifications that results from PRC binding to PCGTs and leads to recruitment of PRC1 through chromodomain-containing components [31]. However, we detected no intermolecular association between SIRT1 and EZH2. The interaction observed in HeLa cells may require the expression of the two proteins at higher levels (as was the case in this previous work, by virtue of expression of recombinant proteins from transgenes) or may be cell-line specific. Moreover, the earlier work showed that SIRT1 is not a component of PRC2 but associates with polycomb proteins in PRC4, a specific PRC containing isoform 2 of EED [34]. EED2 is expressed specifically in undifferentiated pluripotent cells and also cancer cells. The same work showed a direct association between SIRT1 and SUZ12 using purified recombinant proteins. A priority for future work is to determine if SIRT1 interacts directly with other components of the PRCs and to determine by ChIP if SIRT1 binds directly to PCGTs or if its effects on DNA methylation at these sites are indirect.

A factor to consider in interpreting the likely implications of age-associated changes in PCGT DNA methylation and the effects thereon of SIRT1 is the nature of the cell population sampled and/or analysed, specifically whether these be stem cells [3] or, principally, the differentiated progeny [4, 7]. Lack of stemness in stem cells or gain of stem cell-like features in the differentiated progeny could give rise to features of tissue ageing. We propose a speculative model, based on this premise, that can reconcile these observations on DNA methylation changes in ageing cells, including effects at PCGTs, with the observed effects thereon of SIRT1 being a counteracting mechanism (Fig. 6). We propose that the fidelity with which two daughter cells that arise from asymmetric stem cell division acquire the correct pattern of DNA methylation across the genome is compromised in ageing tissue. Viz, the DNA methylation pattern of the retained stem cell becomes more skewed towards that of the differentiated cell and vice versa. Epigenetic drift in a sample of ageing stem cells, therefore, will be towards the epigenome of the differentiated cell. In a mixed cell population, however, where differentiated cells predominate (including intestine [4], as used to compile our list, and leukocytes

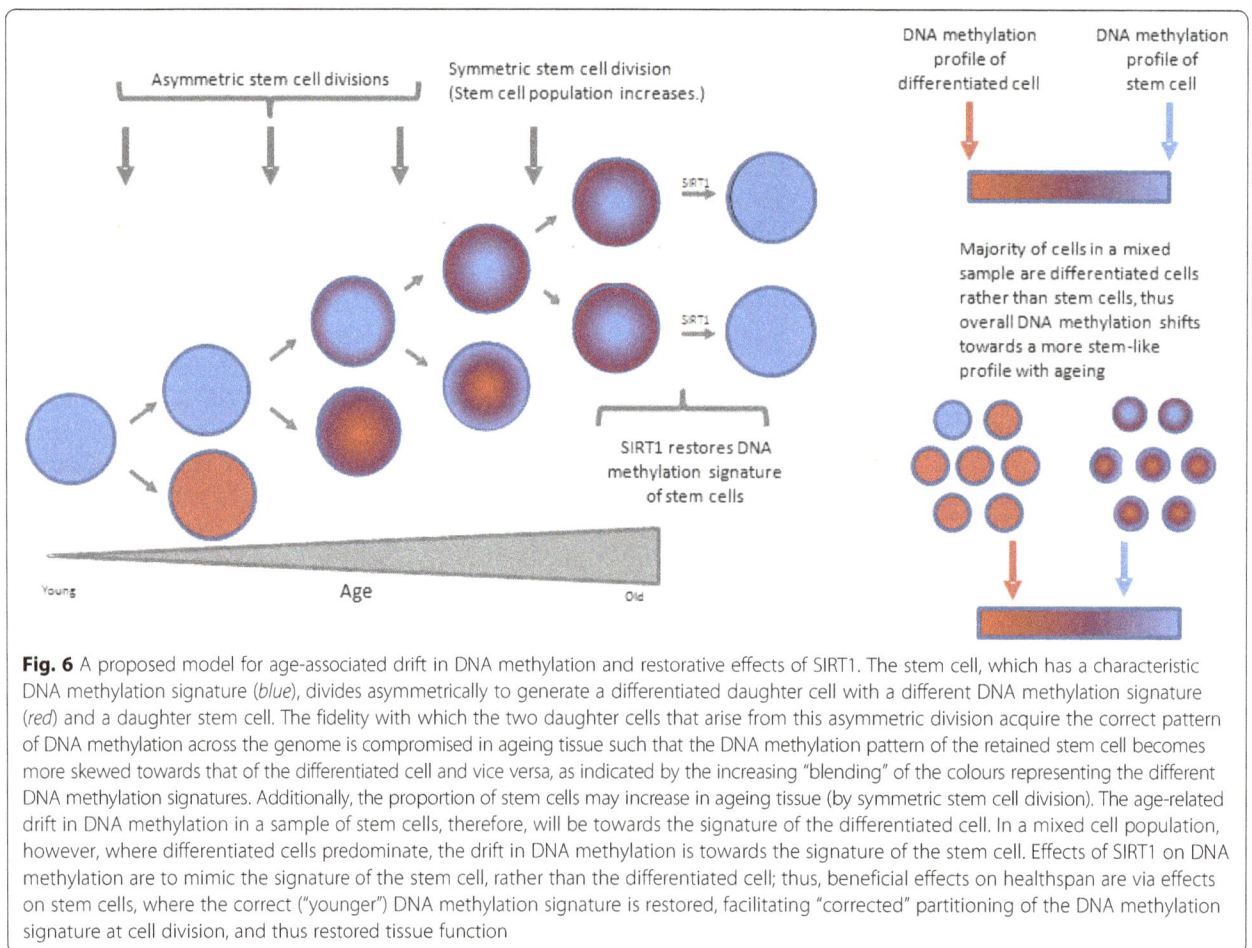

Fig. 6 A proposed model for age-associated drift in DNA methylation and restorative effects of SIRT1. The stem cell, which has a characteristic DNA methylation signature (*blue*), divides asymmetrically to generate a differentiated daughter cell with a different DNA methylation signature (*red*) and a daughter stem cell. The fidelity with which the two daughter cells that arise from this asymmetric division acquire the correct pattern of DNA methylation across the genome is compromised in ageing tissue such that the DNA methylation pattern of the retained stem cell becomes more skewed towards that of the differentiated cell and vice versa, as indicated by the increasing "blending" of the colours representing the different DNA methylation signatures. Additionally, the proportion of stem cells may increase in ageing tissue (by symmetric stem cell division). The age-related drift in DNA methylation in a sample of stem cells, therefore, will be towards the signature of the differentiated cell. In a mixed cell population, however, where differentiated cells predominate, the drift in DNA methylation is towards the signature of the stem cell. Effects of SIRT1 on DNA methylation are to mimic the signature of the stem cell, rather than the differentiated cell; thus, beneficial effects on healthspan are via effects on stem cells, where the correct ("younger") DNA methylation signature is restored, facilitating "corrected" partitioning of the DNA methylation signature at cell division, and thus restored tissue function

[7]), epigenetic drift will be towards the epigenome of the stem cell. The more stem-like DNA methylation signature reflects faltering of the differentiation process such that cells retain some of the signature of their stem progenitors. Additionally, or alternatively, as tissues age, the proportion of differentiated versus stem cells may shift towards there being a larger proportion of stem cells. Indeed, an expanded proliferative zone has been observed in the intestinal mucosa of older, compared with younger, rats and humans [38, 39], and the number of clonogenic cells per intestinal crypt on older mice, detected following irradiation damage, exceeded eightfold the number in younger mice [40]. We make the speculative suggestion that effects of SIRT1 on DNA methylation contribute to improved healthspan by restoring the DNA methylation profile of stem cells, thus enhancing tissue capacity to regenerate and restore function. We propose that SIRT1 has these actions through association with PRCs, which in turn modulates effects of these complexes on DNA methylation. Thus, establishing rigorously which polycomb proteins or combinations thereof co-immunprecipitate with SIRT1 and mapping the sites on the genome where these complexes bind are immediate future priorities.

Conclusions

We show that effects of SIRT1 on DNA methylation cluster at PCGTs. There is already robust evidence that these loci are also hot spots for age-related changes in DNA methylation. The discovery thus advances substantially our understanding of how the pleiotropic effects of SIRT1 may contribute to healthspan. Future research should explore the mechanisms that account for these effects of SIRT1 on DNA methylation at PCGTs and how these actions then affect stem cell biology. Such knowledge will point towards actions of SIRT1 whose mimicry by lifestyle or pharmaceutical interventions may contribute to a longer, healthier life.

Methods
Cell culture

Caco-2 cells were cultured under our standard laboratory conditions as described previously [41]. HuVECs (passage 3) were seeded into 75-cm^2 flasks at a density of approximately 1×10^6 cells per flask and maintained at 37 °C in a humidified atmosphere of 5 % CO_2 in air in EGM™ endothelial growth medium supplemented with EGM™-2 BulletKit™. All tissue culture reagents were supplied by Lonza. The medium was replaced twice weekly. Experiments were carried out at passage 5–6 in EGM™ endothelial growth medium supplemented with 2 % (*v/v*) fetal calf serum (Sigma) and 60 µg/ml gentamycin (Sigma).

Overexpression and knockdown of SIRT1

Overexpression of SIRT1 in both Caco-2 cells and HuVECs was achieved by transient transfection with the plasmid pCMV6-ENTRY-SIRT1 (Origene), and knockdown of SIRT1 was achieved using two different siRNAs and compared with a control siRNA, as described previously for Caco-2 cells [16]. Efficacy of overexpression and knockdown was confirmed by RT-qPCR and Western blotting for both cell lines as described previously for Caco-2 cells [16].

Preparation of DNA and enrichment for the methylated fraction

DNA was extracted from HuVECs and Caco-2 cells using the QIAamp® DNA Mini Kit (Qiagen). Enrichment of HuVEC DNA for the methylated fraction was carried out by NXT-DX, The Netherlands. DNA was fragmented to an average length of 200 bp by sonication and analysed using the Agilent 2100 Bioanalyser (Agilent Technologies). Methylated DNA was captured using the MethylCap kit (Diagenode), based on binding to methylated DNA of a recombinant H6-GST-MBD protein then captured on magnetic beads coated with GSH. Caco-2 DNA was fragmented by sonication and purified from agarose gels to produce DNA fragments ranging from 200–1000 bp in length. Purified fragmented DNA was then spiked with unmethylated and methylated internal controls generated from Lambda phage genomic DNA. To generate these samples, two different fragments of ~500 bp in length were amplified by PCR from Lambda phage dam$^-$ dcm$^-$ genomic DNA (Fermentas) using high specificity HotStar Taq DNA polymerase (Qiagen) and primer pairs AGCAACCAACAAGAAAACACT plus TCATCCTCGGCAAACTCTTT and GTGAGGTGAAT GTGGTGAAGT plus TCGCAGAGATAAAACACGCT. An aliquot of each PCR product was then methylated in vitro using *SssI* DNA methylase (New England Biolabs). The methylation status of the unmethylated and methylated controls was confirmed by digestion with the methylation-sensitive restriction enzyme *AciI* (Fermentas). Spiked DNA samples were denatured at 95 °C then incubated for 4 h at 4 °C with anti-5-methylcytidine (5mC) antibody (Eurogentec) (12 µl per 4.5 µg DNA) in 10 mM Na-Phosphate pH 7.0, 140 mM NaCl, 0.05 % Triton X-100. DNA was then captured using sheep anti-mouse IgG Dynabeads (Invitrogen), following the manufacturer's instructions. Dynabeads were then incubated with Proteinase K (Roche) at 50 °C for 2 h with shaking, and the beads were then removed using a magnetic rack. DNA was then purified using the DNA Clean and Concentrator 5 kit (Zymo Research). Whole genome amplification of input or immunoprecipitated DNA was then performed using the GenomePlex® Complete Whole Genome Amplification kit (Sigma), and retention of fragment size was confirmed by agarose gel electrophoresis.

Analysis of differences in the DNA content of input versus enriched pools

For each condition (SIRT1 overexpressed by transfection with pCMV6-ENTRY-SIRT1, corresponding vector control, each of the siRNAs that target SIRT1 and control siRNA), six biological replicates prepared from HuVECs were pooled for analysis. Sequence reads were mapped to hg19 using Bowtie2 [42] then filtered first to remove unmapped reads using Picard (http://picard.sourceforge.net/). The GENCODE gene set [43] was then used to identify genes, defined as within 2 kb of a TSS or within genes (between the TSS and stop codon). The package MEDIPS (Bioconductor) was used to identify differentially methylated genes, which were then classified as hypomethylated or hypermethylated when SIRT1 expression was increased or reduced.

The methylation profile of DNA from Caco-2 cells was measured using the DNA Methylation Service provided by NimbleGen. Input and methylation-enriched samples were labelled with Cy3 and Cy5, respectively, using the NimbleGen Dual-Colour DNA Labelling Kit then co-hybridised to the Human 3x720K CpG Island Plus RefSeq Promoter Array using the NimbleGen Hybridisation System. The hybridised arrays were then washed with the NimbleGen Wash Buffer Kit and dried with the NimbleGen Microarray Drier, and signal peak intensities for each probe were measured with the NimbleGen MS 200 Microarray Scanner. Data were provided as signal intensity and P value for each probe. These data were used to detect peaks through manual inspection in Excel (Microsoft), accepting positive enrichment being where at least two of three biological replicates of each sample analysed (SIRT1 overexpressed by transfection with pCMV6-ENTRY-SIRT1, corresponding vector control, each of the siRNAs that target SIRT1 and control siRNA) met a >2 P value minimum cutoff.

RNA measurement by RT-qPCR

RNA was prepared using the PureLink RNA MiniKit (Life Technologies) then reverse transcribed using SuperScript III Reverse Transcriptase (Life Technoloiges), following the manufacturer's instructions. Quantitative real-time PCR was performed in a Roche LightCycler 480 with 20 µl reactions set up in a 96-well format containing LightCycler SYBR Green I Master (Roche), 0.5 µM of each primer (see Additional file 2) and 1 µl of cDNA (reverse transcription reaction diluted 1:4). After denaturing for 5 min at 95 °C, 50 cycles were carried out using the following parameters: 95 °C, 10 s; 55 °C, 10 s; and 72 °C, 15 s. Levels of specific RNAs relative to control, and corrected according to levels of reference gene RNAs, were calculated using the $\Delta\Delta C_t$ method. PCR products were sequenced (Eurofins Genomics) to confirm identity.

Immunoprecipitation

SIRT1 or EZH2 and proteins bound to them were co-immunoprecipitated using the Classic IP kit (Pierce). Cells were rinsed with 1× PBS and lysed with lysis/wash buffer in volumes according to cell growth area, as recommended by the manufacturer. Lysate samples (1000 µg each) were pre-cleared with control agarose resin, followed by overnight incubation at 4 °C with either 10 µg SIRT1 antibody (Abcam, ab7343) or 10 µg EZH2 antibody (R&D Systems, AF4767). Negative controls were prepared in an identical manner but omitting antibody. Immune complexes were immobilised on protein A/G Sepharose columns, which were washed three times with lysis/wash buffer then once with conditioning buffer. Protein was then eluted in elution buffer and concentration was determined against BSA standards using Bradford reagent (Bio-Rad Laboratories).

Western blotting

Immunoprecipitated proteins or total cell protein extract prepared as described previously [16] (10 µg) were resolved by SDS-PAGE using 7 % gels then transferred by semi-dry blotting onto PVDF membrane (Amersham Hybond-P, GE Healthcare). Membranes were incubated at room temperature for 1 h with constant shaking first in Odyssey blocking buffer (LI-COR) then with primary antibody (SIRT1 (Abcam, ab7343) or EZH2 (R&D Systems, AF4767)) diluted 1:250 in Odyssey blocking buffer supplemented with 0.1 % (v/v) Tween-20. Membranes were washed five times for 15 min in 1× PBS plus 0.1 % (v/v) Tween-20 then incubated for 1 h at room temperature with secondary antibody (Odyssey LI-COR, IRDye® 680RD donkey anti-rabbit IgG to detect anti-SIRT1 and IRDye® 800CW donkey anti-goat IgG to detect EZH2) diluted 1:2000 in Odyssey blocking buffer supplemented with 0.1 % (v/v) Tween-20. The washing procedure was then repeated before scanning membranes on the Odyssey LI-COR infrared imaging system.

Statistical analysis

Intersections between gene lists were tested for statistical significance by chi-square analysis applying Yates' correction. Data derived by RT-qPCR were analysed by one-way ANOVA then Dunnett's post hoc test.

Availability of supporting data

The data sets supporting the results of this article are available in the Gene Expression Omnibus database (http://www.ncbi.nlm.nih.gov/geo/query/acc.cgi?acc=GSE54072 and http://www.ncbi.nlm.nih.gov/geo/query/acc.cgi?acc=GSE53569).

Abbreviations

MeDIP: methylated DNA immunoprecipitation; DNMT: DNA methyltransferase; PCGT: polycomb group protein target gene; PRC: polycomb repressive complex; RF: representation factor; SIRT1: sirtuin 1; TSS: transcription start site.

Competing interests

The authors declare that they have no competing interests.

Authors' contributions

LAW carried out DNA methylation analysis on HuVECs and supervised most other aspects of the laboratory work. LJI carried out DNA methylation analysis on Caco-2 cells. SME carried out immunoprecipitation experiments, western blotting for EZH2 and several of the RT-qPCR assays. TS, MD and EVH carried out RT-qPCR assays. JAM collaborated and supervised LJI to undertake genome-wide DNA methylation analyses. SJC provided bioinformatics support and analysed sequence data. GJ and LJW contributed to conception of the project and design of experiments. DF conceived and managed the project, analysed data and prepared the manuscript, with input from all other authors. All authors read and approved the final manuscript.

Acknowledgements

This study was funded by BBSRC (BB/E007457/1 and BB/G007993/1 and studentship to LJI), Unilever, SPARC and the Rank Prize Funds (studentship to SME).

Author details

[1]Institute for Cell and Molecular Biosciences, Human Nutrition Research Centre, Newcastle University Medical School, Newcastle upon Tyne NE2 4HH, UK. [2]Faculty of Medical Sciences, Newcastle University Medical School, Newcastle upon Tyne NE2 4HH, UK. [3]Institute of Health and Society, Human Nutrition Research Centre, Newcastle University Medical School, Newcastle upon Tyne NE2 4HH, UK. [4]Unilever R&D, Colworth Discover, Colworth Science Park, Sharnbrook, Bedfordshire MK44 1LQ, UK.

References

1. Bjornsson HT, Sigurdsson MI, Fallin MD, Irizarry RA, Aspelund T, Cui H, et al. Intra-individual change over time in DNA methylation with familial clustering. Jama. 2008;299(24):2877–83.
2. Richardson B. Impact of aging on DNA methylation. Ageing Res Rev. 2003;2(3):245–61.
3. Beerman I, Bock C, Garrison BS, Smith ZD, Gu H, Meissner A, et al. Proliferation-dependent alterations of the DNA methylation landscape underlie hematopoietic stem cell aging. Cell Stem Cell. 2013;12(4):413–25.
4. Maegawa S, Hinkal G, Kim HS, Shen L, Zhang L, Zhang J, et al. Widespread and tissue specific age-related DNA methylation changes in mice. Genome Res; 2010;20(3):332–40.
5. Hannum G, Guinney J, Zhao L, Zhang L, Hughes G, Sadda S, et al. Genome-wide methylation profiles reveal quantitative views of human aging rates. Mol Cell. 2013;49(2):359–67.
6. Heyn H, Li N, Ferreira HJ, Moran S, Pisano DG, Gomez A, et al. Distinct DNA methylomes of newborns and centenarians. Proc Natl Acad Sci U S A. 2012;109(26):10522–7.
7. Teschendorff AE, Menon U, Gentry-Maharaj A, Ramus SJ, Weisenberger DJ, Shen H, et al. Age-dependent DNA methylation of genes that are suppressed in stem cells is a hallmark of cancer. Genome Res. 2010;20(4):440–6.
8. Rakyan VK, Down TA, Maslau S, Andrew T, Yang TP, Beyan H, et al. Human aging-associated DNA hypermethylation occurs preferentially at bivalent chromatin domains. Genome Res. 2010;20(4):434–9.
9. Teschendorff AE, West J, Beck S. Age-associated epigenetic drift: implications, and a case of epigenetic thrift? Hum Mol Genet. 2013;22(R1):R7–15.
10. Rando TA, Chang HY. Aging, rejuvenation, and epigenetic reprogramming: resetting the aging clock. Cell. 2012;148(1–2):46–57.
11. Greer EL, Maures TJ, Ucar D, Hauswirth AG, Mancini E, Lim JP, et al.
Transgenerational epigenetic inheritance of longevity in Caenorhabditis elegans. Nature. 2011;479(7373):365–71.
12. Lee TI, Jenner RG, Boyer LA, Guenther MG, Levine SS, Kumar RM, et al. Control of developmental regulators by Polycomb in human embryonic stem cells. Cell. 2006;125(2):301–13.
13. Widschwendter M, Fiegl H, Egle D, Mueller-Holzner E, Spizzo G, Marth C, et al. Epigenetic stem cell signature in cancer. Nat Genet. 2007;39(2):157–8.
14. Ohm JE, McGarvey KM, Yu X, Cheng L, Schuebel KE, Cope L, et al. A stem cell-like chromatin pattern may predispose tumor suppressor genes to DNA hypermethylation and heritable silencing. Nat Genet. 2007;39(2):237–42.
15. Schlesinger Y, Straussman R, Keshet I, Farkash S, Hecht M, Zimmerman J, et al. Polycomb-mediated methylation on Lys27 of histone H3 pre-marks genes for de novo methylation in cancer. Nat Genet. 2007;39(2):232–6.
16. Ions LJ, Wakeling LA, Bosomworth HJ, Hardyman JE, Escolme SM, Swan DC, et al. Effects of Sirt1 on DNA methylation and expression of genes affected by dietary restriction. Age. 2013;35:1835–49.
17. Guarente L. The many faces of sirtuins: Sirtuins and the Warburg effect. Nat Med. 2014;20(1):24–5.
18. Imai S, Guarente L. NAD+ and sirtuins in aging and disease. Trends Cell Biol. 2014;24(8):464–71.
19. Satoh A, Brace CS, Rensing N, Cliften P, Wozniak DF, Herzog ED, et al. Sirt1 extends life span and delays aging in mice through the regulation of Nk2 homeobox 1 in the DMH and LH. Cell Metab. 2013;18(3):416–30.
20. Burnett C, Valentini S, Cabreiro F, Goss M, Somogyvari M, Piper MD, et al. Absence of effects of Sir2 overexpression on lifespan in C. elegans and Drosophila. Nature. 2011;477(7365):482–5.
21. Banerjee KK, Ayyub C, Ali SZ, Mandot V, Prasad NG, Kolthur-Seetharam U. dSir2 in the adult fat body, but not in muscles, regulates life span in a diet-dependent manner. Cell Reports. 2012;2(6):1485–91.
22. Guarente L. Calorie restriction and sirtuins revisited. Genes Dev. 2013;27(19):2072–85.
23. Guarente L, Picard F. Calorie restriction—the SIR2 connection. Cell. 2005;120(4):473–82.
24. Donmez G, Guarente L. Aging and disease: connections to sirtuins. Aging Cell. 2010;9(2):285–90.
25. Libert S, Guarente L. Metabolic and neuropsychiatric effects of calorie restriction and sirtuins. Annu Rev Physiol. 2013;75:669–84.
26. Gene Expression Omnibus GSE54072 [http://www.ncbi.nlm.nih.gov/geo/query/acc.cgi?acc=GSE54072]
27. Lisanti S, von Zglinicki T, Mathers JC. Standardization and quality controls for the methylated DNA immunoprecipitation technique. Epigenetics Official J DNA Methylation Soc. 2012;7(6):615–25.
28. Gene Expression Omnibus GSE53569 [http://www.ncbi.nlm.nih.gov/geo/query/acc.cgi?acc=GSE53569]
29. Figshare [http://figshare.com/articles/PCGT_DNA_methylation_by_SIRT1/1142611]
30. Boyer LA, Plath K, Zeitlinger J, Brambrink T, Medeiros LA, Lee TI, et al. Polycomb complexes repress developmental regulators in murine embryonic stem cells. Nature. 2006;441(7091):349–53.
31. Schwartz YB, Pirrotta V: A new world of Polycombs: unexpected partnerships and emerging functions. Nature reviews, 14(12):853–864.
32. Vire E, Brenner C, Deplus R, Blanchon L, Fraga M, Didelot C, et al. The Polycomb group protein EZH2 directly controls DNA methylation. Nature. 2006;439(7078):871–4.
33. He J, Shen L, Wan M, Taranova O, Wu H, Zhang Y. Kdm2b maintains murine embryonic stem cell status by recruiting PRC1 complex to CpG islands of developmental genes. Nat Cell Biol. 2013;15(4):373–84.
34. Kuzmichev A, Margueron R, Vaquero A, Preissner TS, Scher M, Kirmizis A, et al. Composition and histone substrates of polycomb repressive group complexes change during cellular differentiation. Proc Natl Acad Sci U S A. 2005;102(6):1859–64.
35. Bock C, Beerman I, Lien WH, Smith ZD, Gu H, Boyle P, et al. DNA methylation dynamics during in vivo differentiation of blood and skin stem cells. Mol Cell. 2012;47(4):633–47.
36. Challen GA, Sun D, Jeong M, Luo M, Jelinek J, Berg JS, et al. Dnmt3a is essential for hematopoietic stem cell differentiation. Nat Genet. 2012;44(1):23–31.
37. Deaton AM, Bird A. CpG islands and the regulation of transcription. Genes Dev. 2011;25(10):1010–22.

38. Holt PR, Yeh KY. Small intestinal crypt cell proliferation rates are increased in senescent rats. J Gerontol. 1989;44(1):B9–14.
39. Roncucci L, Ponz DeLeon M, Scalmati A, Malagoli G, Pratissoli S, Perini M, et al. The influence of age on colonic epithelial cell proliferation. Cancer. 1988;62(11):2373–7.
40. Martin K, Potten CS, Roberts SA, Kirkwood TB. Altered stem cell regeneration in irradiated intestinal crypts of senescent mice. J Cell Sci. 1998;111(Pt 16):2297–303.
41. Cragg RA, Christie GR, Phillips SR, Russi RM, Kury S, Mathers JC, et al. A novel zinc-regulated human zinc transporter, hZTL1, is localized to the enterocyte apical membrane. J Biol Chem. 2002;277(25):22789–97.
42. Langmead B, Salzberg SL. Fast gapped-read alignment with Bowtie 2. Nat Methods. 2012;9(4):357–9.
43. Harrow J, Frankish A, Gonzalez JM, Tapanari E, Diekhans M, Kokocinski F, et al. GENCODE: the reference human genome annotation for The ENCODE Project. Genome Res. 2012;22(9):1760–74.

6

Utility and limitations of exome sequencing as a genetic diagnostic tool for conditions associated with pediatric sudden cardiac arrest/sudden cardiac death

Mindy H. Li[1,3]*, Jenica L. Abrudan[1,3], Matthew C. Dulik[1,3], Ariella Sasson[4], Joshua Brunton[1,3], Vijayakumar Jayaraman[1,3], Noreen Dugan[1,2], Danielle Haley[1,2], Ramakrishnan Rajagopalan[5,6], Sawona Biswas[5], Mahdi Sarmady[4], Elizabeth T. DeChene[5,6], Matthew A. Deardorff[1,3], Alisha Wilkens[1,3], Sarah E. Noon[1,3], Maria I. Scarano[1,3], Avni B. Santani[5,6], Peter S. White[1,4,7,8], Jeffrey Pennington[4], Laura K. Conlin[5,6], Nancy B. Spinner[5,6], Ian D. Krantz[1,3] and Victoria L. Vetter[1,2]

Abstract

Background: Conditions associated with sudden cardiac arrest/death (SCA/D) in youth often have a genetic etiology. While SCA/D is uncommon, a pro-active family screening approach may identify these inherited structural and electrical abnormalities prior to symptomatic events and allow appropriate surveillance and treatment. This study investigated the diagnostic utility of exome sequencing (ES) by evaluating the capture and coverage of genes related to SCA/D.

Methods: Samples from 102 individuals (13 with known molecular etiologies for SCA/D, 30 individuals without known molecular etiologies for SCA/D and 59 with other conditions) were analyzed following exome capture and sequencing at an average read depth of 100X. Reads were mapped to human genome GRCh37 using Novoalign, and post-processing and analysis was done using Picard and GATK. A total of 103 genes (2,190 exons) related to SCA/D were used as a primary filter. An additional 100 random variants within the targeted genes associated with SCA/D were also selected and evaluated for depth of sequencing and coverage. Although the primary objective was to evaluate the adequacy of depth of sequencing and coverage of targeted SCA/D genes and not for primary diagnosis, all patients who had SCA/D (known or unknown molecular etiologies) were evaluated with the project's variant analysis pipeline to determine if the molecular etiologies could be successfully identified.

Results: The majority of exons (97.6 %) were captured and fully covered on average at minimum of 20x sequencing depth. The proportion of unique genomic positions reported within poorly covered exons remained small (4 %). Exonic regions with less coverage reflect the need to enrich these areas to improve coverage. Despite limitations in coverage, we identified 100 % of cases with a prior known molecular etiology for SCA/D, and analysis of an additional 30 individuals with SCA/D but no known molecular etiology revealed a diagnostic answer in 5/30 (17 %). We also demonstrated 95 % of 100 randomly selected reported variants within our targeted genes would have been picked up on ES based on our coverage analysis.

Conclusions: ES is a helpful clinical diagnostic tool for SCA/D given its potential to successfully identify a molecular diagnosis, but clinicians should be aware of limitations of available platforms from technical and diagnostic perspectives.

* Correspondence: lim@email.chop.edu
[1]Department of Pediatrics, Perelman School of Medicine at the University of Pennsylvania, Philadelphia, PA, USA
[3]Division of Human Genetics, The Children's Hospital of Philadelphia, Abramson Research Center, Room 1012G, 3615 Civic Center Blvd, Philadelphia, PA 19104, USA
Full list of author information is available at the end of the article

Background

The rapid development of genomic sequencing and techniques such as massively parallel next-generation sequencing has decreased cost, improved efficiency, and increased the clinical and research use of genetic testing [1, 2]. Exome sequencing (ES), or sequencing the protein-coding portions of a human genome, has become an increasingly utilized approach for investigating Mendelian disorders [3]. Studies report varying diagnostic ES success rates, ranging from 22.8 % [4] to 50 % [5]. As costs continue to decline, it is likely the use of whole genome-sequencing will increase [6]. The application of exome and genome-level sequencing raises many challenges both from a technical execution and diagnostic standpoint [7, 8], and the best use of this testing in clinical practice remains unclear [1]. The role of genetic testing as a tool to investigate cardiovascular disease has had increased focus in recent years [9].

Sudden cardiac arrest/death (SCA/D) is uncommon in the young and occurs in an estimated 2000 individuals under 25 years of age annually in the US [10]. Causes include inherited structural, functional, and electrical cardiac abnormalities [11–13]. There may be no significant previous medical history prior to the occurrence of SCA/D, and standard postmortem analysis may be unrevealing [12, 14, 15] in as many as 10-30 % of cases [16]. A pro-active family screening approach is important to provide life-saving treatment and to help identify other affected members due to the high association of genetic causes [12, 13]. More than 100 genes have been associated with SCA/D [11, 17]. Guidelines for genetic testing for channelopathies and cardiomyopathies, were published in 2011 [18].

Increased accessibility to ES warrants its examination as a possible front line diagnostic tool for inherited conditions associated with SCA/D. Compared with targeted gene sequencing and comprehensive panels specific for disease, which are currently available [11], there are important differences to consider with ES. The workflow and challenges of completing ES are described in detail by Bamshad et al. [3]. Importantly, only 1-2 % of the human genome contains protein–coding sequences [3, 6, 19]; thus, these regions must undergo an exon-targeting "*capture*" process before being sequenced. Traditional sequencing methods for individual genes or panels do not require this step.

A second consideration is determining the "*coverage*" of these captured regions. Coverage (also known as "*depth*" of sequencing) refers to how many times a nucleotide meeting criteria for being a high-quality base is represented in a random collection of raw sequences [20]. This helps differentiate sequencing errors from a true sequence variant; the higher the coverage, the more likely the captured base is accurate and not a false read due to technical errors. For example, a captured base "T" with 20x coverage means that base is represented at least 20 times at that position in multiple raw sequences. Of note, "coverage", in addition to meaning *depth* of sequencing, may also refer to the general *proportion* of bases covered in genomic sequence at a specific depth [21]. For example, an exon with 90 % coverage at 20x means 90 % of the bases in that exon are represented at least 20 times on multiple raw sequences. In our manuscript, *depth* of sequencing will refer to the number of times a nucleotide is represented, and *coverage* will refer to the general proportion of genomic sequence covered unless otherwise specified.

Figure 1 outlines a basic schematic of a hypothetical analysis of patients carrying pathogenic SCA/D variants via ES and the steps that potentially can lead to missed calls. Though there is much debate regarding the use and implications of ES in a clinical setting, there is little information available regarding the capture and coverage of cardiac genes commonly found within comprehensive panels, which has implications for diagnostics and ability to find potential pathogenic mutations. The main objective of this study was to investigate the utility of ES as a potential diagnostic tool by investigating capture and depth of coverage of a group of targeted genes related to SCA/D.

Methods

Study population

This research study was approved by The Children's Hospital of Philadelphia (CHOP) Institutional Review Board (IRB). Samples from 102 pediatric individuals were enrolled under an IRB approved protocol of informed consent at The Children's Hospital of Philadelphia and de-identified. Detailed demographic information of individuals was not readily available due to the de-identification process. Of these 102 patients, thirteen had known molecular diagnosis for SCA/D, 59 had known molecular etiologies for other conditions including hearing loss, intellectual disability and mitochondrial disease (all known diagnoses were identified by Clinical Laboratory Improvement Amendments certified laboratories), and 30 individuals did not have known molecular etiologies for SCA/D.

Exome capture, sequencing, and bioinformatics

Peripheral blood from patients was collected in sterile EDTA tubes (BD vacutainer) at Phlebotomy, Children's Hospital of Philadelphia. Blood tubes were stored immediately at 4 °C and Genomic DNA (gDNA) was manually extracted using standard procedures with the Gentra Puregene Blood Kit Plus (Qiagen, 158489). gDNA quality was assessed on an agarose gel, Nanodrop spectrophotometer and quantified via the Qubit system. 3-6ug of gDNA from each sample were prepared and sent to the Beijing Genomics Institute (BGI) facility at CHOP. Exome capture was done with Agilent SureSelect V4, and whole-exome sequencing was completed on Illumina Hi-Seq

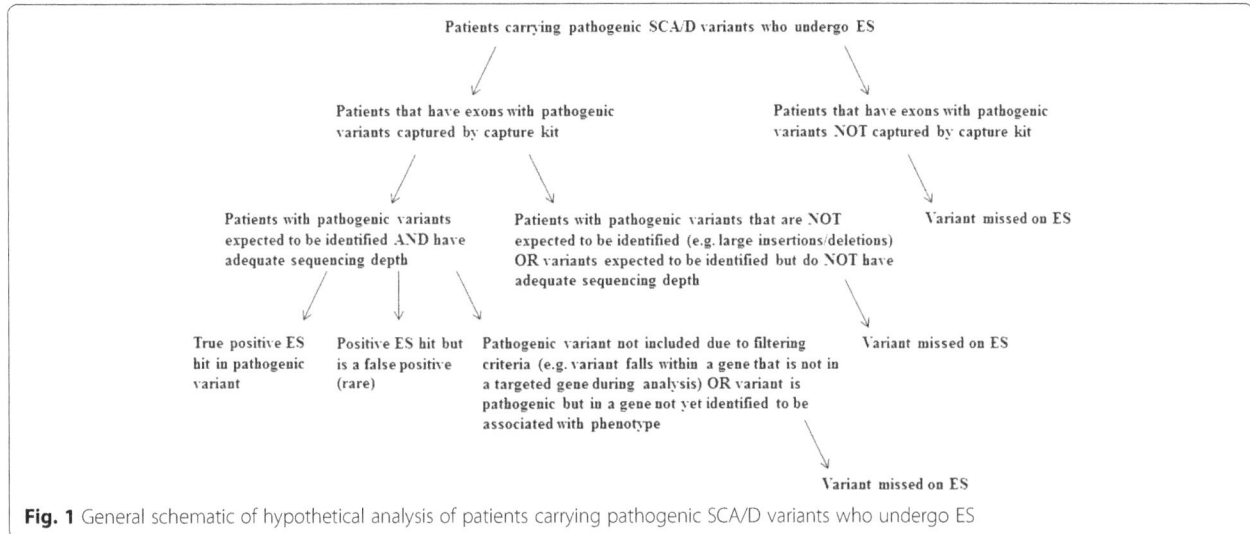

Fig. 1 General schematic of hypothetical analysis of patients carrying pathogenic SCA/D variants who undergo ES

2000 sequencers at an average coverage depth of 100X. Sequencing reads were obtained in FASTQ format and were examined via the Pediatric Genetic Sequencing Project (PediSeq) exome sequence coverage analysis pipeline.

The sequence reads were mapped to human genome assembly, GRCh37.p10, using Novoalign (V3.00.02) (www.novocraft.com), which has been shown to optimize alignment [22]. Coverage statistics per exons in the SCA/D genes bed file were generated using GATK Depth of Coverage tool version 2.2. Quality control steps to filter out poorly mapped reads included minBaseQuality 20/minMapping Quality 20 settings during variant calling and removal of variants with a minimum depth of coverage of less than 20 reads. Multi-sample variant calling was not done as patients are clinically evaluated and analyzed independently. Refer to Additional file 1 for further details regarding the sequencing protocol and data processing.

The primary aim of this study was to evaluate the adequacy of depth of sequencing, coverage of targeted SCA/D genes, and platform efficacy, not to identify individual molecular diagnoses. Therefore, the main *coverage* analysis was completed on the first 72 samples, regardless of their underlying clinical findings and molecular etiologies. Of these 72 samples, additional analysis (described at the end of the methods section) was completed on the 13 cases with known molecular causes for SCA/D. In addition, *diagnostic* analysis was performed on 30 patients without known causes for their SCA/D. 15 of these 30 patients (50 %) had variable prior genetic workup that was non-diagnostic.

A list of 103 genes associated with SCA/D (Table 1) was manually curated and used as a primary filter for analysis. To determine the full scope of variants that could be missed using ES, the first step was to determine the total number of exons present within the targeted genes, and within those exons, determine how many variants have

been previously reported and suspected to be pathogenic. Information regarding reported variants can be found in the Human Gene Mutation Database (HGMD), a comprehensive database that aims to compile nuclear gene germline mutations that have been associated with human disease [23]. We recognize not all variants reported in HGMD may be considered pathogenic, but for the purposes of this study any variant reported in HGMD was considered a potentially disease causing change that would require further review if picked up on ES.

We then evaluated the proportion of different variant types present within the 2,190 exons. This was critical as changes such as large insertions or deletions do not have specific genomic coordinates, and locating these changes can be problematic due to current limitations of ES technology. As large insertions and deletions were unlikely to be picked up with ES without separate and additional computational analyses, they were not included in the final analysis.

The next round of analysis focused on three main aspects: 1) Examining how well the exons in the targeted genes were *captured* on the Agilent SureSelect V4 platform, 2) Of the exons that were captured, how adequate was the *depth* of sequencing of these exons, using 20x (when a nucleotide on average is represented at least 20 times in a group of random raw sequences) as our standard for defining adequate depth of sequencing, and 3) Of the captured exons, what proportion of the exons met criteria for adequate sequencing *coverage* (percentage of bases within the exons that are sequenced at an appropriate read depth, which in this case was 20x). Coverage scores of all 2,190 exons were obtained for each sample individually, and then data for each exon was averaged across all samples.

Although the primary objective of this study was to evaluate the adequacy of depth of sequencing and

Table 1 Curated List of 103 Genes Associated with SCA/D

ABCC9	CAV3	ELN	KCNE3	MYH6	RBM20	TCAP
ACTA2	CBS	EMD	KCNH2	MYH7	RYR2	TGFB3
ACTC1	COL3A1	EYA4	KCNJ2	MYL2	SCN1B	TGFBR1
ACTN2	COL5A1	FBN1	KCNJ5	MYL3	SCN3B	TGFBR2
AKAP9	COL5A2	FBN2	KCNJ8	MYLK2	SCN4B	TMEM43
ANK2	CRYAB	FHL2	KCNQ1	MYOZ2	SCN5A	TMPO
ANKRD1	CSRP3	FKTN	KRAS	NEXN	SDHA	TNNC1
BAG3	CTF1	GATAD1	LAMA4	NRAS	SGCD	TNNI3
BRAF	DES	GLA	LAMP2	PKP2	SHOC2	TNNT2
CACNA1B	DMD	GPD1L	RPSA	PLN	SLC25A4	TPM1
CACNA1C	DPP6	HRAS	LDB3	PRKAG2	SLC2A10	TTN
CACNA2D1	DSC2	JPH2	LMNA	PSEN1	SMAD3	TTR
CACNB2	DSG2	JUP	MAP2K1	PSEN2	SNTA1	VCL
CALR3	DSP	KCNE1	MYBPC3	PTPN11	SOS1	
CASQ2	DTNA	KCNE2	MYH11	RAF1	TAZ	

coverage of targeted SCA/D genes and not for primary diagnosis, all patients enrolled in the study who had known molecular etiologies for SCA/D were evaluated with the project's variant analysis pipeline to determine if the molecular etiologies could be successfully identified. Project members completing this analysis were blinded to the known molecular diagnosis of the patients to avoid bias during the evaluation process. As this group of patients was relatively small ($n = 13$), additional variants within the targeted genes associated with SCA/D were selected and evaluated for depth of sequencing and coverage to determine how well a random number of potentially disease related mutations would be picked up on ES. One hundred variants reported in the Human Gene Mutation Database (HGMD) that were within the 103 genes associated with SCA/D were randomly selected for analysis to ensure a varied distribution. Statistics regarding the capture and depth of sequencing on those specific variants were generated using GATK Depth of Coverage tool version 2.2.

Beyond the 72 individuals used in the primary coverage analysis, the additional 30 patients with a history of SCA/D but no known molecular causes were analyzed using the same variant analysis pipeline to determine a diagnostic yield. Results were deemed "positive" if there were variants in genes related to SCA/D categorized as likely pathogenic or pathogenic, "uncertain" if there were only variants in genes related to SCA/D categorized as variants of uncertain significance (VUS), and "negative" if there were no VUS, suspected pathogenic, or pathogenic variants identified in genes related to SCA/D.

Results

Within the 103 targeted genes associated with SCA/D, there were 2,190 exons present (Fig. 2a). Within these exons, the total number of reported variants suspected to be disease causing in HGMD was 11,452 (Fig. 3). Of the 11,452 variants reported in HGMD in our targeted exons, 1,896 (16.6 %) were large deletions or insertions, and 9,556 (83.4 %) were variants with a reported genomic position (Fig. 3). Of these variants with a coordinate, 8,578 (89.8 %) were associated with a unique genomic position (e.g., base "T" at position "X"), and 978 (10.2 %) represented the number of positions with multiple base pair changes reported at the same position (e.g., both base "T" and base "A" changes reported at position "X") (Fig. 3). Using our specific capture kit and sequencing technology, 2,138 out of 2,190 exons were captured (97.6 %), and 52 out of the 2,190 exons (2.4 %) were not captured (Fig. 2a) during the sequencing process. Within the captured exons, there were 8538 genomic positions (99.5 % of 8578 total genomic positions) published HGMD variants (Fig. 2b).

We were also interested in the proportion of unique genomic positions falling within exons that were captured but had poor *depth* of sequencing and thus potentially poor *coverage*. We considered an exon to have inadequate coverage ("not covered" or "no coverage") when less than 40 % of the bases within the exon met criteria for having sequence depth of at least 20x. Within the 72 samples, the number of unique genomic positions falling within captured exons that fell in this category ranged from 44 to 587, with a median of 374. Averaged across all 72 samples, there were approximately 344/8538 (4 %) unique genomic positions falling within captured exons that had less than 40 % of bases sequenced at 20x depth ("no coverage").

In contrast, Fig. 4 delineates the proportion of exons that met criteria for "adequate" sequencing coverage in the captured exons. The exon proportions were separated into different categories based on what percentage of bases within the exons had adequate depth of sequencing at 20x. We considered an exon "fully covered" if 100 % of the bases in that exon were covered at 20x, "well covered" if ≥90 % to less than 100 % of bases in the exon were covered at 20x, "mostly covered" if ≥70 % to less than 90 % of the bases in the exon were covered at 20x, lightly covered if ≥40 % to less than 70 % of the bases in the exon were covered at 20x, and "not covered" if <40 % of the bases in the exon were covered at 20x. Proportions for each of the coverage categories were obtained for samples individually, and then they were averaged across all samples. On average, 81 % of the total exons were fully covered, 5.04 % of exons were well covered, 4.91 % of exons were mostly covered, 3.39 % of exons were lightly covered, and 5.66 % of exons were not covered.

All patients enrolled in the study who had known molecular etiologies for SCA/D ($n = 13$) were also evaluated

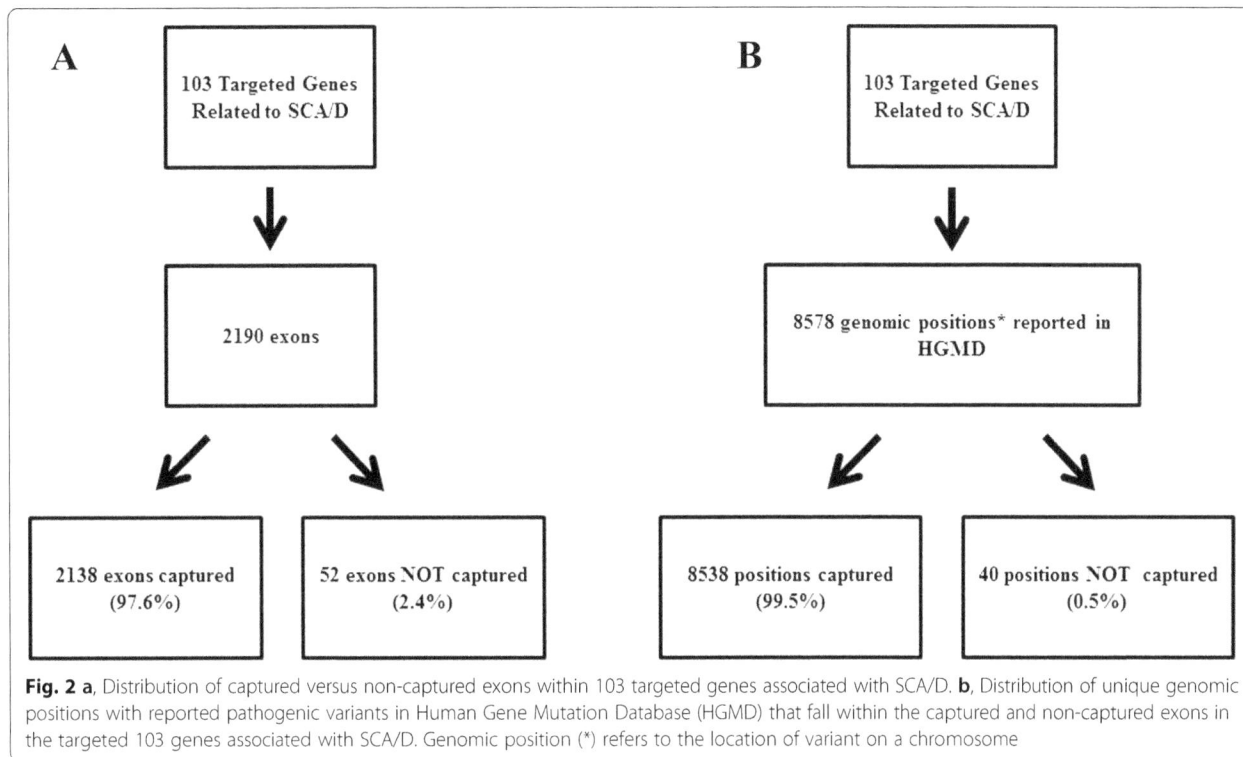

Fig. 2 a, Distribution of captured versus non-captured exons within 103 targeted genes associated with SCA/D. **b**, Distribution of unique genomic positions with reported pathogenic variants in Human Gene Mutation Database (HGMD) that fall within the captured and non-captured exons in the targeted 103 genes associated with SCA/D. Genomic position (*) refers to the location of variant on a chromosome

to see if the causal molecular etiologies could be identified using the project's variant analysis ES pipeline. Of these patients, 13/13 pathogenic variants previously called by CLIA certified laboratories were accurately identified. Due to a small sample size, additional coverage analysis was performed on 100 random variants within the 103 SCA/D genes reported in HGMD to evaluate how well these would be picked up on ES (see Additional file 2 for variant list). All 100 variants were captured by the capture kit (Fig. 5). The next step was to address adequate depth of sequencing. Using 20x as the standard for defining adequate depth of sequencing, 95/100 variants met this criteria, and the remaining 5/100

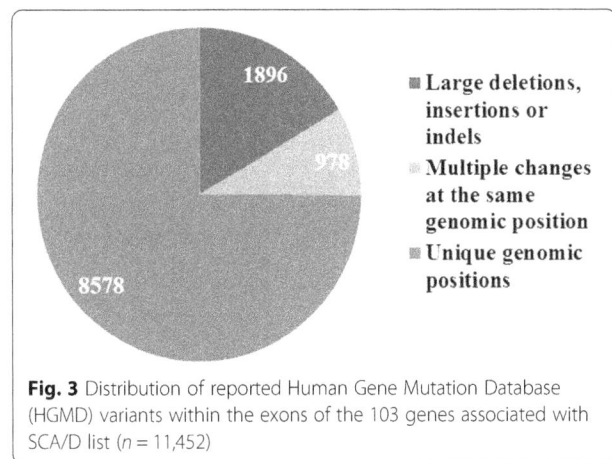

Fig. 3 Distribution of reported Human Gene Mutation Database (HGMD) variants within the exons of the 103 genes associated with SCA/D list (*n* = 11,452)

variants had a sequencing depth of less than 20x (Fig. 5; Table 2). Of these variants, 3 had sequencing depths of 10x-19x, and 2 had a sequencing depth of less than 5x (Fig. 5; Table 2).

An additional 30 patients with history of SCA/D and no known molecular etiology were also analyzed using the same variant analysis pipeline. Results revealed a positive causative finding in 5/30 cases (17 %, Table 3), uncertain results in 16/30 cases (53 %), and negative results in 9/30 (30 %).

Discussion

Our results revealed a number of findings demonstrating the strengths and limitations of using ES as a diagnostic tool. First, there was a fair percentage (16.6 %) of variants within the targeted genes reported in HGMD (large deletions or insertions without genomic positions) that would not be expected to be seen with ES due to the limitations of current technology. Though the ability to identify such changes will likely improve with better technology, clinicians should be aware of the types of platforms being used to capture exonic sequence as well as the limitations of sequence and variant calling technologies to successfully sequence and identify certain types of mutations. In our analysis, amongst the variants reported that have a genomic position, the majority (89.8 %) were unique genomic positions. We were primarily interested in the number of unique genomic positions since presumably a position that has good depth of

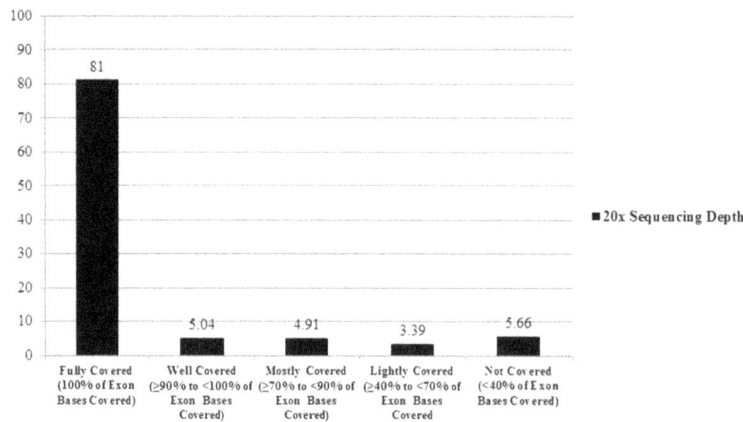

Fig. 4 Average Percentage of Exon Coverage Categories of Targeted SCA/SCD Genes at 20x sequencing depth

sequencing would be covered regardless of the base pair change at that location.

Second, the results demonstrate that the majority of exons (97.6 %) within the targeted SCA/SCD genes were captured with our specific capture kit. The remaining small portion of exons (2.4 %) were not captured primarily because our capture kit did not target these exons, so these areas would not be expected to be picked up even on subsequent sample runs. There are regions of DNA that can be difficult to capture due to the inherent sequence/structure (e.g., repetitive and GC rich regions) resulting in a technical inability to target and capture every exon in the human genome. Thus, the proportion of unique genomic positions reported in HGMD falling within these non-captured regions would be potentially

missed. However, within the non-captured exons, there were only 40 genomic positions (0.5 % of 8578 total genomic positions) with published HGMD variants that would not be expected to be picked up (Fig. 2b). This reflects the process was able to capture the majority of exons with minimal reported HGMD positions missed due to capture issues alone. It is important to be mindful that the goal of individual capture kits is to obtain consistent coverage on the desired targets, but capture and coverage of the non-targeted regions will vary depending on the run due to the limitations of technology.

Aside from the initial capture process, variants could also be missed due to inadequate depth of sequencing. For example, 2 out of 2,190 exons were targeted for capture across all 72 samples but had zero depth of sequencing

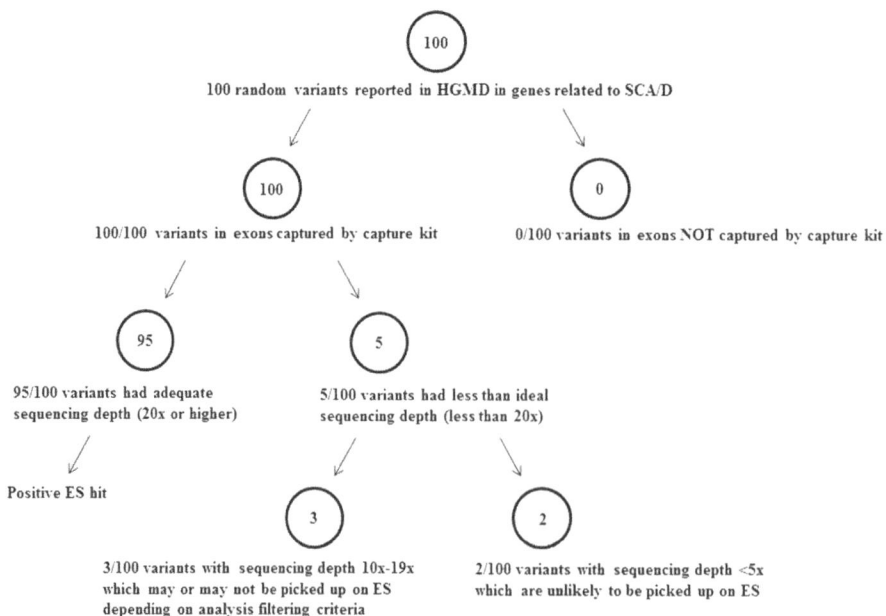

Fig. 5 Schematic analysis of 100 random pathogenic variants, suspected pathogenic variants, or variants of uncertain significance (VUS) reported in Human Gene Mutation Database (HGMD) in genes related to SCA/D

Table 2 Randomly selected variants potentially associated with disease within targeted SCA/D genes with less than 20x sequencing depth

Gene	OMIM[a] #	Phenotype	Variant	Sequencing depth
COL3A1	120180	Ehlers-Danlos syndrome IV	c.3230G > T	12
CTF1	600435	Cardiomyopathy, dilated	c.274G > A	2
PKP2	602861	Arrhythmogenic right ventricular dysplasia	c.1237C > T	11
SDHA	600857	Complex II deficiency & Dilated cardiomyopathy, 1GG	c.1664G > A	1
TMPO	188380	Cardiomyopathy, dilated	c.2068C > T	13

[a]Online Mendelian Inheritance in Man

(nucleotides in those exons were represented zero times on sequences). Within captured exons of individual samples, we were most interested in the proportion of reported HGMD unique genomic positions falling within exons meeting criteria for "no coverage" because these have the potential to be missed on ES analysis. There was a wide range within individual samples, but overall the average proportion of positions in non-covered regions was relatively small (4 %). When examining the proportion of the total number of exons that fell under different coverage categories at 20x sequencing depth, there was a wide distribution with the majority (81 %) of exons being fully covered and a minority (5.66 %) of exons not covered at all. The middle categories of well covered, mostly covered, and lightly covered (13.34 % of exon proportions in aggregate) still made up a fair portion of exons without complete coverage in our targeted gene list. Our results highlight the variability that exists with coverage using ES, and poorly covered regions, as well as moderately covered regions would likely benefit from improving coverage enrichment of kits in these areas. In spite of these limitations in coverage, it is notable that all of the known pathogenic mutations in the 13 samples with confirmed molecular etiologies for SCA/D were identified by using ES (Table 4), and causal findings were identified in 17 % of an additional set of 30 SCA/D patients without known molecular etiologies referred for primary analysis (Table 3). While the main purpose of this study was not to identify diagnostic yields of a particular cohort, these results are reflective of how a primary diagnosis can successfully be made using ES.

We took an additional step to evaluate 100 random variants that have been potentially associated with disease in our targeted SCA/D genes across all individual samples (Additional file 2). The capture of these variants was ideal with all 100 variants successfully captured by the capture kit (Fig. 5). In terms of coverage, the majority (95/100) of the variants were sequenced adequately at 20x and thus would have been picked up on ES. Of the remaining variants with sequencing depths less than 20x, only 2 had sequencing depths less than 5x and would be likely to be missed on ES (Fig. 5; Table 2). Although 20x is often the ideal standard for sequencing depth, many analysis pipelines include variants with lower cutoffs; thus, depending on what analysis protocols are used, up to 5 % of these particular variants may not have been picked up on ES due to sequencing depth. In sum, although at this time ES does not fully cover every base pair at 20x within our targeted genes, the likelihood of a missed variant due to coverage issues remains small.

Limitations

There are a number of limitations to this study that should be considered. First, coverage data from this analysis was limited to the focused gene list that was curated by our team. It is recognized that all genes associated with SDA/D are not included in this list and that new genes are frequently discovered as updated information becomes available. Since the completion of this study, we have added additional genes associated with SCA/D to our list. A pathogenic variant may exist in a patient, but it will not be picked up on ES if the gene has not yet been associated with that particular phenotype and/or human disease. Genes are uniquely different in terms of genomic location, size, number of exons, repeat regions and GC rich regions, and other

Table 3 Pathogenic or likely pathogenic variants identified on ES in samples without prior known molecular diagnosis

Gene	Phenotype of gene	Variant	Protein change	Zygosity
SCN5A	BrS[a], DCM[b], Familial atrial fib, Long QT	c.4867C > T	p.Arg1623*	Heterozygous
DSP	ARVD[c], DCM	c.928dupG	p.Glu310Glyfs*13	Heterozygous
KCNE1	Long QT	c.226G > A	p.Asp76Asn	Heterozygous
KCNH2	Long QT, Short QT	c.1750G > A	p.Gly584Ser	Heterozygous
KCNQ1	Familial atrial fib, Long QT, Short QT	c.513C > A	p.Tyr171*	Heterozygous

[a]Brugada syndrome, [b]Dilated cardiomyopathy, [c]Arrhythmogenic right ventricular dysplasia

Table 4 Pathogenic variants identified on ES in samples with known molecular diagnosis

#	Gene	Phenotype of gene	Variant	Protein change	Zygosity	Change found
1	KCNH2	Long QT, Short QT	c.1882G > A	p.Gly628Ser	Heterozygous	Yes
2	KCNQ1	Familial atrial fib, Long QT, Short QT	c.1552 C > T	p.Arg518[a]	Heterozygous	Yes
3	KCNH2	Long QT, Short QT	c.1838C > T	p.Thr613Met	Heterozygous	Yes
4	SCN5A	BrS[a], DCM[b], Familial atrial fib, Long QT	c.4978A > G	p.Ile1660Val	Heterozygous	Yes
5	KCNQ1	Familial atrial fib, Long QT, Short QT	c.704 T > A	p.Ile235Asn	Heterozygous	Yes
6	MYL2	HCM[c]	c.173G > A	p.Arg58Gln	Heterozygous	Yes
7	KCNE1	Long QT	c.226G > A	p.Asp76Asn	Heterozygous	Yes
8	KCNQ1	Familial atrial fib, Long QT, Short QT	c.1140G > T	p.Arg380Ser	Heterozygous	Yes
9	MYH7	DCM, HCM, LVNC[d]	c.2572C > T	p.Arg858Cys	Heterozygous	Yes
10	MYL2	HCM	c.173G > A	p.Arg58Gln	Heterozygous	Yes
11	TTN	DCM, HCM	c.59216 T > A; c.94578delT	p.Val19664Glu; p.Thr31451Thrfs[a]9	Compound Heterozygous	Yes
12	FBN1	Marfan syndrome	c.2347 A > C	Asn783His	Heterozygous	Yes
13	ACTC; TTN	DCM, HCM, LVNC; DCM, HCM	ACTC: c.806 T > C; TTN: c.11323 G > A	Ile269Thr; Ala3775Thr	Heterozygous	Yes

[a]Brugada syndrome, [b]Dilated cardiomyopathy, [c]Hypertrophic cardiomyopathy, [d]Left ventricular noncompaction

characteristics that can potentially affect the ability to capture exons and have appropriate coverage. As new genes related to SCA/SCD are discovered, it is important to consider how these might be best captured and what is necessary to improve coverage. Sims et al. reviewed in depth issues of sequencing coverage [21]. Definitions for what constitutes specific coverage "levels" also vary among institutions and should be taken into account when performing ES analysis. Additionally, available exome capture kits differ between vendors, though functionality has been found to be generally equal [3], and reproducibility varies with each use, even when using the same kits. Factors such as level of enrichment, genomic library detail, and consistency of captured targets play a role [3].

Finally, as the goal of ES is to identify those variants that may be potentially pathogenic and disease causing, it is equally important to have efficient strategies and appropriate variant analysis pipelines. Without a solid analysis pipeline, consistent capture and high coverage alone is not adequate to detect meaningful variants. A pathogenic variant may be present, but if it is not targeted in analysis it may not be found depending on the filtering parameters. This consideration will continue to be of importance even as the use of genome level sequencing potentially increases beyond ES in the future. Genome sequencing would allow changes beyond the coding regions to be identified, and it would not face the technical challenges seen in ES such as exon capture

and coverage. However, the number of potential variants to analyze would increase tremendously and would be require more sophisticated analysis pipelines to filter and identify disease-causing changes.

Conclusions

Given the high genetic heterogeneity of conditions leading to SCA/SCD, genomic sequencing has the potential to provide invaluable clinical information to high-risk families and clinicians and to help identify at-risk individuals in whom management can help to prevent future SCA/SCD. Our results revealed both the abilities and limitations of using ES as a tool to evaluate genes related to conditions associated with SCA/SCD. Although ES is not fully comprehensive for our targeted genes at this time compared to traditional single or multi-gene panels, the majority of exons were still captured with commercially available kits and were also fully covered on average at 20x sequencing depth. Also, the proportion of HGMD unique genomic positions reported within poorly covered exons remained small. Exonic regions with less coverage reflect the need to enrich these regions to improve coverage. Despite limitations in coverage, our results show ES has a strong potential to pick up molecular changes as we were able to identify 100% of cases with known molecular etiologies for SCA/D in our small cohort. Additionally, in a cohort of 30 patients without a known molecular etiology for their SCA/D we were able to identify a likely etiology in 17 %. We were also able to demonstrate at least 95 % of a number of randomly selected HGMD reported variants would have been picked up on ES as well based on coverage analysis. Overall, ES is a helpful genetic diagnostic tool for SCA/SCD in the clinical setting given its potential to successfully reveal a molecular diagnosis, but clinicians should be aware of limitations of currently available platforms from both a technical and diagnostic perspective.

Competing interests

The authors declare that they have no competing interests.

Authors' contributions

MHL participated in study conception, design and coordination, performed statistical analysis, data interpretation and drafted the manuscript. JLA, MCD, AS, JB and VJ participated in study design, statistical analysis, data interpretation, and contributed to the manuscript. ND, DH, SB, ETD, AW, SN, MIS and MAD participated in study coordination. RR, MS, ABS, PSW, and JP participated in bioinformatics development and analysis support. LKC participated in data interpretation. NBS participated in study design and helped to draft the manuscript. IDK and VLV participated in study conception, design and coordination, data interpretation and helped to draft the manuscript. All authors approved of the final manuscript.

Acknowledgements

We would like to acknowledge the patients and their families who participated in this study.
Funding for this work was supported by the NIH/NHGRI UO1-HG006546 as part of the Clinical Exploratory Sequencing Consortium (CSER) (IDK, NBS, VLV, JP, PSW) and National Institute of General Medical Sciences (NIGMS) 5T32GM008638 (MHL). The funding body did not participate in the study design, collection, analysis and interpretation of data, in the writing of the manuscript, or in the decision to submit the manuscript for publication.

Author details

[1]Department of Pediatrics, Perelman School of Medicine at the University of Pennsylvania, Philadelphia, PA, USA. [2]Division of Cardiology, The Children's Hospital of Philadelphia, Philadelphia, PA, USA. [3]Division of Human Genetics, The Children's Hospital of Philadelphia, Abramson Research Center, Room 1012G, 3615 Civic Center Blvd, Philadelphia, PA 19104, USA. [4]Department of Biomedical and Health Informatics, The Children's Hospital of Philadelphia, Philadelphia, PA, USA. [5]Department of Pathology & Laboratory Medicine, Perelman School of Medicine at the University of Pennsylvania, Philadelphia, PA, USA. [6]Division of Genomic Diagnostics, The Children's Hospital of Philadelphia, Philadelphia, PA, USA. [7]Division of Oncology, The Children's Hospital of Philadelphia, Philadelphia, PA, USA. [8]Present address: Department of Pediatrics, Cincinnati Children's Hospital and Medical Center, and Department of Biomedical Informatics, College of Medicine, University of Cincinnati, Cincinnati, OH, USA.

References

1. Jamal SM, Yu J, Chong JX, Dent KM, Conta JH, Tabor HK, Bamshad MJ. Practices and Policies of Clinical Exome Sequencing Providers: Analysis and Implications. Am J Med Genet A. 2013;161A:n/a-n/a.
2. Interpreting Secondary Cardiac Disease Variants in an Exome Cohort. Circ Cardiovasc Genet. 2013;6:337–46.
3. Bamshad MJ, Ng SB, Bigham AW, Tabor HK, Emond MJ, Nickerson DA, et al. Exome sequencing as a tool for Mendelian disease gene discovery. Nat Rev Genet. 2011;12:745–55.
4. Atwal PS, Brennan M, Cox R, Niaki M, Platt J, Homeyer M, et al. Clinical whole-exome sequencing: are we there yet? Genet Med. 2014;16:717–9.
5. Clinical application of exome sequencing in undiagnosed genetic conditions. J Med Genet. 2012;49:353–61.
6. Wang Z, Liu X, Yang B, Gelernter J. The Role and Challenges of Exome Sequencing in Studies of Human Diseases. Front Genet. 2013;4:160.
7. Yang Y, Muzny DM, Reid JG, Bainbridge MN, Willis A, Ward PA, et al. Clinical Whole-Exome Sequencing for the Diagnosis of Mendelian Disorders. N Engl J Med. 2013;369:1502–11.
8. Ng SB, Turner EH, Robertson PD, Flygare SD, Bigham AW, Lee C, et al. Targeted capture and massively parallel sequencing of 12 human exomes. Nature. 2009;461:272–6.
9. Arndt A, MacRae CA. Genetic testing in cardiovascular diseases. Curr Opin Cardiol. 2014;29:235–40.
10. Pediatric Sudden Cardiac Arrest. Pediatrics. 2012;129:e1094-e1102.
11. Wilde AAM, Behr ER. Genetic testing for inherited cardiac disease. Nat Rev Cardiol. 2013;10:571–83.
12. Contribution Of Inherited Heart Disease To Sudden Cardiac Death In Childhood. Pediatrics. 2007;120:E967-E973.
13. Yield of Molecular and Clinical Testing for Arrhythmia Syndromes: Report of 15 Years' Experience. Circulation. 2013;128:1513–21.
14. Kauferstein S, Kiehne N, Jenewein T, Biel S, Kopp M, König R, et al. Genetic analysis of sudden unexplained death: A multidisciplinary approach. Forensic Sci Int. 2013;229:122–7.
15. Kumar S, Peters S, Thompson T, Morgan N, Maccicoca I, Trainer A, et al. Familial cardiological and targeted genetic evaluation: Low yield in sudden unexplained death and high yield in unexplained cardiac arrest syndromes. Heart Rhythm. 2013;10:1653–60.
16. Tester DJ, Ackerman MJ. Postmortem Long QT Syndrome Genetic Testing for Sudden Unexplained Death in the Young. J Am Coll Cardiol. 2007;49:240–6.
17. OMIM. Online Mendelian Inheritance in Man, OMIM®. Johns Hopkins University, Baltimore, MD. World Wide Web URL: http://www.omim.org.
18. HRS/EHRA Expert Consensus Statement on the State of Genetic Testing for the Channelopathies and Cardiomyopathies: This document was developed as a partnership between the Heart Rhythm Society (HRS) and the European Heart Rhythm Association (EHRA). Europace. 2011;13:1077–109.
19. Krawitz P, Mundlos S. Strategies for exome and genome sequence data analysis in disease-gene discovery projects. Clin Genet. 2011;80:127–32.
20. Raymond C, Raymond C, Aravind L. Initial sequencing and analysis of the human genome. Nature. 2001;409:860–921.
21. Sims D, Sudbery I, Ilott NE, Heger A, Ponting CP. Sequencing depth and coverage: key considerations in genomic analyses. Nat Rev Genet. 2014;15:121–32.
22. Li H. Aligning Sequence Reads, Clone Sequences And Assembly Contigs With BWA-MEM. 03 2013. 1303.
23. Stenson PD, Mort M, Ball EV, Shaw K, Phillips AD, Cooper DN. The Human Gene Mutation Database: building a comprehensive mutation repository for clinical and molecular genetics, diagnostic testing and personalized genomic medicine. Hum Genet. 2013;133:1–9.

7

Epigenetic inheritance and the missing heritability

Marco Trerotola[1], Valeria Relli[1], Pasquale Simeone[1] and Saverio Alberti[1,2]*

Abstract

Genome-wide association studies of complex physiological traits and diseases consistently found that associated genetic factors, such as allelic polymorphisms or DNA mutations, only explained a minority of the expected heritable fraction. This discrepancy is known as "missing heritability", and its underlying factors and molecular mechanisms are not established. Epigenetic programs may account for a significant fraction of the "missing heritability." Epigenetic modifications, such as DNA methylation and chromatin assembly states, reflect the high plasticity of the genome and contribute to stably alter gene expression without modifying genomic DNA sequences. Consistent components of complex traits, such as those linked to human stature/height, fertility, and food metabolism or to hereditary defects, have been shown to respond to environmental or nutritional condition and to be epigenetically inherited. The knowledge acquired from epigenetic genome reprogramming during development, stem cell differentiation/de-differentiation, and model organisms is today shedding light on the mechanisms of (a) mitotic inheritance of epigenetic traits from cell to cell, (b) meiotic epigenetic inheritance from generation to generation, and (c) true transgenerational inheritance. Such mechanisms have been shown to include incomplete erasure of DNA methylation, parental effects, transmission of distinct RNA types (mRNA, non-coding RNA, miRNA, siRNA, piRNA), and persistence of subsets of histone marks.

Keywords: DNA methylation, Missing heredity, Transgenerational inheritance

Introduction

Patterns of heritable traits within the human population determine body phenotypes, through a deeply inter-twined interaction between genetic components and the environment. Specific genetic/DNA sequence variants are typically inherited transgenerationally as Mendelian alleles and are supposed to carry with them all the genetic information that acts as inheritable determinant [1]. Genome-wide association studies (GWAS) have recently demonstrated that multiple genomic loci are linked to complex traits, such as body development and height ([2, 3] and references therein). Several common disorders, such as type 2 diabetes, Crohn's disease, and rheumatoid arthritis, were also shown to possess significant genetic components, as provided by multiple polymorphic loci [4–6]. These findings led to postulate models whereby numerous genetic factors provide small,

independent contributions to complex phenotypes, such effects being essentially additive [5]. However, simple models of additive effects of ever-smaller components have remained as yet unproven [4–7]. On the other hand, identified genetic factors associated with complex diseases have been found to confer far less disease risk than expected from empirical estimates of heritability and typically explain only a minority of the heritable traits. As a consequence, "pure" genetic models are prone to underestimate the interactions among loci [5], globally designated as epistasis. Epistatic components need to be integrated by estimates of the contribution of non-genetic factors, globally designated as the "missing heritability" [7, 8]. Hence, the issue remained open, whereby identified genetic factors associated with complex diseases conferred far less disease risk than expected from empirical estimates of heritability. As an example, Crohn's disease is a recessive disorder which shows about 80 % heritability. However, the genetic components identified to date only explain 20 % of this heritable fraction [9]. An additional example is that of

* Correspondence: s.alberti@unich.it
[1]Unit of Cancer Pathology, CeSI, Foundation University 'G. d'Annunzio', Chieti, Italy
[2]Department of Neuroscience, Imaging and Clinical Sciences, Unit of Physiology and Physiopathology, 'G. d'Annunzio' University, Chieti, Italy

human population stature [2, 3]. A significant fraction of height-determining genes has been identified by GWAS analysis [10–12]. Most of these genes were demonstrated to be largely overlapping in Caucasian and non-Caucasian populations [13, 14], consistent with an actual identification of the most relevant height-associated determinants. However, the identified polymorphisms were found to account only for a minor fraction of stature heritability. Although dedicated procedures for SNP-associated analyses have significantly increased their combined predictive power [15], a large amount of heritable height-associated factors remains undetectable by conventional GWAS, suggesting that such non-DNA sequence-linked information may be associated to epigenetic heredity.

Epigenetic heritability

Epigenetic modifications, such as DNA methylation, can contribute to alter gene expression in heritable manner without affecting the underlying genomic sequences. Such epigenetic contribution would be systematically missed by conventional DNA sequence-based analyses. A model of epigenetic inheritance, as additional to Mendelian heredity of polymorphic DNA sequences, would thus efficiently explain the lack of detection in conventional GWAS as "missing heritability". It would also help explaining the cases of rapid, heritable adaptations to changing environmental conditions, such as for human stature [2, 3], and the occurrence of hereditary epistatic effects. Support for this model is provided by the evidence that phenotypic plasticity can emerge over rapid time scales, at rates that are orders of magnitude higher than the processes of natural selection [16, 17].

However, to be tenable, such a model of epigenetic inheritance poses rigorous requirements: (a) mitotic inheritance of epigenetic traits across cell generations (see discussion on DNA methylation maintenance through mitotic cycles); (b) epigenetic inheritance across successive meiotic divisions (see the paragraphs describing gamete generation and the development of primordial germ cells (PGC); and (c) true transgenerational inheritance, which requires proof of heritability beyond the first generation that has not been unexposed to the causal epigenetic modifiers (see the paragraphs describing transgenerational inheritance of DNA methylation and of chromatin states).

Evidence is now accumulating that provides insight in the mechanisms that underlie epigenetic transmission. These include the following: (1) DNA methylation, (2) histone modifications and chromatin remodeling, (3) inheritance of specific mRNAs, long non-coding RNAs (ncRNAs) and siRNAs/miRNAs, (4) feedback loops through which mRNA or protein products of a gene can stimulate its own transcription and enable "heritable states" of gene expression, and (5) the activity of chaperones such as Hsp90 that plays an important role in chromatin remodeling and can mediate epigenetic transgenerational variation. The most relevant mechanisms are described in Table 1.

Table 1 Molecular mechanisms of epigenetic transgenerational heredity

Steps	Molecular mechanisms
DNA sequence-invariant heritable traits	DNA methylation/histone post-translational modifications
DNA methylation maintenance across cell division cycles	Hemimethylated DNA-guided, DNMT1-mediated CpG methylation pattern maintenance
DNA demethylation	Passive DNA demethylation
	5-mC to 5-hmC conversion
	Active DNA demethylation
	Glycosylase-mediated base removal and base excision repair mechanisms
Histone code	Condensed chromatin
	HAT inactivation
	HMT activation
	Relaxed chromatin
	HAT activation
	HMT inactivation
Epigenetic modulation of mother-to-fetus transmission	Maternal nutrition status
	Maternal exposure to environmental toxins and food contaminants
	BPA
	Phthalates
	Dioxins
	Tobacco smoke
Cell differentiation and body development	Epigenetic signature reprogramming
	Erasure/reprogramming in the zygote (mitotic transmission)
	Erasure/reprogramming in PGCs (meiotic transmission)
	Gamete-carried transmission
	DNA methylation profiles in sperm and oocytes
	H3K4 and H3K27 histone methylation in sperm cells
	RNA molecules carried by sperm cells (mRNA, non-coding RNA, miRNA, siRNA, piRNA)
Stem cell reprogramming	Epigenetic signature of induced pluripotency
	Decreased TETs/decreased hydroxymethylation at ES gene promoters
	Reprogramming-resistant regions enriched for H3K9me3

DNA methylation

Approximately 60–80 % of the 28 million CpG dinucleotides in the human genome are methylated [18, 19]. DNA methyltransferases (DNMT) recognize hemimethylated sites (maintenance methylation) or specific unmethylated sequences (de novo methylation) (Fig. 1). Maintenance DNA methylation occurs during DNA replication and is predominantly dependent on DNMT1, whereas de novo DNA methylation is carried out by DNMT3A, DNMT3B [20], and DNMT3L [21, 22]. Both DNMT3A and DNMT3B localize to methylated, CpG-dense regions and preferentially bind to the bodies of transcribed genes but are excluded from regions of active promoters and enhancers [23].

DNA methylation at CpG islands modulates gene transcription and is involved in alternative promoter usage and regulation of short and long ncRNA processing and of enhancer activity [24–26]. DNA methylation also affects determinants of higher-order DNA structure, e.g. in X chromosome inactivation, imprinting control regions (ICR) [27], heterochromatin folding and maintenance of genomic stability [28, 29].

DNA demethylation

The dynamic on/off switching of gene expression requires a balanced action of DNA methylation versus DNA demethylation. Both active and passive DNA demethylation have indeed been shown to occur.

Active DNA demethylation, i.e. replication-independent enzymatic removal of 5-methyl cytosine (5-mC), occurs through the processing of 5-mC to 5-hydroxymethyl cytosine (5-hmC) catalyzed by the ten-eleven translocation (TET) dioxygenases [30–32]. 5-hmC is then converted to 5-formylcytosine (5-fC) and 5-carboxylcytosine (5-caC) [30]. The thymine DNA glycosylase (TDG) efficiently excises both 5-fC and 5-caC [33, 34]. This leaves an abasic site that is subsequently processed through the base excision repair (BER) pathway [33] (Fig. 1).

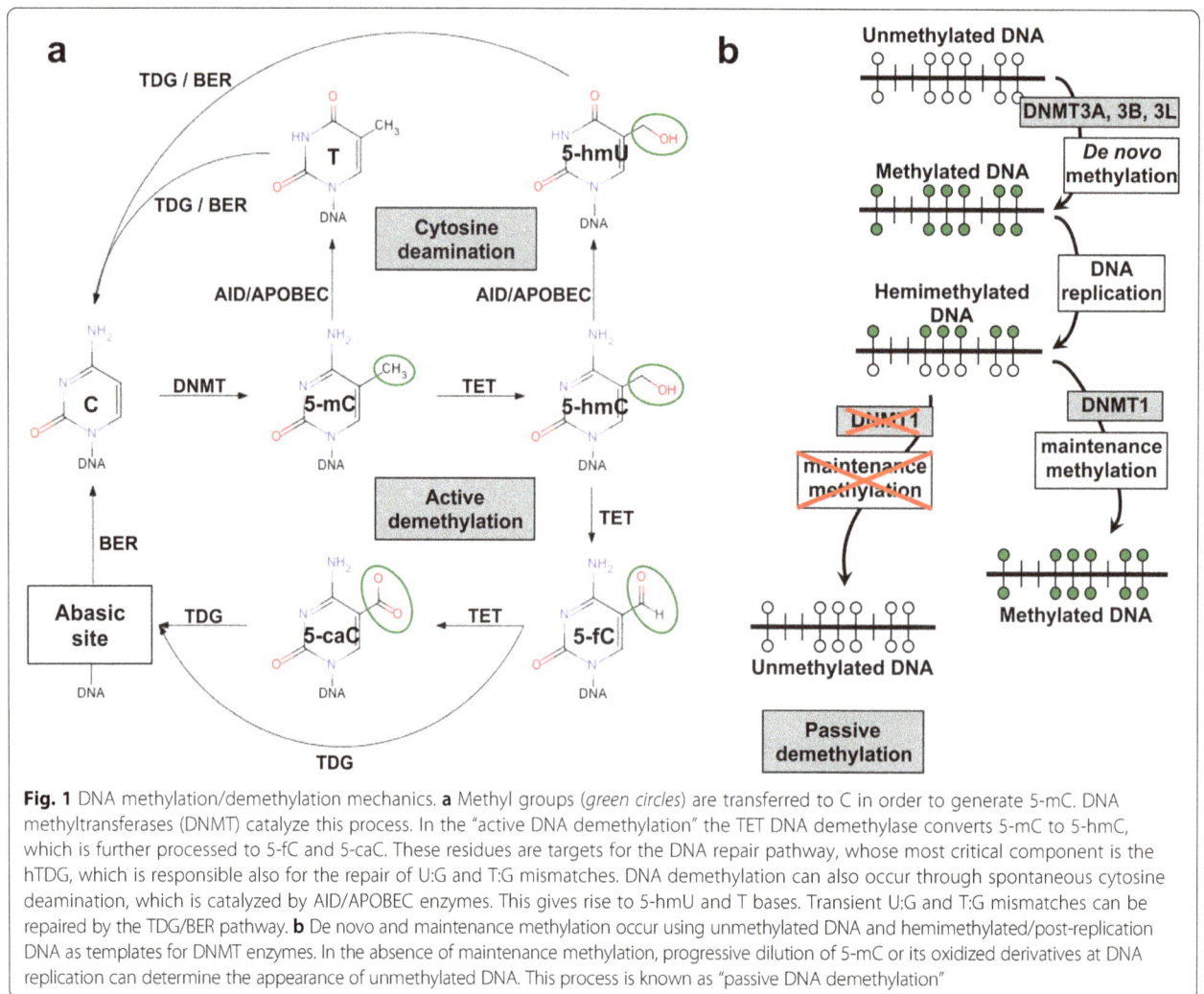

Fig. 1 DNA methylation/demethylation mechanics. **a** Methyl groups (*green circles*) are transferred to C in order to generate 5-mC. DNA methyltransferases (DNMT) catalyze this process. In the "active DNA demethylation" the TET DNA demethylase converts 5-mC to 5-hmC, which is further processed to 5-fC and 5-caC. These residues are targets for the DNA repair pathway, whose most critical component is the hTDG, which is responsible also for the repair of U:G and T:G mismatches. DNA demethylation can also occur through spontaneous cytosine deamination, which is catalyzed by AID/APOBEC enzymes. This gives rise to 5-hmU and T bases. Transient U:G and T:G mismatches can be repaired by the TDG/BER pathway. **b** De novo and maintenance methylation occur using unmethylated DNA and hemimethylated/post-replication DNA as templates for DNMT enzymes. In the absence of maintenance methylation, progressive dilution of 5-mC or its oxidized derivatives at DNA replication can determine the appearance of unmethylated DNA. This process is known as "passive DNA demethylation"

Passive DNA demethylation can also proceed through the processing of 5-mC to 5-hmC [30–32] (Fig. 1). Neither maintenance methylase DNMT1 nor de novo methylases DNMT3A and DNMT3B recognize 5-hmC, and no mechanisms for 5-hmC maintenance have been as yet identified [32]. Hence, 5-hmC is inevitably lost at each replication cycle [35]. This appears to have a functional impact on PGC, which progressively lose 5-mC between embryonic day (E) E9.5 and E10.5, until 5-mC becomes undetectable by E11.5 [31]. This progressive loss of 5-mC occurs concurrently with an enrichment of 5-hmC, suggesting a genome-scale conversion of 5-mC to 5-hmC for the epigenomic reprogramming of these cells [31].

An additional path of passive DNA demethylation proceeds through destabilization of DNMT1, which is then followed by "loss by dilution" of 5-mC through successive replication cycles. Overexpression of SET7 leads to decreased DNMT1 levels via induction of proteasome-mediated degradation [36]. SET7 directly interacts with DNMT1 and specifically monomethylates Lys-142 of DNMT1 [36]. On the opposite side, AKT1 phosphorylates DNMT1 at Ser143 and stabilizes the protein [37]. Phosphorylation of DNMT1 at Ser143 interferes with monomethylation of the nearby Lys142 [37], making these two modifications mutually exclusive. *Rb* and *ATM* also affect the stability of DNMT1 [38]. The inactivation of pRB promotes a Tip60 (acetyltransferase)-dependent ATM activation. This allows activated ATM to physically bind to DNMT1, forming a complex with Tip60 and the E3 ligase ubiquitin-like containing PHD and RING finger domain protein 1 (Uhrf1) and accelerates the DNMT1 ubiquitination driven by Tip60-dependent acetylation [38, 39]. In contrast, histone deacetylase 1 (HDAC1) and the deubiquitinase HAUSP (herpes virus-associated ubiquitin-specific protease) stabilize DNMT1 [39].

Of note, 5-mC frequently undergoes deamination (Fig. 1). Hence, a DNA methylation-dependent modification can end up in a permanently fixed DNA sequence change. Physiological enzymes are involved, raising the intriguing issue of "guided" mutagenesis of Mendelian traits. The AID and APOBEC enzyme families catalyze the cytosine processing which leads to cytosine deamination. This occurs predominantly on 5-hmC and 5-mC residues, giving rise to formation of 5-hydromethyluracil (5-hmU) and thymine (T) bases, respectively [40] (Fig. 1). Consequently, transient U:G and T:G mismatches can be generated, though most of these mutations can be efficiently repaired by the TDG/BER pathway [40]. Notably, dysregulated APOBEC3B-catalyzed deamination can provide a chronic source of DNA damage, with consequent TP53 inactivation; this was shown to lead to development of breast cancer [41].

Reversible changes of epigenetic patterns

Epigenetic reprogramming through the mechanisms described above has been demonstrated in mammals over distinct, key developmental stages:

Erasure of DNA methylation patterns in the zygote Erasure of DNA methylation patterns of the gametes (oocyte and sperm [42]) in the zygote was shown to occur immediately after fertilization. This process has been traditionally considered as a mechanism for resetting epigenetic marks between generations, to ensure the totipotency of the zygote after fertilization. Recent evidence from genome-scale DNA methylation analysis of human development confirmed a transient, highly dynamic state of global hypomethylation that affects most CpGs [43]. However, neither histone codes nor DNA methylation patterns are completely erased and are carried over through zygote divisions and generation of PGCs, thus, providing some of the means for transgenerational inheritance.

Erasure and reconstitution of DNA methylation patterns in PGCs Epigenetic marks have been shown to undergo reprogramming across meiotic divisions of PGCs during gametogenesis. However, genome-wide DNA methylation profiling in PGCs revealed that, although the bulk of the genome becomes demethylated [44, 45], several loci escape this epigenetic erasure [46]. This leads to preserving the methylation status of more than 40 % of all 5-mC [16]. Substantial numbers of genes have been found to retain parental DNA methylation patterns in sperm and oocytes as a result of epigenetic transmission from PGCs [47]. It was recently reported that 5-hmC and 5-fC do exist in both maternal and paternal genomes and that 5-mC or its oxidized derivatives can be converted to unmodified cytosines through active demethylation rather than by passive dilution during embryonic development [48].

Erasure and reconstitution of epigenetic signatures during early body development Reprogramming/re-establishment of epigenetic signatures was also shown to be necessary for proper development of a mature organism [49]. It has been shown that during early mitotic divisions of a mammalian embryo, daughter cells derived from the zygote have a globally hypomethylated genome [50] and a transcriptionally active chromatin due to histone H4 acetylation [51]. Therefore, they are an epigenetically homogeneous cell population [52, 53]. At the blastocyst stage, peripheral cells (that will become the extraembryonic tissue) have low levels of DNA methylation and are epigenetically different from cells of the inner cell mass (ICM, which will form the embryo) that have already undergone re-establishment of some methylation patterns [52, 53]. A major epigenetic switch then occurs during implantation at the transition from the blastocyst to the post-implantation epiblast [47].

Erasure and reconstitution of DNA methylation patterns in adult stem cells Somatic cell nuclear transfer has been utilized as one of the procedures to obtain reprogramming of somatic cells toward a totipotent state, with the generation of induced pluripotent stem cells [54, 55]. This, and the possibility to isolate embryonic stem (ES) cells from the blastocyst ICM, provided unprecedented opportunities to investigate the mechanics of erasure and re-establishment of epigenetic patterns and to define the molecular components involved in these processes. Declining levels of TETs during differentiation were shown to be associated with decreased hydroxymethylation levels at the promoters of ES cell-specific genes [32]. Thus, the balance between hydroxymethylation and methylation in the genome appears to be linked with the balance between pluripotency and lineage commitment [32]. Moreover, reprogramming-resistant regions strongly enriched for H3K9me3 were demonstrated to be critical barriers for efficient reprogramming [56]. Hence, modulation of epigenetic inheritance appears to play a key role in stem cell differentiation/de-differentiation.

Gamete-carried epigenetic traits
Nucleosomes are largely replaced by protamines in mature human sperm, thus erasing most chromatin patterns. However, not all histones in sperm are replaced by protamines, and epigenetic marks such as H3K4me2 and H3K27me3 have been detected in sperm [57, 58]. These retained nucleosomes are significantly enriched at loci of developmental importance, including imprinted gene clusters, microRNA clusters, HOX gene clusters, and the promoters of stand-alone developmental transcription and signalling factors [57]. H3K4me2 was found to be enriched at promoters of genes coding for developmental transcription factors, whereas H3K4me3 was found predominantly localized at promoters of genes important for spermatogenesis and rearrangement of nuclear architecture and presumably active during the gametogenesis [57].

Our findings indicated that sperm methylation pattern of the CD5/Leu1 and CD8/Leu2 genes is incompletely erased. This had a heavy impact on gene function and tightly prevented the expression of the CD5 gene, though not of CD8 [59]. Genes encoding olfactory receptors, in cases where mice associated specific odors with fearful experience, were also found differentially methylated in sperm, and this methylation pattern was transmitted to F1 and F2 generations [60].

RNA molecules packaged in sperm represent an additional contributor to transgenerational transmission of epigenetic traits and have been shown to profoundly affect offspring phenotypes [61, 62]. Injecting sperm RNA from traumatized males into fertilized wild type oocytes reproduced the altered behavior in the offspring. Moreover, miRNA-mediated signals can change DNA methylation patterns in the F2 sperm, and this signature can be maintained and replicated through subsequent mitotic and meiotic cycles [63].

Notably, RNA expressed in somatic cells can be transferred to gametes via extracellular vesicles [64]. Subcutaneous injection of human melanoma cells stably expressing enhanced green fluorescent protein (EGFP) led to the transfer of EGFP mRNA in murine sperm heads, likely through exosomes-mediated transport [65]. Furthermore, in *C. elegans*, it has recently been shown that neurons can transmit double-stranded RNA (dsRNA) to the germ cells to initiate transgenerational silencing of their target genes [16, 66]. Thus, extracellular vesicles have been revealed as an important route for transgenerational inheritance of epigenetic signatures.

Transgenerational inheritance of DNA methylation patterns
Transgenerational inheritance of epigenetic patterns [16, 66] is key to a model of "epigenetic missing heredity". In this regard, it should be noted that purely parental effects, such as the impact of direct in utero exposure to particular nutritional, hormonal, or stress/toxin environments, do not represent true transgenerational heredity [67]. In addition, F1 gametes are potentially exposed in utero to maternal experiences, and this may subsequently affect F2 offsprings. Hence, proof of transgenerational transmission of ancestral memory requires demonstrating the passing of the epigenetic trait through unexposed F2 gametes to F3 offsprings [16].

Physiological traits, such as body stature, were shown to rapidly and progressively adaptate to changing environments, e.g., nutritional status [2]. The epigenetic inactivation of height-associated genes (e.g., *BMP2, BMP6, CABLES1, DLEU7, GNAS, GNASAS, HHIP, MOS, PLAGL1*) was shown to be functionally equivalent to Mendelian physical loss of the corresponding alleles. Moreover, distinct epigenetic defects, such as in Beckwith–Wiedemann, Prader–Willi, Angelman's, Rett, and Silver–Russell syndromes ([2] and references therein), were correspondingly shown to cause hereditary growth anomalies, indicating that body stature is under a heritable epigenetic control.

Correspondingly, the appearance of several pathological conditions with heritable components, such as diabetes and Crohn's disease, is affected by interaction with dynamic environmental factors, such as host pathogens or nutritional status [68]. Exposure of parents to distinct diet regimens, stress, drugs, or endocrine/metabolic dysfunctions, was shown to additionally affect the transgenerational transmission of altered DNA methylation patterns [69–75]. As an example, F2 generation offspring (i.e., the grandchildren) of alcohol-abusing women have a higher tendency to show fetal alcohol syndrome than the F2 progeny of control women [76, 77].

Epigenetic modifications were also shown to be caused by exposure to environmental toxins, including metals (cadmium, arsenic, nickel, chromium, and methylmercury), solvents (trichloroethylene), air pollutants (black carbon, benzene), food-chain contaminants (dioxins), and tobacco smoke (nicotine, benzo(a)pyrene) [78, 79]. In

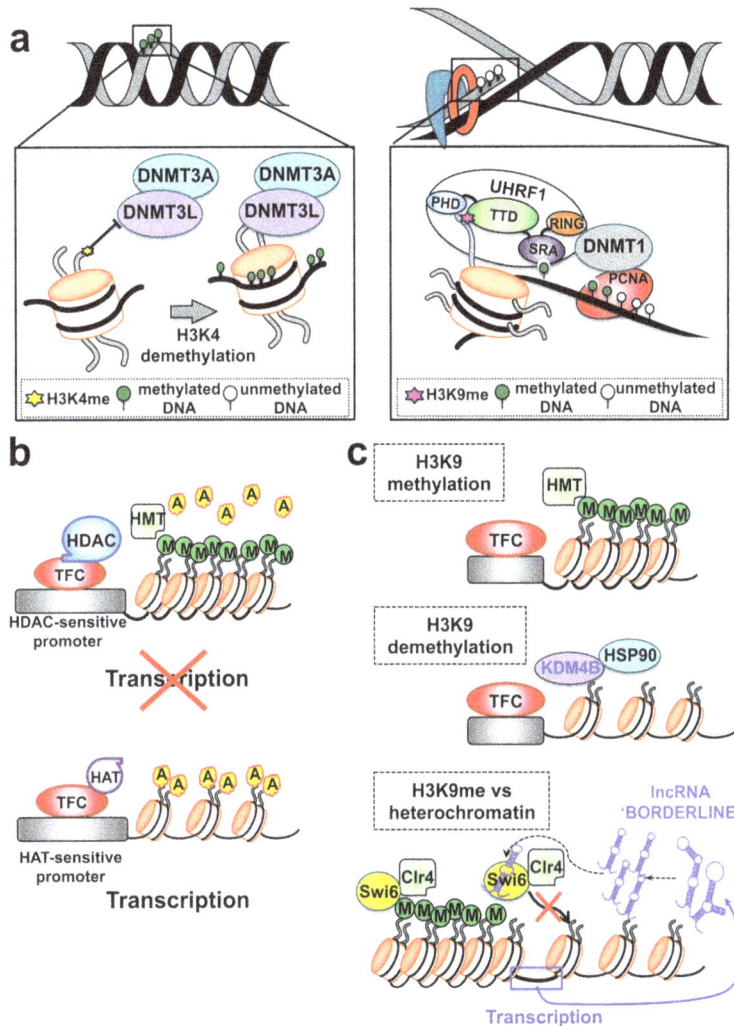

Fig. 2 Histone modifications and DNA cooperate in re-shaping chromatin organization and regulating gene expression. **a** (*left*) De novo DNA methylation occurs on unmethylated DNA. It is catalyzed by the DNMT3, whose subunits can be positioned in proximity of their target sites through physical interaction with unmethylated H3K4. **a** (*right*) Maintenance DNA methylation occurs during the DNA replication and is catalyzed by the DNMT1. Uhrf1 and proliferating cell nuclear antigen (PCNA) associates to DNMT1 and recruit it to the replication fork, concentrating its activity on hemimethylated DNA. The Uhrf1 TTD domain interacts with H3K9me. This binding allows a faithful propagation of DNA methylation patterns throughout mitosis. **b** (*top*) HDAC and the transcription factor complex (TFC) can be recruited on sensitive promoters, leading to histone deacetylation. HMT-driven methylation of the histone tails causes tight wrapping of DNA around nucleosome cores and inhibition of gene expression. **b** (*bottom*) The accumulation of HAT-driven histone acetylation determines DNA relaxation around the nucleosomes surrounding HAT-sensitive promoters; this leads to increased transcription and gene expression. **c** Methylation of H3K9 plays a central role in non-DNA-dependent mechanisms of regulation of gene activity (*top*). Hsp90 has a strong effect on the histone code via stabilization of KDM4B, which demethylases H3K9 (*middle*). Non-coding RNAs alter the histone code through siRNA-dependent mechanisms that lead to direct competition between BORDERLINE ncRNAs and H3K9me for binding to the HP1 proteins, such as Swi6. This occurs at heterochromatin/euchromatin boundary sites and counteracts the spreading of heterochromatin into neighboring euchromatin (*bottom*). A acetyl groups, M methyl groups, HAT histone acetyltransferase, HDAC histone deacetylase, HMT histone methyltransferase

utero and neonatal exposure to low doses of Bisphenol A and/or phthalates causes epigenetic alterations [71], such as differential methylation at CpG islands, histone modifications, and altered expression of ncRNAs and miRNAs [80, 81]. DNA methylation was shown to be affected by periconceptional maternal plasma concentrations of micronutrients involved in one-carbon metabolism, such as folate, B2 vitamin, methionine, and betaine [70]. The first evidence in mammals for a true transgenerational transmission of exposure-determined epigenetic traits was obtained in rats exposed to endocrine disruptors vinclozolin or methoxychlor during gestation. This induced in the F1 generation an adult phenotype of decreased spermatogenic capacity and of increased incidence of male infertility. Remarkably, transgenerational transmission of these effects through the male germ line was then observed in F1 to F4 generations [82]. DNMT3L has been reported to be necessary for maternal methylation imprinting [22]. DNMT3L is enzymatically inactive but acts as a stimulatory factor for de novo methylation

by DNMT3A [21]. DNA methylation signatures driving altered behavioral/metabolic phenotypes (from exposure to maternal stress) were subsequently shown to be transgenerationally transmitted to the offspring [83], through miRNA delivery by sperm cells to the oocyte [63, 83].

Transgenerational inheritance of chromatin states

Distinct types of histone post-translational modifications (PTMs) play a critical role in the nucleosome-dependent regulation of gene transcription. The largest body of knowledge has been gathered on histone methylation and acetylation [84], which cooperates in re-shaping chromatin organization. This histone code interacts closely with DNA methylation: unmethylated Lys-4 on H3 histone (H3K4) acts as a docking site for DNMT3A, which is recruited on nucleosomes and methylates associated target nucleotides [21, 22]. Recent findings showed that in the absence of histone H3, DNMT3A exists in an autoinhibitory form, in which the ATRX–DNMT3–DNMT3L (ADD) domain binds to

Table 2 Epigenetic hereditary traits contribution to developmental diseases[a]

Non-cancerous syndromes	Phenotypes/clinical features	Molecular defects
ATR-X	Upswept frontal hair line; hypertelorism; epicanthic folds; flat nasal bridge; small triangular upturned nose; tented upper lip; everted lower lip; hypotonic facies	Mutations in *ATRX* gene, hypomethylation of specific repeat and satellite sequences
Fragile X	Mild to severe intellectual disabilities; elongated face; large or protruding ears; macroorchidism; stereotypic movements (e.g., hand-flapping); social anxiety	Expansion and methylation of CGG repeat in *FMR1* 5'UTR, promoter methylation
ICF	Hypertelorism; low-set ears; epicanthal folds; macroglossia	*DNMT3B* mutations, DNA hypomethylation
Angelman	Severe intellectual and developmental disabilities; sleep disturbance; seizures; jerky movements (e.g., hand-flapping); frequent laughter or smiling; a happy behavior	Deregulation of one or more imprinted genes at 15q11–13 (maternal)
Prader–Willi	Low muscle tone; short stature; incomplete sexual development; cognitive disabilities; chronic feeling of hunger leading to excessive eating and life-threatening obesity	Deregulation of one or more imprinted genes at 15q11–13 (paternal)
Beckwith–Wiedemann	Macroglossia; macrosomia; midline abdominal wall defects; ear creases or ear pits; neonatal hypoglycemia	Deregulation of one or more imprinted genes at 11p15.5 (e.g., *IGF2*)
Rett	Small hands and feet; decelerated rate of head growth; repetitive stereotyped hand movements (e.g., wringing and/or repeatedly putting hands into the mouth); gastrointestinal disorders; seizures; no verbal skills; scoliosis; growth failure; constipation	*MeCP2* mutations
Rubinstein–Taybi	Short stature; moderate to severe learning difficulties; broad thumbs and first toes; increased risk of developing benign and malignant tumors, leukemia, and lymphoma	Mutation in CREB-binding protein (histone acetylation)
Coffin–Lowry	Abnormal growth; cardiac defects; kyphoscoliosis; auditory and visual abnormalities	Mutation in Rsk-2 (histone phosphorylation)
Silver–Russel	Feeding problems; hypoglycemia; excessive sweating; triangular shaped face with a small jaw and a pointed chin that tends to lessen slightly with age; curved down mouth; blue tinge to the whites of the eyes in younger children; normal size of head circumference, disproportionate to a small body size; wide and late-closing fontanelle; clinodactyly; body asymmetric growth; precocious puberty; low muscle tone; gastroesophageal reflux disease; lack of subcutaneous fat; late closing of the opening between the heart hemispheres; constipation	Loss of methylation on the *ICR1* paternal allele at the *H19/IGF2* locus (11p15)

[a]Phenotype-genotype correlations were extracted from the OMIM databank (www.ncbi.nlm.nih.gov/omim)

the catalytic domain and hinders its DNA-binding capacity [85]. Once the DNMT3A–DNMT3L complex is recruited to the nucleosome, unmethylated H3K4 binds to the ADD domain and stimulates DNMT3A to undergo a significant conformational change from an autoinhibitory form to an active form that can bind DNA and exert DNA methylation activity [85]. Moreover, interaction of Dppa3 with histone H3K9me2 blocks the activity of TET3, favoring the maintenance of DNA methylation [86]. Less common histone PTMs have been recently identified [87, 88]. Among them, histone H1 arginine 54 citrullination (H1R54ci) determines histone displacement from chromatin and chromatin decondensation. Histone PTMs were further shown to be a critical mechanism for maintaining stem cell pluripotency [89].

Histone PTMs and corresponding DNA methylation patterns can affect imprinting in mammalian cells [90, 91] through the selective recruitment of effector proteins, known as "readers" (e.g. the bromodomain motif that docks onto acetylated lysines [92]), which drive chromatin packaging around nucleosomes [88, 91] (Fig. 2).

Histone PTM-driven heritable silencing of gene expression is also affected by various chromatin remodeling factors. Among them, the polycomb-group (PcG) proteins recognize specific histone modifications such as the H3K27me3 and participate in maintenance of repressed chromatin domains [93]. PcG proteins take part to the regulation of X-chromosome inactivation and maintenance of stem cell identity [94]. Correspondingly, the molecular chaperone Hsp90 was shown to alter the histone code via interaction and stabilization of KDM4B, which demethylates H3K9. Pharmacological inhibition of Hsp90 results in ubiquitin-dependent proteasomal degradation of KDM4B, which is accompanied by increased methylation of H3K9 [95] (Fig. 2).

Although the mechanisms through which DNA methylation states can propagate across cell divisions have been studied in depth, it is still unclear how the histone code (e.g., the histone PTM levels) is restored through multiple rounds of DNA replication. During DNA replication, nucleosomes are disrupted ahead of the replication machinery and reassembled on the two newly synthesized DNA strands. Histones PTMs are transmitted with high efficiency at replication forks, indicating a specific recycling of old histones. However, incorporation of new, largely unmodified histones also occurs [96–98]. Reincorporated parental, modified histones may serve as a blueprint to modify neighboring new histones, limiting the possible "dilution" of the corresponding code. Recent findings suggested that most PTM levels are maintained according to the simple paradigm that new histones acquire modifications to become identical to the old ones [99]. These and other findings also demonstrated that H3K9me3 and H3K27me3 are central marks for cellular memory [97, 100], and are propagated by continuous modification of both new and old histones through the generations. Epigenetic heritability of H3K9 methylation was recently investigated in fission yeast and demonstrated to involve the activity of a single H3K9 methyltransferase, Clr4, that directs all H3K9 methylation and heterochromatin through a "read-write" mechanism [101, 102]. Hence, histones act as carriers of epigenetic information, and the kinetics of PTM restoration appears to play a critical role in epigenetic inheritance [90, 98].

Epigenetic inheritance in model organisms

Non-DNA methylation-mediated epigenetic heritability has been demonstrated also in non-mammalian species, such as *C. elegans* [1], *D. melanogaster* [103], and *S. pombe* [104].

C. elegans lacks DNA methylation and has evolved histone methylation/demethylation pathways to regulate the

Table 3 Epigenetic heredity of cancer-causing genes

Cancers	
Bladder	Aberrant methylation of *TWIST*, *NID2*, and *RUNX3*
Brain	Aberrant methylation of *RASSF1A* and *MGMT*
Breast	Aberrant methylation of *BRCA1*, *Sat-2*, *IGF2*, *ATM*, *RASSF1A*, and other genes
Cervix	Hypermethylation of *CDKN2A/p16*
Colon-Rectum	Aberrant methylation of *MHL1*, *SEPT9*, *IGF2*, *THBD*, *C9orf50*, and other genes
Esophagus	Aberrant methylation of *CDH1*, *PIGR*, *RIN2*, and other genes
Head/Neck	Hypermethylation of *CDKN2A/p16* and *MGMT*
Kidney	Hypermethylation of *TIMP-3*
Leukemia	Hypermethylation of *p15* and chromosomal translocations involving HATs and HMTs
Liver	Aberrant methylation of multiple genes
Lung	Hypermethylation of *CDKN2A/p16*, *p73*, *RARb*, *RASSF1A*, *GSTP1*, *MGMT*, and other genes
Lymphoma/Myeloma	Hypermethylation of *DAPK*
Ovary	Hypermethylation of *BRCA1* and hypomethylation of *SAT2*
Pancreas	Hypermethylation of *APC* and hypomethylation of other genes
Prostate	Hypermethylation of *BRCA2*
Rhabdomyosarcoma	Hypermethylation of *PAX3*
Stomach	Hypomethylation of *Cyclin D2*
Thymus	Hypomethylation of *POMC*
Urothelial	Hypomethylation of Satellite DNA
Uterus	Hypermethylation of *hMLH1* leading to Microsatellite instability

transmission of epigenetic information through multiple generations [1, 105]. In *C. elegans*, transgenerational inheritance has been shown to be mediated by transmission of piRNAs. piRNAs induce a highly-stable, long-term gene silencing, which persists at least through 20 generations. The inheritance of the phenotype then becomes independent of the original piRNA, as is taken over by siRNAs [106]. These siRNAs act by modulating transcriptional gene silencing of histone methyltransferases, with consequent rearrangement of the chromatin structure [106]. Such heterochromatin-like configuration is required for stable silencing [16].

D. melanogaster shows "paramutations," a form of epigenetic inheritance whereby one allele at a gene locus is capable of inducing a structural modification in the paired allele, in the absence of DNA sequence changes, which is then inherited through meiotic divisions [103]. The paramutated allele itself becomes paramutagenic and is capable to epigenetically convert a new paramutable allele. The paramutation has been shown to occur without any chromosome pairing between the paramutagenic and the paramutated loci and is mediated by maternal inheritance of piRNAs.

In the yeast *S. pombe*, transgenerational inheritance has been demonstrated to be mediated by multiple long ncRNAs termed BORDERLINEs, which act in a sequence-independent but locus-dependent manner [104] (Fig. 2). BORDERLINE ncRNAs are processed by Dicer into short RNAs referred to as brdrRNAs [104]. brdrRNAs then compete with H3K9me for binding to the HP1 protein Swi6. This prevents spreading of the HP1 protein Swi6 and histone H3K9 methylation beyond pericentromeric repeat regions and leads to Swi6 removal from chromatin, which counteracts the spreading of heterochromatin into neighboring euchromatin by preventing the spreading of H3K9me [104].

Phenotypic impact of epigenetic heredity

Epigenetic alterations can have strong impact on hereditary disease phenotypes. Altered balance of epigenetic networks has been reported to cause major pathologies, including complex phenotype syndromes and cancer (Tables 2 and 3).

Most hereditary diseases linked to defect of epigenetic control present with multi-organ abnormalities and overall developmental defects. Several of these syndromes are characterized by mental retardation and other central nervous system defects. Bone and cartilage growth abnormalities are also frequent, consistent with a key role of epigenetic regulation in body development [2, 3].

Notably, several epigenetic, hereditary syndromes are also characterized by gross chromosomal anomalies. This is consistent with the role that epigenetic mechanisms that play in the regulation of chromosome architecture and

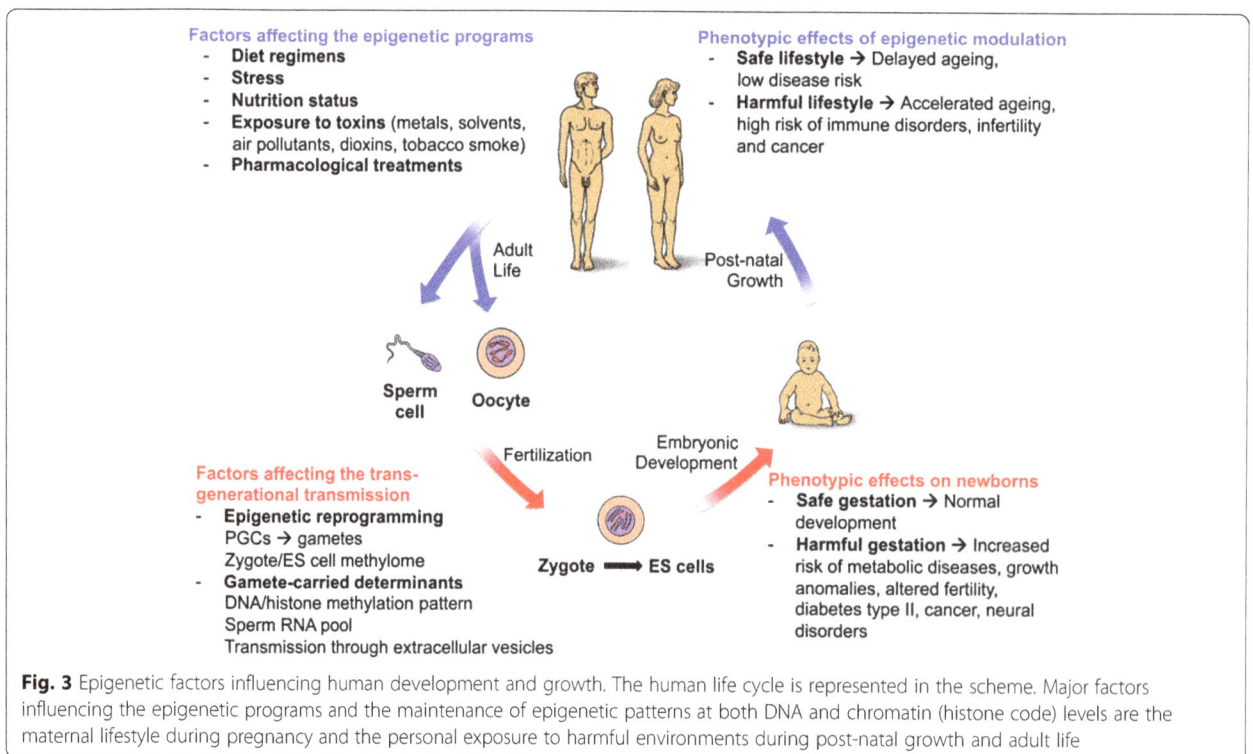

Fig. 3 Epigenetic factors influencing human development and growth. The human life cycle is represented in the scheme. Major factors influencing the epigenetic programs and the maintenance of epigenetic patterns at both DNA and chromatin (histone code) levels are the maternal lifestyle during pregnancy and the personal exposure to harmful environments during post-natal growth and adult life

maintenance of genomic stability [28, 29], as well as in modulation of regulatory networks that involve p53.

Epigenetic changes can also have a major role in the development of cancer [107]. Most studied examples include patients with sporadic colorectal cancer with a microsatellite instability phenotype that shows methylation and silencing of the gene encoding MLH1 (Table 3), indicating that epigenetic silencing can result directly in genomic instability in transformed cells [108, 109]. However, epigenetic regulation of key oncogenes/tumor suppressor genes appears much more widespread than commonly appreciated, as major targets include cyclins, cyclin inhibitors, APC, BRCA1, retinoic acid receptors, protease modulators, IGF2, and transcription factors associated with epithelial-mesenchymal transition, e.g. Twist.

Conclusions

The genomes of eukaryotic organisms are adaptable to non-genetic/environmental-driven changes through epigenetic modulation of gene expression across generation cycles (Fig. 3). Epigenetic modifications include DNA methylation and histone modifications. These have the ability to alter gene expression patterns without affecting the nucleotide sequence of the underlying genome, the only exception being deamination of 5-mC to thymine. Tight regulation of the activity of DNA methyltransferases as well as of demethylases plays a mechanistic role in the establishment, maintenance and transient erasure of DNA methylation patterns. Correspondingly, post-translational modifications of histones and of regulatory proteins were shown to play a role in hereditary transmission of chromatin composition and configuration states, both in mammals and in model organisms. Thus, epigenetic programs contribute to the transgenerational inheritance of complex traits, which may help accounting for the "missing heritability" in current GWAS studies.

Abbreviations

5-caC: 5-carboxylcytosine; 5-fC: 5-formylcytosine; 5-hmC: 5-hydroxymethyl cytosine; 5-hmU: 5-hydromethyluracil; 5-mC: 5-methyl cytosine; BER: base excision repair; DNMT: DNA methyltransferase; dsRNA: double-stranded RNA; EGFP: enhanced green fluorescent protein; ES: embryonic stem; GWAS: genome-wide association studies; HDAC: histone deacetylase; ICM: inner cell mass; ICR: imprinting control regions; ncRNA: non-coding RNA; PcG: polycomb-group; PGC: primordial germ cells; PTMs: post-translational modifications; T: thymine; TDG: thymine DNA glycosylase; TET: ten-eleven translocation.

Competing interests

The authors declare that they have no competing interests.

Authors' contributions

MT, PS, and SA wrote the manuscript. MT and VR generated the figure panels. All authors contributed to discussing the content of the article. All authors read and approved the final manuscript.

Acknowledgements

This study was supported by grants by the Fondazione of the Cassa di Risparmio della Provincia di Chieti, the Oncoxx Biotech, the Italian Ministry of Health (RicOncol RF-EMR-2006-361866), the Compagnia di San Paolo (Grant 2489IT), and the Ministry of Development—Made in Italy (contract MI0100424).

References

1. Lim JP, Brunet A. Bridging the transgenerational gap with epigenetic memory. Trends Genet. 2013;29:176–86.
2. Tripaldi R, Stuppia L, Alberti S. Human height genes and cancer. BBA Reviews Cancer. 1836;2013:27–41.
3. Simeone P, Alberti S. Epigenetic heredity of human height. Physiological reports. 2014;2:e12047.
4. Marian AJ. Elements of "missing heritability". Curr Opin Cardiol. 2012;27:197–201.
5. Zuk O, Hechter E, Sunyaev SR, Lander ES. The mystery of missing heritability: Genetic interactions create phantom heritability. Proc Natl Acad Sci U S A. 2012;109:1193–8.
6. Zuk O, Schaffner SF, Samocha K, Do R, Hechter E, Kathiresan S, et al. Searching for missing heritability: designing rare variant association studies. Proc Natl Acad Sci U S A. 2014;111:E455–464.
7. Koch L. Epigenetics: an epigenetic twist on the missing heritability of complex traits. Nat Rev Genet. 2014;15:218.
8. Manolio TA, Collins FS, Cox NJ, Goldstein DB, Hindorff LA, Hunter DJ, et al. Finding the missing heritability of complex diseases. Nature. 2009;461:747–53.
9. Park JH, Wacholder S, Gail MH, Peters U, Jacobs KB, Chanock SJ, et al. Estimation of effect size distribution from genome-wide association studies and implications for future discoveries. Nat Genet. 2010;42:570–5.
10. Gudbjartsson DF, Walters GB, Thorleifsson G, Stefansson H, Halldorsson BV, Zusmanovich P, et al. Many sequence variants affecting diversity of adult human height. Nat Genet. 2008;40:609–15.
11. Lettre G, Jackson AU, Gieger C, Schumacher FR, Berndt SI, Sanna S, et al. Identification of ten loci associated with height highlights new biological pathways in human growth. Nat Genet. 2008;40:584–91.
12. Weedon MN, Lango H, Lindgren CM, Wallace C, Evans DM, Mangino M, et al. Genome-wide association analysis identifies 20 loci that influence adult height. Nat Genet. 2008;40:575–83.
13. Cho YS, Go MJ, Kim YJ, Heo JY, Oh JH, Ban HJ, et al. A large-scale genome-wide association study of Asian populations uncovers genetic factors influencing eight quantitative traits. Nat Genet. 2009;41:527–34.
14. Okada Y, Kamatani Y, Takahashi A, Matsuda K, Hosono N, Ohmiya H, et al. A genome-wide association study in 19 633 Japanese subjects identified LHX3-QSOX2 and IGF1 as adult height loci. Hum Mol Genet. 2010;19:2303–12.
15. Yang J, Benyamin B, McEvoy BP, Gordon S, Henders AK, Nyholt DR, et al. Common SNPs explain a large proportion of the heritability for human height. Nat Genet. 2010;42:565–9.
16. Szyf M. Nongenetic inheritance and transgenerational epigenetics. Trends Mol Med. 2015;21:134–44.
17. Laforsch C, Tollrian R. Embryological aspects of inducible morphological defenses in Daphnia. Journal of Morphology. 2004;262:701–7.
18. Ziller MJ, Gu H, Muller F, Donaghey J, Tsai LT, Kohlbacher O, et al. Charting a dynamic DNA methylation landscape of the human genome. Nature. 2013;500:477–81.
19. Rivera CM, Ren B. Mapping human epigenomes. Cell. 2013;155:39–55.
20. Okano M, Bell DW, Haber DA, Li E. DNA methyltransferases Dnmt3a and Dnmt3b are essential for de novo methylation and mammalian development. Cell. 1999;99:247–57.
21. Chen ZX, Mann JR, Hsieh CL, Riggs AD, Chedin F. Physical and functional interactions between the human DNMT3L protein and members of the de novo methyltransferase family. J Cell Biochem. 2005;95:902–17.
22. Suetake I, Shinozaki F, Miyagawa J, Takeshima H, Tajima S. DNMT3L stimulates the DNA methylation activity of Dnmt3a and Dnmt3b through a direct interaction. J Biol Chem. 2004;279:27816–23.

23. Baubec T, Colombo DF, Wirbelauer C, Schmidt J, Burger L, Krebs AR, et al. Genomic profiling of DNA methyltransferases reveals a role for DNMT3B in genic methylation. Nature. 2015;520(7546):243–7.

24. Kulis M, Queiros AC, Beekman R, Martin-Subero JI. Intragenic DNA methylation in transcriptional regulation, normal differentiation and cancer. Biochim Biophys Acta. 1829;2013:1161–74.

25. Lujambio A, Portela A, Liz J, Melo SA, Rossi S, Spizzo R, et al. CpG island hypermethylation-associated silencing of non-coding RNAs transcribed from ultraconserved regions in human cancer. Oncogene. 2010;29(48):6390–401.

26. Ehrlich M, Lacey M. DNA methylation and differentiation: silencing, upregulation and modulation of gene expression. Epigenomics. 2013;5:553–68.

27. Henckel A, Nakabayashi K, Sanz LA, Feil R, Hata K, Arnaud P. Histone methylation is mechanistically linked to DNA methylation at imprinting control regions in mammals. Hum Mol Genet. 2009;18(18):3375–83.

28. Alberti S, Nutini M, Herzenberg LA. DNA methylation prevents the amplification of TROP1, a tumor associated cell surface antigen gene. Proc Natl Acad Sci USA. 1994;91:5833–7.

29. Nasr AF, Nutini M, Palombo B, Guerra E, Alberti S. Mutations of TP53 induce loss of DNA methylation and amplification of the TROP1 gene. Oncogene. 2003;22:1668–77.

30. Tahiliani M, Koh KP, Shen Y, Pastor WA, Bandukwala H, Brudno Y, et al. Conversion of 5-methylcytosine to 5-hydroxymethylcytosine in mammalian DNA by MLL partner TET1. Science. 2009;324:930–5.

31. Hackett JA, Sengupta R, Zylicz JJ, Murakami K, Lee C, Down TA, et al. Germline DNA demethylation dynamics and imprint erasure through 5-hydroxymethylcytosine. Science. 2013;339:448–52.

32. Ficz G, Branco MR, Seisenberger S, Santos F, Krueger F, Hore TA, et al. Dynamic regulation of 5-hydroxymethylcytosine in mouse ES cells and during differentiation. Nature. 2011;473:398–402.

33. He YF, Li BZ, Li Z, Liu P, Wang Y, Tang Q, et al. Tet-mediated formation of 5-carboxylcytosine and its excision by TDG in mammalian DNA. Science. 2011;333:1303–7.

34. Song CX, Szulwach KE, Dai Q, Fu Y, Mao SQ, Lin L, et al. Genome-wide profiling of 5-formylcytosine reveals its roles in epigenetic priming. Cell. 2013;153:678–91.

35. Sarkar DK. Male germline transmits fetal alcohol epigenetic marks for multiple generations: a review. Addiction Biology. 2015. doi:10.1111/adb.12186.

36. Esteve PO, Chin HG, Benner J, Feehery GR, Samaranayake M, Horwitz GA, et al. Regulation of DNMT1 stability through SET7-mediated lysine methylation in mammalian cells. Proc Natl Acad Sci U S A. 2009;106:5076–81.

37. Esteve PO, Chang Y, Samaranayake M, Upadhyay AK, Horton JR, Feehery GR, et al. A methylation and phosphorylation switch between an adjacent lysine and serine determines human DNMT1 stability. Nat Struct Mol Biol. 2011;18:42–8.

38. Shamma A, Suzuki M, Hayashi N, Kobayashi M, Sasaki N, Nishiuchi T, et al. ATM mediates pRB function to control DNMT1 protein stability and DNA methylation. Mol Cell Biol. 2013;33:3113–24.

39. Du Z, Song J, Wang Y, Zhao Y, Guda K, Yang S, et al. DNMT1 stability is regulated by proteins coordinating deubiquitination and acetylation-driven ubiquitination. Sci Signal. 2010;3:ra80.

40. Pastor WA, Aravind L, Rao A. TETonic shift: biological roles of TET proteins in DNA demethylation and transcription. Nat Rev Mol Cell Biol. 2013;14:341–56.

41. Burns MB, Lackey L, Carpenter MA, Rathore A, Land AM, Leonard B, et al. APOBEC3B is an enzymatic source of mutation in breast cancer. Nature. 2013;494:366–70.

42. Gannon JR, Emery BR, Jenkins TG, Carrell DT. The sperm epigenome: implications for the embryo. Adv Exp Med Biol. 2014;791:53–66.

43. Smith ZD, Chan MM, Humm KC, Karnik R, Mekhoubad S, Regev A, et al. DNA methylation dynamics of the human preimplantation embryo. Nature. 2014;511:611–5.

44. Guo F, Yan L, Guo H, Li L, Hu B, Zhao Y, et al. The transcriptome and DNA methylome landscapes of human primordial germ cells. Cell. 2015;161:1437–52.

45. Tang WW, Dietmann S, Irie N, Leitch HG, Floros VI, Bradshaw CR, et al. A unique gene regulatory network resets the human germline epigenome for development. Cell. 2015;161:1453–67.

46. Gkountela S, Zhang KX, Shafiq TA, Liao WW, Hargan-Calvopina J, Chen PY, et al. DNA demethylation dynamics in the human prenatal germline. Cell. 2015;161:1425–36.

47. Borgel J, Guibert S, Li Y, Chiba H, Schubeler D, Sasaki H, et al. Targets and dynamics of promoter DNA methylation during early mouse development. Nat Genet. 2010;42:1093–100.

48. Wang L, Zhang J, Duan J, Gao X, Zhu W, Lu X, et al. Programming and inheritance of parental DNA methylomes in mammals. Cell. 2014;157:979–91.

49. Seisenberger S, Peat JR, Hore TA, Santos F, Dean W, Reik W. Reprogramming DNA methylation in the mammalian life cycle: building and breaking epigenetic barriers. Philos Trans R Soc Lond B Biol Sci. 2013;368:20110330.

50. Meissner A, Mikkelsen TS, Gu H, Wernig M, Hanna J, Sivachenko A, et al. Genome-scale DNA methylation maps of pluripotent and differentiated cells. Nature. 2008;454:766–70.

51. Reik W. Stability and flexibility of epigenetic gene regulation in mammalian development. Nature. 2007;447:425–32.

52. Santos F, Hendrich B, Reik W, Dean W. Dynamic reprogramming of DNA methylation in the early mouse embryo. Dev Biol. 2002;241:172–82.

53. Santos J, Pereira CF, Di-Gregorio A, Spruce T, Alder O, Rodriguez T, et al. Differences in the epigenetic and reprogramming properties of pluripotent and extra-embryonic stem cells implicate chromatin remodelling as an important early event in the developing mouse embryo. Epigenetics Chromatin. 2010;3:1.

54. Krishnakumar R, Blelloch RH. Epigenetics of cellular reprogramming. Curr Opin Genet Dev. 2013;23:548–55.

55. Papp B, Plath K. Epigenetics of reprogramming to induced pluripotency. Cell. 2013;152:1324–43.

56. Matoba S, Liu Y, Lu F, Iwabuchi KA, Shen L, Inoue A, et al. Embryonic development following somatic cell nuclear transfer impeded by persisting histone methylation. Cell. 2014;159:884–95.

57. Hammoud SS, Nix DA, Zhang H, Purwar J, Carrell DT, Cairns BR. Distinctive chromatin in human sperm packages genes for embryo development. Nature. 2009;460(7254):473–8.

58. Brykczynska U, Hisano M, Erkek S, Ramos L, Oakeley EJ, Roloff TC, et al. Repressive and active histone methylation mark distinct promoters in human and mouse spermatozoa. Nat Struct Mol Biol. 2010;17:679–87.

59. Alberti S, Herzenberg LA. DNA methylation prevents transfection of genes for specific surface antigens. Proc Natl Acad Sci USA. 1988;85:8391–4.

60. Dias BG, Ressler KJ. Parental olfactory experience influences behavior and neural structure in subsequent generations. Nat Neurosci. 2014;17:89–96.

61. Carone BR, Fauquier L, Habib N, Shea JM, Hart CE, Li R, et al. Paternally induced transgenerational environmental reprogramming of metabolic gene expression in mammals. Cell. 2010;143:1084–96.

62. Rassoulzadegan M, Grandjean V, Gounon P, Vincent S, Gillot I, Cuzin F. RNA-mediated non-mendelian inheritance of an epigenetic change in the mouse. Nature. 2006;441:469–74.

63. Liebers R, Rassoulzadegan M, Lyko F. Epigenetic regulation by heritable RNA. PLoS Genet. 2014;10:e1004296.

64. Johnson GD, Mackie P, Jodar M, Moskovtsev S, Krawetz SA. Chromatin and extracellular vesicle associated sperm RNAs. Nucleic Acids Res. 2015 [Epub ahead of print].

65. Cossetti C, Lugini L, Astrologo L, Saggio I, Fais S, Spadafora C. Soma-to-germline transmission of RNA in mice xenografted with human tumour cells: possible transport by exosomes. PLoS One. 2014;9:e101629.

66. Sharma A. Transgenerational epigenetic inheritance requires a much deeper analysis. Trends Mol Med. 2015;21:269–70.

67. Heard E, Martiensssen RA. Transgenerational epigenetic inheritance: myths and mechanisms. Cell. 2014;157:95–109.

68. Silventoinen K, Kaprio J, Lahelma E, Koskenvuo M. Relative effect of genetic and environmental factors on body height: differences across birth cohorts among Finnish men and women. Am J Public Health. 2000;90:627–30.

69. Waterland RA, Kellermayer R, Laritsky E, Rayco-Solon P, Harris RA, Travisano M, et al. Season of conception in rural gambia affects DNA methylation at putative human metastable epialleles. PLoS Genet. 2010;6:e1001252.

70. Dominguez-Salas P, Moore SE, Baker MS, Bergen AW, Cox SE, Dyer RA, et al. Maternal nutrition at conception modulates DNA methylation of human metastable epialleles. Nat Commun. 2014;5:3746.

71. Dolinoy DC, Huang D, Jirtle RL. Maternal nutrient supplementation counteracts bisphenol A-induced DNA hypomethylation in early development. Proc Natl Acad Sci U S A. 2007;104:13056–61.

72. Davis AP, Murphy CG, Saraceni-Richards CA, Rosenstein MC, Wiegers TC, Mattingly CJ. Comparative Toxicogenomics Database: a knowledgebase and

72. discovery tool for chemical-gene-disease networks. Nucleic Acids Res. 2009;37:D786–792.

73. Singh S, Li SS. Phthalates: toxicogenomics and inferred human diseases. Genomics. 2011;97:148–57.

74. Singh S, Li SS. Bisphenol A and phthalates exhibit similar toxicogenomics and health effects. Gene. 2012;494:85–91.

75. Berdasco M, Esteller M. Aberrant epigenetic landscape in cancer: how cellular identity goes awry. Dev Cell. 2010;19:698–711.

76. Kvigne VL, Leonardson GR, Borzelleca J, Brock E, Neff-Smith M, Welty TK. Alcohol use, injuries, and prenatal visits during three successive pregnancies among American Indian women on the Northern Plains who have children with fetal alcohol syndrome or incomplete fetal alcohol syndrome. Maternal and Child Health Journal. 2008;12 Suppl 1:37–45.

77. Kvigne VL, Leonardson GR, Borzelleca J, Welty TK. Characteristics of grandmothers who have grandchildren with fetal alcohol syndrome or incomplete fetal alcohol syndrome. Maternal and Child Health Journal. 2008;12:760–5.

78. Baccarelli A, Bollati V. Epigenetics and environmental chemicals. Current Opinion in Pediatrics. 2009;21:243–51.

79. Maccani MA, Avissar-Whiting M, Banister CE, McGonnigal B, Padbury JF, Marsit CJ. Maternal cigarette smoking during pregnancy is associated with downregulation of miR-16, miR-21, and miR-146a in the placenta. Epigenetics. 2010;5:583–9.

80. Avissar-Whiting M, Veiga KR, Uhl KM, Maccani MA, Gagne LA, Moen EL, et al. Bisphenol A exposure leads to specific microRNA alterations in placental cells. Reprod Toxicol. 2010;29:401–6.

81. Esteller M. Epigenetics in cancer. N Engl J Med. 2008;358:1148–59.

82. Anway MD, Cupp AS, Uzumcu M, Skinner MK. Epigenetic transgenerational actions of endocrine disruptors and male fertility. Science. 2005;308:1466–9.

83. Gapp K, Jawaid A, Sarkies P, Bohacek J, Pelczar P, Prados J, et al. Implication of sperm RNAs in transgenerational inheritance of the effects of early trauma in mice. Nat Neurosci. 2014;17:667–9.

84. Rose NR, Klose RJ. Understanding the relationship between DNA methylation and histone lysine methylation. Biochim Biophys Acta. 2014;1839(12):1362–72.

85. Guo X, Wang L, Li J, Ding Z, Xiao J, Yin X, et al. Structural insight into autoinhibition and histone H3-induced activation of DNMT3A. Nature. 2014;517(7536):640–4.

86. Nakamura T, Liu YJ, Nakashima H, Umehara H, Inoue K, Matoba S, et al. PGC7 binds histone H3K9me2 to protect against conversion of 5mC to 5hmC in early embryos. Nature. 2012;486:415–9.

87. Arnaudo AM, Garcia BA. Proteomic characterization of novel histone post-translational modifications. Epigenetics Chromatin. 2013;6:24.

88. Rothbart SB, Strahl BD. Interpreting the language of histone and DNA modifications. Biochim Biophys Acta. 1839;2014:627–43.

89. Christophorou MA, Castelo-Branco G, Halley-Stott RP, Oliveira CS, Loos R, Radzisheuskaya A, et al. Citrullination regulates pluripotency and histone H1 binding to chromatin. Nature. 2014;507:104–8.

90. Campos EI, Stafford JM, Reinberg D. Epigenetic inheritance: histone bookmarks across generations. Trends Cell Biol. 2014;24(11):664–74.

91. Strahl BD, Allis CD. The language of covalent histone modifications. Nature. 2000;403:41–5.

92. Dhalluin C, Carlson JE, Zeng L, He C, Aggarwal AK, Zhou MM. Structure and ligand of a histone acetyltransferase bromodomain. Nature. 1999;399:491–6.

93. Fouse SD, Shen Y, Pellegrini M, Cole S, Meissner A, Van Neste L, et al. Promoter CpG methylation contributes to ES cell gene regulation in parallel with Oct4/Nanog, PcG complex, and histone H3 K4/K27 trimethylation. Cell Stem Cell. 2008;2:160–9.

94. Yu Q. Cancer gene silencing without DNA hypermethylation. Epigenetics. 2008;3:315–7.

95. Ipenberg I, Guttmann-Raviv N, Khoury HP, Kupershmit I, Ayoub N. Heat shock protein 90 (Hsp90) selectively regulates the stability of KDM4B/JMJD2B histone demethylase. J Biol Chem. 2013;288:14681–7.

96. Annunziato AT. Split decision: what happens to nucleosomes during DNA replication? J Biol Chem. 2005;280:12065–8.

97. Probst AV, Dunleavy E, Almouzni G. Epigenetic inheritance during the cell cycle. Nat Rev Mol Cell Biol. 2009;10:192–206.

98. Alabert C, Groth A. Chromatin replication and epigenome maintenance. Nat Rev Mol Cell Biol. 2012;13:153–67.

99. Alabert C, Barth TK, Reveron-Gomez N, Sidoli S, Schmidt A, Jensen ON, et al. Two distinct modes for propagation of histone PTMs across the cell cycle. Genes Dev. 2015;29:585–90.

100. Apostolou E, Hochedlinger K. Chromatin dynamics during cellular reprogramming. Nature. 2013;502:462–71.

101. Ragunathan K, Jih G, Moazed D. Epigenetics. Epigenetic inheritance uncoupled from sequence-specific recruitment Science. 2015;348:1258699.

102. Audergon PN, Catania S, Kagansky A, Tong P, Shukla M, Pidoux AL, et al. Epigenetics. Restricted epigenetic inheritance of H3K9 methylation. Science. 2015;348:132–5.

103. de Vanssay A, Bouge AL, Boivin A, Hermant C, Teysset L, Delmarre V, et al. Paramutation in Drosophila linked to emergence of a piRNA-producing locus. Nature. 2012;490:112–5.

104. Keller C, Kulasegaran-Shylini R, Shimada Y, Hotz HR, Buhler M. Noncoding RNAs prevent spreading of a repressive histone mark. Nat Struct Mol Biol. 2013;20:994–1000.

105. Greer EL, Beese-Sims SE, Brookes E, Spadafora R, Zhu Y, Rothbart SB, et al. A histone methylation network regulates transgenerational epigenetic memory in C. elegans. Cell Rep. 2014;7:113–26.

106. Castel SE, Martienssen RA. RNA interference in the nucleus: roles for small RNAs in transcription, epigenetics and beyond. Nat Rev Genet. 2013;14:100–12.

107. Esteller M, Corn PG, Baylin SB, Herman JG. A gene hypermethylation profile of human cancer. Cancer Res. 2001;61:3225–9.

108. Arnold CN, Goel A, Compton C, Marcus V, Niedzwiecki D, Dowell JM, et al. Evaluation of Microsatellite Instability, hMLH1 Expression and hMLH1 Promoter Hypermethylation in Defining the MSI Phenotype of Colorectal Cancer. Cancer Biol Ther. 2004;3.

109. Hawkins NJ, Ward RL. Sporadic colorectal cancers with microsatellite instability and their possible origin in hyperplastic polyps and serrated adenomas. J Natl Cancer Inst. 2001;93:1307–13.

8

A survey of computational tools for downstream analysis of proteomic and other omic datasets

Anis Karimpour-Fard[1*], L. Elaine Epperson[2] and Lawrence E. Hunter[1]

Abstract

Proteomics is an expanding area of research into biological systems with significance for biomedical and therapeutic applications ranging from understanding the molecular basis of diseases to testing new treatments, studying the toxicity of drugs, or biotechnological improvements in agriculture. Progress in proteomic technologies and growing interest has resulted in rapid accumulation of proteomic data, and consequently, a great number of tools have become available. In this paper, we review the well-known and ready-to-use tools for classification, clustering and validation, interpretation, and generation of biological information from experimental data. We suggest some rules of thumb for the reader on choosing the best suitable learning method for a particular dataset and conclude with pathway and functional analysis and then provide information about submitting final results to a repository.

Keywords: Proteomics, Machine learning, Random forests, PLS, PCA, SVM, Proteomics repository

Introduction

Proteomics, the assessment and quantitation of protein expression changes in a given type of biological sample, contributes heavily to current views in modern biology, genetics, biochemistry, and environmental sciences. Expression proteomics studies investigate the presence or absence patterns of proteins in disease compared to normal using a mass spectrometry approach often preceded by gel separation methods. Proteomics is a science that focuses on the study of proteins: their roles, their structures, their localization, their interactions, and other factors. Proteomics has emerged as a powerful tool in many different fields and is a technique widely used across biology, mainly applied in disease [1–3], agriculture, and food microbiology. Proteomics is becoming increasingly important for the study of many different aspects of plant functions. For example, it is used to help identify candidate proteins involved in the defensive response of plants to herbivorous insects [4, 5]. In agriculture, a proteomic approach was used to investigate population growth and the effect of global climate changes on crop production [6]. In food technology, proteomics is utilized for characterization and standardization of raw materials, process development, and detection of batch-to-batch variations and quality control of the final product, in particular to food safety in terms of microbial content and the use of genetically modified foods [7]. The study of interactions between microbial pathogens and their hosts is called "infectomics" and comprises a growing area of interest in proteomics [8].

A protein may exist in multiple forms within a cell or cell type. These protein isoforms derive from transcriptional, post-transcriptional, translational, post-translational, regulatory, and degrading and preserving processes that affect protein structure, localization, function, and turnover. The field has thus evolved to include a variety of methods for separation of complex protein samples followed by identification using mass spectrometry. It is inherently a systems science that considers not only protein abundances in a cell but also the interplay of proteins, protein complexes, signaling pathways, and networks. To address the relevant challenges, we categorize the analytical tools into three types: (1) basic traditional statistical analysis, (2) machine learning approaches, and (3) assignment of functional and biological information to describe and understand protein interaction networks.

Traditional statistics is used as a critical first pass to identify the "low-hanging fruit" in the dataset. Methods such as t test and its nonparametric equivalent, the Wilcoxon test, univariate, or analysis of variance (ANOVA)

* Correspondence: anis.karimpour-fard@ucdenver.edu
[1]Department of Pharmacology, University of Colorado School of Medicine, Aurora, CO 80045, USA
Full list of author information is available at the end of the article

are applied to identify the significant proteins. Due to inherent variability, statistics alone is often insufficient to discover most of the biologically relevant information in a proteomic dataset but is an important first step of every analysis. For the purposes of this review, we focus mainly on approaches that are more specific to proteomic and other "omic" data. But statistically significant results are very useful as seed data or bait in the machine learning approaches.

Machine learning classification complements traditional statistics as it allows for consideration of many variables at once and also removes much of the researcher bias. Dataset complexity is reduced as correlations, and trends are identified that may not withstand statistical scrutiny or may be undetectable using traditional statistics, e.g., clustering using iterative subsampling. Machine learning also bypasses researcher bias by revealing patterns within the data that may not relate to the original hypothesis or that relate in an unanticipated manner. The researcher is then able to examine the clustering or classification results for new biological features that were not initially predicted. Thus, in addition to being potentially inconsistent with the hypotheses of any particular researcher, machine learning and network tools enable hypothesis generation as they uncover the real biology of the system in question. Swan et al. [9] discussed the benefit of machine learning methods for application to proteomic data and show that machine learning methods give an overall view of data and also offer a large potential for identifying relevant information within data.

Pathway analysis following statistical analysis and classification and clustering can help organize a long list of proteins onto a short list of pathway knowledge maps, easing interpretation of the molecular mechanisms underlying altered proteins or their expressions [10].

Here we primarily review tools for machine learning and clustering of omic data. The machine learning section of this review will introduce the concept of supervised and unsupervised classification for seven types of machine learners: principal component analysis (PCA), independent component analysis (ICA), K-means, hierarchical clustering, partial least square (PLS), random forests (RF), and support vector machines (SVM). These methods are also summarized and compared in Table 1, which provides an overview of different machine learning and clustering tools and how to select a method most likely to be effective for a specific dataset. We include a brief discussion of experimental design and feature selection, i.e., the selection of significant attributes for reduction of datasets, with the aim to increase the accuracy of classification models that are applied to the selected features. The machine learning and clustering section is followed by a brief summary of tools for

analysis of longitudinal (time series) data. Next, we discuss tools that can achieve automated learning of pathway modules and features and those that help perform integrated network visual analytics. Finally, we provide information for public repository of proteomics data.

Experimental design

Although the purpose of this review is to discuss tools that are useful for data analysis after completion of a proteomic experiment, we want to recognize the essential nature of thoughtful upfront experimental design. Sample groups should be as large and reproducible as possible, representing a consistent proteomic phenotype in the harvested sample for a particular sample group. Even when the researcher is not establishing a study in a prospective manner, samples and sample groups should be chosen to reflect this insofar as is possible for the researcher. For example, if the experimental purpose is to find changes in the mouse hypothalamus with respect to circadian rhythm, the surgeries should—ideally—be performed by the same researcher at precise times of the day until a minimum of five or six samples, preferably more, are collected for every treatment group in question. The power of the experiment increases with each additional sample. Treatment groups should ideally be similar in size. Consistent collection, storage, and sample handling during the experiment will greatly increase chances of high-quality omic data. Furthermore, reduction of a sample to fewer or a specific cell type will increase the quality of proteomic or RNA data. Gene expression is a cell-type-specific phenomenon so that, in order to increase the signal-to-noise ratio for a gene expression study, the experimental design should consider tissue and sample complexity. A protein extract from liver, for example, primarily comprises hepatocyte proteins, whereas the brain contains cells that express hugely variable mRNA and protein signatures. We encourage the researcher to plan carefully regarding experimental design, as this investment will yield greatly improved resulting data. For review of experimental design, see [11, 12].

Guidelines for analyzing a large dataset

The following guidelines are listed as sequential steps, but they are meant to be more of a frame for thought rather than rigid steps in a series. For example, steps one and three may overlap and provide answers to the main questions of the experiment. Step two may obviate the need to perform extensive machine learning. Our hope is to relieve the distress of inheriting or creating an enormous mass of data that seems impenetrable.

Step one: Observe your data, quality control
Observe your data by creating plots and descriptive statistics to assess data distribution, overall variation, and

Table 1 Summary and comparison of classification and clustering methods

	Classification					Clustering	
	PCA	ICA	RF	PLS	SVM	K-means	Hierarchical
What does it do?	Separates features into groups based on commonality and reports the weight of each component's contribution to the separation	Separates features into groups by eliminating correlation and reports the weight of each component's contribution to the separation	Separates features into groups based on commonality; identifies important predictors	Separates features into groups based on maximal covariation and reports the contribution of each variable	Uses a user-specified kernel function to quantify the similarity between any pair of instances and create a classifier	Separates features into clusters of similar expression patterns	Clusters treatment groups, features, or samples into a dendrogram
By what mechanism?	Orthogonal transformation; transfers a set of correlated variables into a new set of uncorrelated variables	Nonlinear, non-orthogonal transformation; standardizes each variable to a unit variance and zero mean	Uses an ensemble classifier that consists of many decision trees	Multivariate regression	Finds a decision boundary maximizing the distance to nearby positive and negative examples	Compares and groups magnitudes of changes in the means into K clusters where K is defined by the user	Compares all samples using either agglomerative or divisive algorithms with distance and linkage functions
Strengths	Unsupervised, nonparametric, useful for reducing dimensions before using supervision	Works well when other approaches do not because data are not normally distributed	Robust to outliers and noise; gives useful internal estimates of error; resistant to overtraining	Diverse experiments that have the same features are made comparable; variables can outnumber features	Robust to outliers, gives useful internal estimates of error, can exploit knowledge of the domain if using appropriate kernel functions	Easily visualized and intuitive; greatly reduces complexity; performs well when distance information between data points is important to clustering	Unsupervised; easily visualized and intuitive
Weaknesses	Number of features must exceed number of treatment groups	Features are assumed to be independent when they actually may be dependent	Does not allow missing data (requires imputation to replace missing values)	Fails to deal with data containing outliers	Selection of an inappropriate kernel yields poor results	Sensitive to initial conditions and specified number of clusters (K)	Does not provide feature contributions; not iterative, therefore, sensitive to cluster distance measures and noise/outliers
More information			Performance depends on number of trees and varies among experiments	Supervised; requires training and testing; groups pre-defined	Supervised; requires training and testing; many good kernel functions have been described, e.g, based on structural alignment	Tools are available to determine the optimal cluster count (K)	User does not define the number of clusters
Sample size/data characteristics	Unlimited sample size, data normally distributed	Unlimited sample size; data non-normally distributed	Performs well on small sample size and is resistant to over-fitting	Unlimited sample size; sensitive to outliers	Performs well on small sample size and resistant to over-fitting	Performs best with a limited dataset, i.e., ~20 to 300 features	Performs best with limited dataset, i.e., ~20 to 300 features or samples

variability within each treatment group. Compare means and variability from those means. Look for any anomalies that could cause a problem in the analysis. Plotting the data is effectively the first unsupervised clustering step. How do the data cluster? Are the data normally distributed? Most parametric statistical approaches assume normality, so if data are not normally distributed, they may need to be transformed or analyzed using nonparametric methods. Curves, scatter plots, and boxplots are useful for observing comparability of different groups or whether two different datasets can be combined. Is there a batch effect? If so, the data must be normalized or corrected for this effect. If using unsupervised approaches such as hierarchical clustering or principal component analysis, do the subjects partition according to predicted treatment groups? Correlation plots can be used to compare treatment groups. Are the correlations as expected?

Step two: Traditional statistics
Groups identified by the researcher either during experimental design or during the data observation step can be compared here using Student's t test, analysis of variance (ANOVA), and their nonparametric equivalents such as Kruskal-Wallis, in addition to regression modeling and other tests of traditional statistics. Many tests done simultaneously should be corrected using a multiple test correction such as the Benjamini-Hochberg correction algorithm [13]. If these tests yield an abundance of significant data, the machine learning methods of step three can be used to reduce dimensionality. These lists of significant features can be used directly for pathway analysis. Or alternately, these significant features can be used as a seed or paradigm for training the supervised machine learning methods in step three to retrieve interesting data that were not found to be significant by traditional statistical methods.

For example, suppose we identify 100 significant features (proteins, transcripts, etc.) after multiple test correction. These 100 can be tested internally for correlation, for pattern recurrence, and for pathway analysis (DAVID, GO, Ingenuity, etc., Table 2). Suppose we used K-means to look for ten patterns, and one of the ten patterns happens to contain five features whose expression profiles appear to match what we know about their biology based on previous experiments or established literature. This is the step we might call "kicking the tires" of this dataset. If gene expression for a few proteins or transcripts follows known patterns, the entire dataset becomes more credible; other significant data can thus be relied upon as informative for further analysis and for interrogating the rest of the data.

From these lists, one can transition directly to pathway analysis (step four), or these data can be used for classification of the rest of the dataset using machine learning methods.

Step three: Dimension reduction with machine learning
The "curse of dimensionality" is inherent to large datasets. At the beginning of any large dataset analysis, the dimension count and the feature count are the same. The purpose of machine learning is to reduce the dimensions such that multiple features (or data points) are contained within a single dimension so that a dataset with 5000 features may contain 500 groups of ten features each where those ten features have something in common as determined by the classifier such as PCA, RF, and K-means. Thus, machine learning allows the data to partition according to the biology of the experiment, and it allows the researcher to better comprehend the data and the potential biological processes that drive the experimental question.

Many machine learning tools are available including Weka [14], Scikit-learn (Machine Learning in Python) [15], and SHOGUN [16]. R has an enormous number of machine learning algorithms with advanced implementations as well that were written by the developers of the algorithm [17].

If performed independently, machine learning and traditional statistics ought to reveal the same results in the data. They confirm each other. As stated in Table 1, different tools for machine learning are appropriate for different datasets. The observation of data in step one will help the researcher to identify which statistics and machine learning approaches might prove to be most effective in partitioning the data in question. For example, if data are not normally distributed and transformation of the data is not desirable, one should start by using nonparametric statistical analyses and independent component analysis.

Step four: Pathway analysis
Genes and features of interest are entered into pathway analysis software and tools, which are rapidly increasing in sophistication. Still, we have found that computational tools for pathway analysis should always be supplemented with individual manual research into relevant literature and textbook information for real biological insights. Only when the individual researcher or team is able to absorb the biological implications of the new data will the true understanding take place. The computational tools enable new connections to be established, but the biological story still requires concept synthesis on the part of the researcher.

Machine learning and clustering methods
It is reasonable to assume on biological grounds that the proteins present in the proteomic profile are not fully

Table 2 Summary of functional and network tools

Name	Description	Link	References	Function
KEGG	Kyoto Encyclopedia of Genes and Genomes	http://www.genome.jp/kegg/	Kanehisa and Goto (2000) [76]	Pathway
DAVID	The Database for Annotation, Visualization and Integrated Discovery	http://david.abcc.ncifcrf.gov/	Dennis et al. (2003) [96]	Pathway and functional annotation using GO
PID	Pathway Interaction Database	http://pid.nci.nih.gov/	Schaefer et al. (2009) [97]	Pathway interaction
IPA	Ingenuity Pathway Analysis	http://www.ingenuity.com/		Pathway and functional annotation
Cytoscape	An open source platform for complex network analysis and visualization	http://www.cytoscape.org/	Shannon et al. (2003) [98]	Network visualization
HAPPI	Human Annotated and Predicted Protein Interaction Database	http://bio.informatics.iupui.edu/HAPPI	Chen et al. (2009) [99]	Protein interaction
GSEA	Gene Set Enrichment Analysis	http://www.broadinstitute.org/gsea/	Subramanian et al. (2005) [77]	Pathway analysis and functional annotation
Reactome	Curated database of pathways and reactions (pathway steps)	http://www.reactome.org/	Matthews et al. (2009) [100]	Pathway
BioCarta	Pathway database	http://www.biocarta.com/	Nishimura (2001) [101]	Pathway
HPD	Integrated Human Pathway Database	http://discovery.informatics.iupui.edu/HPD/	Chowbina et al. (2009) [102]	Pathway
PAGED	Pathway and Gene Enrichment Database	http://omictools.com/paged-s3492.html	Huang et al. (2012) [103]	Pathway, functional annotation
HPRDB	Human Protein Reference Database	http://www.hprd.org/	Keshava Prasad, T. S. et al. (2009) [104]	Annotation
DrugBank	Drug Bank	http://www.drugbank.ca/		Combines drug data with drug target
CPDB	Consensus Path DB	http://consensuspathdb.org/	Kamburov, A. et al. (2013) [105]	Interaction networks (protein-protein, genetic, metabolic, signaling, gene regulatory, and drug-target)
BINGO	Biological Network Gene Ontology Tool	http://www.psb.ugent.be/cbd/papers/BiNGO/Home.html	Maere S, Heymans K, and Kuiper M (2005) [106]	Biological network gene ontology
GATHER	Gene Annotation Tool to Help Explain Relationships	http://gather.genome.duke.edu	Chang JT, and Nevins JR. (2006) [84]	Gene annotation tool

independent of each other in vivo. For this reason, a multivariate approach to analysis is preferred because it can address the correlations among variables. Dimension reduction methods project a large number of genes or proteins onto a smaller and more manageable number of features. The art of machine learning starts with the design of appropriate data representations, and better performance is often achieved using features derived from the original input and experimental design of the researcher. Building a feature representation is an opportunity to incorporate domain knowledge into the data and can be very application-specific. Nonetheless, there are a number of generic feature construction methods, including the following: clustering, basic linear transforms of the input variables (PCA/ICA/PLS), more sophisticated linear transforms like spectral transforms (Fourier, Hadamard), convolutions and kernels, and applying simple functions to subsets of variables. Among these techniques, some of the most important approaches include (i) dimensionality reduction, (ii) feature selection, and (iii) feature extraction.

There are many benefits regarding the dimensionality reduction when the datasets have a large number of features. Machine learning algorithms work best when the dimensionality is lower (curse of dimensionality). Additionally, the reduction of dimensionality can eliminate irrelevant features, reduce noise, and produce more robust learning models due to the involvement of fewer features. In general, the dimensionality reduction by selecting new features which are a subset of the old ones is known as feature selection. Three main approaches exist for feature selection, namely the following: embedded, filter, and wrapper approaches [18]. In the case of feature extraction, a new set of features can be created from the initial set that captures all the significant information in a dataset. The creation of new sets of features allows for gathering the described benefits of dimensionality reduction.

Sometimes classifications or clustering decisions are susceptible to high bias (under-fitting) or high variance and low bias (over-fitting). If there is under-fitting that results in a high error rate in both training and test, it might help to (1) add more features, (2) use a more sophisticated model, or (3) employ fewer samples. If the dataset has a high variance and low bias (over-fitting) that results in a low error rate in training but high error rate in the test case, it might help to (1) use fewer features or (2) use more training samples. Over-fitting is usually a more common problem in classification than under-fitting. Over-fitting the data causes the model to fit the noise rather than the actual underlying behavior.

The application of different feature selection techniques usually produces different predictive feature lists, presumably because each method captures different features from the data or the small number of samples.

Classification methods have been used extensively for visualization and classification of high-throughput data. These algorithms group objects based on a similarity metric that is computed for features. There are several issues that can affect the outcome of the methods, including (1) a large number of features, (2) mean of the groups, (3) variance and (4) correlation among groups, (5) distribution of the data, and (6) outliers. Thus, exploiting the hidden structure within a dataset is critical for improving classification selection and accuracy and speed of prediction systems. No free lunch (NFL) theorems previously showed that any two optimization algorithms are equivalent when their performance is averaged across all possible problems [19, 20]. Here we emphasize the importance of the hidden structure of the data in order to achieve superior performance of learning systems.

Supervised machine learning involves training a model based on data samples that have known class labels associated with them. This is in contrast with unsupervised classification, or clustering, where no samples have associated class labels, and instead, samples with similar attribute profiles are grouped together.

Each of the supervised classification methods described can make errors, either by incorrectly identifying an instance as a member of a class (a "false positive") or by incorrectly failing to identify an instance as a member of a class (a "false negative"). The rates of both types of errors can be estimated; the proportion of false positive results is reported using *specificity* and the proportion of false negatives using *sensitivity*. There is often a trade-off between these types of errors; increases in specificity (fewer false positives) often lead to decreases in sensitivity (more false negatives) and vice versa. Some classification methods always treat these types of errors as equally important, but others allow the user to set an explicit trade-off ratio, e.g., telling the classifier that sensitivity is twice as important as specificity or vice versa.

Methods that have adjustable sensitivity/specificity trade-offs are noted in Table 1. There are no "one size fits all" tests in classification or clustering methods, and different datasets can make errors which are specific to that dataset (i.e., the no free lunch theorem).

Unsupervised classification and clustering
Principal component
The principal component analysis (PCA) [21] is a mathematical procedure that transforms a number of possibly correlated variables into a smaller number of uncorrelated variables, which are then ordered by reducing variability. These variables are called principal components. The first principal component accounts for as much of the variability in the data as possible, and each succeeding component accounts for as much of the remaining variability as possible. PCA is an unsupervised analysis tool since samples are classified without including disease status in the training algorithm and best if the variables are standardized, and in most of the implementation, this is done by default. PCA is not only useful as a visualization tool [22]. It also helps to detect outliers and perform quality control. PCA has been widely used in analysis of high-throughput data including proteomic data, e.g., [23–25].

Independent component
Independent component analysis (ICA) [26] is a method for finding underlying factors or components from multidimensional data. ICA is also known as blind signal separation (BSS). PCA and ICA have very different goals, and naturally, they may give quite different results. PCA finds directions of maximal variance (using second-order statistics) while ICA finds directions that maximize independence (using higher order statistics) [27]. ICA maximizes non-Gaussianity and makes the assumption of combinatorial linearity of components, satisfied by removing the correlated data. In contrast to PCA, ICA analysis seeks not a set of orthogonal components but a set of independent components. Two components are independent if any knowledge about one implies nothing about the other, such that independent components (IC) represent different non-overlapping information. Since the number of components can be very high, it is relatively easy for the ICA estimation to over-fit the data.

Safavi et al. used ICA to separate groups of proteins that may be differentially expressed across treatment groups [28]. They also showed that the univariate ANOVA technique with false discovery rate (FDR) correction is very sensitive to the FDR-derived p value, whereas ICA is able to identify and separate differential expression into the correct factors without any p value threshold. Other studies have applied ICA to MS data and have shown that ICA represents a powerful unsupervised technique [29, 30].

K-means

K-means [31, 32] is a popular partitioning method due to its ease of programming, allowing a good trade-off between achieved performance and computational complexity. It performs well when the distance information between data points is important to the clustering. K-means requires the analyst to specify the number of clusters to extract, and there are tools available to determine the appropriate number of clusters [33]. Although this is a widely used technique, it suffers from several drawbacks: K-means does not scale well with high dimensional datasets and is prone to local minima problems. It is sensitive to initial conditions, does not remove undesirable features for clustering, and it is best but even then it is prone to local maxima. In spite of the weaknesses, with thoughtful application, the K-means algorithm is very useful in analysis of proteomics data due to its simple algorithmic assumptions and intuitively clear and interpretable visualization [34, 35].

Hierarchical clustering

Hierarchical clustering outputs a dendrogram tree representation of the data. Leaves are the input patterns and non-leaf nodes represent a hierarchy of groupings. This method comes in two flavors: agglomerative and divisive. Agglomerative algorithms work from the bottom up, with each pattern in a separate cluster. Clusters are then iteratively merged according to some criterion. Conversely, divisive algorithms start from the whole dataset in a single cluster and work top down by iteratively dividing each cluster into two components until all clusters are singletons. Hierarchical clustering suffers from the disadvantage of any merging/division decision being irreversible and any errors being dragged through the rest of the hierarchy (in another word, established mergers cannot be undone). Thus, hierarchical clustering analysis and principal component analysis can be used to identify subgroups on the basis of similarities between the proteins' expression profile. Hierarchical clustering methodologies commonly used in transcriptomic studies have also been performed on proteomic data [36, 37]. The different methods will shed light on different aspects of the data [38, 39].

Supervised classification
Partial least squares

Partial least squares (PLS) [40] is a method of dimensionality reduction that maximizes the covariance between groups. PLS constructs a set of orthogonal components that maximize the sample covariance between the response and the linear combination of the predictor variables. It generalizes and combines the features of PCA and multilinear regression [41, 42]. Through maximizing the covariance of dependent and independent variables,

PLS searches for the components that capture the majority of the information contained in independent variables as well as in the relations between dependent and independent variables. PLS regression is particularly useful when users have a very large set of predictors that are highly collinear. In case of over-fitting, the PLS will (1) reduce the predictors to a smaller set of uncorrelated components—these components are mapped in a new space—and (2) perform least squares regression on the new set of components. Although PLS regression was not originally designed for classification and discrimination problems, it has often been used for this purpose [23, 25, 43–49].

Random forests

Random forests (RF) [50] are another classifier method that consists of many decision trees and can be either supervised or unsupervised. It is a popular method that has gained recognition for its ability to construct robust classifiers and select discriminant variables in proteomics [34, 35, 51–54].

RF is an extension to bagging and uses *de-correlated* trees; it is capable of minimizing the number of selected features. For a given decision tree, a subset of samples is selected to build the tree; the remaining samples are predicted from this tree. Bagging (bootstrap aggregating) can be used as an ensemble method [55]. To see which variables contribute the most to the separation, "importance" measures are computed, e.g., the "mean decrease accuracy" and the Gini index [50].

Principal component analyses are used for dimension reduction, but the reduction is valid only when the number of components (i.e., subjects in a study) is less than the number of features (i.e., measured entities in the experiment). In contrast, random forests can be used when the number of features (metabolites, genes, or proteins) is smaller than the number of subjects. A random forest tends to be resistant to over-fitting and also not very sensitive to outliers. A random forest does not handle missing data, and missing values either need to be eliminated or imputation of missing data is needed.

Support vector machine

Support vector machine (SVM) [56] is a supervised learning method that constructs a hyperplane or set of hyperplanes in a high-dimension or infinite dimensional space. A good separation is achieved when the hyperplane has the largest distance to the nearest training data point of any class (the so-called functional margin).

SVM can be applied to different data types by designing the kernel function for such data; selection of a specific kernel and parameters is usually a trial and error process. A kernel function is one that corresponds to an inner product in some expanded feature space. Kernel methods are a kernel class of algorithms for pattern analysis. Since SVM is using regularization, it is highly resistant to over-fitting, even in cases where the number of attributes is greater than the number of observations. In practice, this depends on the careful choice of a C and kernel parameter. A C parameter is an optimization or regularization parameter which is chosen by the user to allow the SVM to best classify the training set. For larger C, the optimization will choose a smaller margin hyperplane if that does a better job of getting all the training points classified correctly. For a very small value of C, this will cause the optimizer to look for a larger margin-separating hyperplane even if that hyperplane misclassifies more points. SVM has been used in various fields to identify biomarkers including proteomics datasets [57–60].

Longitudinal or time-series data

Several software tools are available that specifically address the problems associated with time-series data. TimeClust is a stand-alone tool which is available for different platforms and allows the clustering of gene expression data collected over time with distance-based, model-based, and template-based methods [61]. There are also several other packages available in R such as maSigPro [62], timecourse [63], BAT [64], betr [65], fpca [66], timeclip [67], rnits [68], and STEM [69].

Python probabilistic graphical query language (pGQL) [70] allows its user to interactively define linear HMM queries on time-course data using rectangular graphical widgets called probabilistic time boxes. The analysis is fully interactive, and the graphical display shows the time courses along with the graphical query. In JAVA, PESTS [71] and OPTricluster [72] both of which are stand-alone with a GUI interface are useful for the clustering of short time-series data in MATLAB. DynamiteC is a dynamic modeling and clustering algorithm which interleaves clustering time-course gene expression data with estimation of dynamic models of their response by biologically meaningful parameters [73].

Pathway analysis

After statistical and/or machine learning analysis, the next challenge is how to extract functional and biological information from a long list of proteins identified or discovered from high-throughput proteomic experiments. In order to provide biological insights into the underlying molecular mechanisms of different conditions [10] or changes involved during the progression of disease as well as identification of potential drug targets [74–76], pathway and network analysis techniques can help to address the challenges of interpretation. We categorize these tools into three types: (1) tools with basic functional information (e.g., GO category analysis),

(2) tools with rich functional information and topological features (e.g., GSEA [77], IPA [78]), and (3) tools with topological features (e.g., Cytoscape [79]).

For pathway analysis, we refer to data analysis that aims to identify activated pathways or pathway modules from functional proteomic data. For network analysis, we refer to data analysis that builds, overlays, visualizes, and infers protein interaction networks from functional proteomics and other systems biology data. It is at this stage that metabolomic and proteomic data intersect to reveal active biological processes in a particular system.

Pathway Commons [80] is publicly available and has pathway information for multiple organisms. Pathways include biochemical interactions, complex assembly, transport and catalysis events, physical interactions involving proteins, DNA, RNA, small molecules and complexes, genetic interactions, and co-expression relationships. HumanCyc plus Pathway Tools [81] provides another set of options. HumanCyc contains well-curated content on human metabolic pathways. The associated Pathway Tools software will let you paint gene expression, proteomics, or metabolomics data onto the HumanCyc pathway map, and Pathway Tools will also perform enrichment analysis. PathVisio [82] is a publicly available pathway editor and visualization and analysis software. 3Omics [83] is a web-based systems biology visualization tool for integrating human transcriptomic, proteomic, and metabolomic data. It covers and connects cascades from transcripts, proteins, and metabolites and provides five commonly used analyses including correlation network, co-expression, phenotype generation, KEGG/HumanCyc pathway enrichment, and GO enrichment. For these tools, the user uploads transcriptome and proteome expression data. The metabolome is inferred using KEGG Pathway. 3Omics derives the relationship between the proteome and the metabolome from the literature.

GSEA [77] enables molecular-signature-based statistical significance testing, which integrates protein functional category information effectively with statistical testing of functional genomics or proteomics results. GATHER [84] is a functional enrichment tool (for KEGG pathways) along with several other categories which provides information for a list of genes/proteins in the context of genes, GO terms, predicted miRNAs, pathways, or diseases. The Protein ANalysis THrough Evolutionary Relationships (PANTHER) [85] classification system is designed to classify proteins (and their genes) to support high-throughput analysis. It combines human curation with gene ontology and utilizes other sources for high-level analysis of protein lists.

A number of visualization tools and plug-ins are available for Cytoscape [79] which can be used for biological network construction.

Ultimately, future tools must support elucidation of complex molecular mechanisms suggested from multi-scale network data and molecular signature data. However, there are still significant challenges in designing next-generation network/pathway analysis tools. Network analysis and pathway analysis have been extensively applied to proteomic datasets, e.g., [75, 86, 87]. Some of the pathway and network analysis tools that have become available in the last decade are listed in Table 2. Although the content of most of these tools is based on knowledge and is freely available, a user might not be able to reproduce the same result using a different selection of tools. These tools integrate information from different sources; they obtain pathway information from the literature and by computational prediction.

Proteomics data repositories

There has been great progress in the last few years in making raw proteomic data publicly available, which provides a considerable value to the community. Currently, several repositories compile proteomic data. The PRoteomics IDEntifications (PRIDE) [88] database at the EBI is a public repository that includes protein and peptide identifications, post-translational modifications, and supporting spectral evidence. The PeptideAtlas database [89] from ISB's Proteome Center accepts only the raw output of mass spectrometers, and all raw data are processed through a uniform pipeline of search software plus validation with the Trans-Proteomic Pipeline (TPP) [90]. The results of this processing are coalesced and made available to the community through a series of builds for different organisms or sample types.

The *Mass* spectrometry *Interactive Virtual Environment* (MassIVE) is a community resource developed by the NIH-funded Center for Computational Mass Spectrometry to promote the global, free exchange of mass spectrometry data [91]. The MassIVE can be run with UCSD proteomics [92]. Chorus is a simple web application for storing, sharing, visualizing, and analyzing spectrometry files [93]. A user can upload experiment files along with the metadata, analyze them, and also make them available to collaborators. The Global Proteome Machine Database (GPMDB) collects spectra and identifications that have been uploaded by researchers to a GPM analysis engine and presents the summarized results back to the community [94].

To make the process of data submission easier for the user, the ProteomeXchange consortium is set up to provide a single point of submission to proteomics repositories [95]. Once the data are submitted to the ProteomeXchange entry point, they can be automatically distributed to all other repositories (PRIDE, MassIVE, and PeptideAtlas).

Discussion and conclusion

Machine learning and clustering approaches have been applied to proteomic and mass spectrometric data from many different biological disciplines in order to identify biomarkers for normal phenotypic characterization [38] and for diagnosis, prognosis, and treatment of specific disease [48, 57]. The bioinformatics tools that are currently available for omic data analysis span a large panel of very diverse applications ranging from simple tools to sophisticated software for large-scale analysis. Technical advances and growing interest in the field have given rise to a great number of specialized tools and software to derive biologically meaningful information. These computational approaches assist in generating hypotheses to be tested in orthogonal experiments.

Machine learning and its methods have increasingly gained attention in bioinformatics research. With the availability of different types of classification methods, it is common for researchers to apply these tools to classify and mine their data. But one should keep in mind that no matter how sophisticated the bioinformatics tools, the quality of the results they produce is directly dependent on the quality of input data they are given. In addition, new experimental methods are likely to require newly adapted bioinformatics tools as mass spectrometers become more powerful and as novel experimental design results in more complex datasets. One area of rapidly expanding complexity is at the integration of the fronts of metabolomic and proteomic data. Each software tool has some advantage and disadvantage, so it benefits the user to employ a combination of tools to examine one dataset rather than a single software tool. Each dataset contains its own quirks, positive and negative, and it is up to the end users and analysts to decide the most effective approach for assessing the biology that is taking place within their experiment.

Competing interests

The authors declare that they have no competing interests.

Authors' contributions

AKF conceived of the project and drafted the manuscript. LEE participated in its design. AKF, LEE, and LEH wrote the manuscript. All authors read and approved the final manuscript.

Acknowledgements

This study was supported by NIH 2R01LM009254 and 2R01LM008111 for AKF and LEH.

Author details

[1]Department of Pharmacology, University of Colorado School of Medicine, Aurora, CO 80045, USA. [2]Integrated Center for Genes, Environment, and Health, National Jewish Health, Denver, CO 80206, USA.

References

1. Hanash S. Disease proteomics. Nature. 2003;422(6928):226–32.

2. Fliser D, Novak J, Thongboonkerd V, Argilés A, Jankowski V, Girolami MA, et al. Advances in urinary proteome analysis and biomarker discovery. J Am Soc Nephrol. 2007;18:1057–71.

3. McGregor E, Dunn MJ. Proteomics of the heart: unraveling disease. Circ Res. 2006;98:309–21.

4. Wang H, Wu K, Liu Y, Wu Y, Wang X. Integrative proteomics to understand the transmission mechanism of Barley yellow dwarf virus-GPV by its insect vector Rhopalosiphum padi. Sci Rep. 2015;5:10971.

5. Liu W, Gray S, Huo Y, Li L, Wei T, Wang X. Proteomic analysis of interaction between a plant virus and its vector insect reveals new functions of hemipteran cuticular protein. Mol Cell Proteomics. 2015;14:2229–42.

6. Komatsu S, Mock H-P, Yang P, Svensson B. Application of proteomics for improving crop protection/artificial regulation. Front Plant Sci. 2013;4:522.

7. Dajana G-S, Kova S, JosiC D. Application of proteomics in food technology and food biotechnology: process development, quality control and product safety.

8. Huang S-H, Triche T, Jong AY. Infectomics: genomics and proteomics of microbial infections. Funct Integr Genomics. 2002;1:331–44.

9. Swan AL, Mobasheri A, Allaway D, Liddell S, Bacardit J. Application of machine learning to proteomics data: classification and biomarker identification in postgenomics biology. Omics. 2013;17(12):595–610.

10. Khatri P, Sirota M, Butte AJ. Ten years of pathway analysis: current approaches and outstanding challenges. PLoS Comput Biol. 2012;8(2):e1002375.

11. Epperson LE, Martin SL. Proteomic strategies to investigate adaptive processes. In: Eckersall PD, Whitfield PD, editors. Methods in animal proteomics. Oxford: Wiley-Blackwell; 2011.

12. González-Fernández R, Jorrín-Novo JV. Proteomics of fungal plant pathogens: the case of Botrytis cinerea. In. Current research, technology and education topics in applied microbiology and microbial biotechnology. 2010.

13. Benjamini Y, Hochberg Y. Controlling the false discovery rate: a practical and powerful approach to multiple testing. J R Stat Soc Ser B. 1995;57:289–300.

14. Hall M, Frank E, Holmes G, Pfahringer B, Reutemann P, Witten IH. The WEKA data mining software. ACM SIGKDD Explor Newsl. 2009;11:10.

15. scikit-learn. [http://scikit-learn.org/stable/]

16. Sonnenburg S, Rätsch G, Henschel S, Widmer C, Behr J, Zien A, et al. The SHOGUN machine learning toolbox. J Mach Learn Res. 2010;11:1799–802.

17. The R project for statistical computing. [https://www.r-project.org/]

18. Tan P-N, Steinbach M, Kumar V: Introduction to data mining. 1996.

19. Wolpert DH, Macready WG. Coevolutionary free lunches. IEEE Trans Evol Comput. 2005;9:721–35.

20. Wolpert DH. The lack of a priori distinctions between learning algorithms. Neural Comput. 1996;8:1341–90.

21. Jolliffe IT. Principal component analysis, second edition. Encycl Stat Behav Sci. 2002;30:487.

22. Wen X, Fuhrman S, Michaels GS, Carr DB, Smith S, Barker JL, et al. Large-scale temporal gene expression mapping of central nervous system development. Proc Natl Acad Sci. 1998;95:334–9.

23. Purohit PV, Rocke DM. Discriminant models for high-throughput proteomics mass spectrometer data. Proteomics. 2003;3:1699–703.

24. Fearn T. Principal component discriminant analysis. Stat Appl Genet Mol Biol. 2008;7:Article6.

25. Hoefsloot HCJ, Smit S, Smilde AK. A classification model for the Leiden proteomics competition. Stat Appl Genet Mol Biol. 2008;7:Article8.

26. Jutten C, Herault J. Blind separation of sources, part I: an adaptive algorithm based on neuromimetic architecture. Signal Process. 1991;24:1–10.

27. Comon P. Independent component analysis, a new concept? Signal Process. 1994;36:287–314.

28. Safavi H, Correa N, Xiong W, Roy A, Adali T, Korostyshevskiy VR, et al. Independent component analysis of 2-D electrophoresis gels. Electrophoresis. 2008;29:4017–26.

29. Hilario M, Kalousis A, Pellegrini C, Müller M. Processing and classification of protein mass spectra. Mass Spectrom Rev. 2006;25:409–49.

30. Rodríguez-Piñeiro AM, Carvajal-Rodríguez A, Rolán-Alvarez E, Rodríguez-Berrocal FJ, Martínez-Fernández M, De Páez La Cadena M. Application of relative warp analysis to the evaluation of two-dimensional gels in proteomics: studying isoelectric point and relative molecular mass variation. J Proteome Res. 2005;4:1318–23.

31. Jain AK, Dubes RC. Algorithms for clustering data. 1988.

32. MacQueen J. Some methods for classification and analysis of multivariate observations. In Proceedings of the Fifth Berkeley Symposium on Mathematical Statistics and Probability, Volume 1: Statistics. The Regents of the University of California. 1967.

33. Pham DT, Dimov SSNC. Selection of k in K-means clustering. Mech Eng Sci. 2004;219:103–19.

34. Hindle AG, Karimpour-Fard A, Epperson LE, Hunter LE, Martin SL. Skeletal muscle proteomics: carbohydrate metabolism oscillates with seasonal and torpor-arousal physiology of hibernation. Am J Physiol Regul Integr Comp Physiol. 2011;301:R1440–52.

35. Jani A, Orlicky DJ, Karimpour-Fard A, Epperson LE, Russell RL, Hunter LE, et al. Kidney proteome changes provide evidence for a dynamic metabolism and regional redistribution of plasma proteins during torpor-arousal cycles of hibernation. Physiol Genomics. 2012;44:717–27.

36. Meunier B, Dumas E, Piec I, Béchet D, Hébraud M, Hocquette JF. Assessment of hierarchical clustering methodologies for proteomic data mining. J Proteome Res. 2007;6:358–66.

37. Laville E, Sayd T, Morzel M, Blinet S, Chambon C, Lepetit J, et al. Proteome changes during meat aging in tough and tender beef suggest the importance of apoptosis and protein solubility for beef aging and tenderization. J Agric Food Chem. 2009;57:10755–64.

38. Jacobsen S, Grove H, Jensen KN, Sørensen HA, Jessen F, Hollung K, et al. Multivariate analysis of 2-DE protein patterns - practical approaches. Electrophoresis. 2007;28:1289–99.

39. Maurer MH, Feldmann RE, Brömme JO, Kalenka A. Comparison of statistical approaches for the analysis of proteome expression data of differentiating neural stem cells. J Proteome Res. 2005;4:96–100.

40. Wold S, Albano C, Dunn III WJ, Edlund U, Esbensen K, Geladi P, et al. Chemometrics. Netherlands: Springer; 1984.

41. Helland IS. Partial least squares regression and statistical models. Scandinavian Journal of Statistics. Wiley. 1990;17(2):97–114.

42. Helland IS. On the structure of partial least squares regression. Commun Stat - Simul Comput. 1988;17:581–607.

43. Nguyen DV, Rocke DM. Partial least squares proportional hazard regression for application to DNA microarray survival data. Bioinformatics. 2002;18:1625–32.

44. Tan Y, Shi L, Tong W, Hwang GTG, Wang C. Multi-class tumor classification by discriminant partial least squares using microarray gene expression data and assessment of classification models. Comput Biol Chem. 2004;28:235–44.

45. Boulesteix A-L, Porzelius C, Daumer M. Microarray-based classification and clinical predictors: on combined classifiers and additional predictive value. Bioinformatics. 2008;24:1698–706.

46. Rajalahti T, Arneberg R, Kroksveen AC, Berle M, Myhr KM, Kvalheim OM. Discriminating variable test and selectivity ratio plot: quantitative tools for interpretation and variable (biomarker) selection in complex spectral or chromatographic profiles. Anal Chem. 2009;81:2581–90.

47. Karp NA, Griffin JL, Lilley KS. Application of partial least squares discriminant analysis to two-dimensional difference gel studies in expression proteomics. Proteomics. 2005;5:81–90.

48. Rosenberg LH, Franzén B, Auer G, Lehtiö J, Forshed J. Multivariate meta-analysis of proteomics data from human prostate and colon tumours. BMC Bioinformatics. 2010;11:468.

49. Azimi A, Pernemalm M, Frostvik Stolt M, Hansson J, Lehtiö J, Egyházi Brage S, et al. Proteomics analysis of melanoma metastases: association between S100A13 expression and chemotherapy resistance. Br J Cancer. 2014;110(10):2489–95.

50. Breiman L. Random Forests. Mach Learn. 2001; 45(1):5–32.

51. Izmirlian G. Application of the random forest classification algorithm to a SELDI-TOF proteomics study in the setting of a cancer prevention trial. Ann N Y Acad Sci. 2004;1020:154–74.

52. Barrett JH, Cairns DA. Application of the random forest classification method to peaks detected from mass spectrometric proteomic profiles of cancer patients and controls. Stat Appl Genet Mol Biol. 2008;7:Article4.

53. Hindle AG, Grabek KR, Epperson LE, Karimpour-Fard A, Martin SL. Metabolic changes associated with the long winter fast dominate the liver proteome in 13-lined ground squirrels. Physiol Genomics. 2014;46:348–61.

54. Epperson LE, Karimpour-Fard A, Hunter LE, Martin SL. Metabolic cycles in a circannual hibernator. Physiol Genomics. 2011;43:799–807.

55. Breiman L. Bagging predictors. Mach Learn. 1996;24:123–40.

56. Cortes C, Vapnik V. Support-vector networks. Mach Learn. 1995;20:273–97.

57. Zhang X, Lu X, Shi Q, Xu X-Q, Leung H-CE, Harris LN, et al. Recursive SVM feature selection and sample classification for mass-spectrometry and microarray data. BMC Bioinformatics. 2006;7:197.

58. Smith FM, Gallagher WM, Fox E, Stephens RB, Rexhepaj E, Petricoin EF, et al. Combination of SELDI-TOF-MS and data mining provides early-stage response prediction for rectal tumors undergoing multimodal neoadjuvant therapy. Ann Surg. 2007;245:259–66.

59. Hart TC, Corby PM, Hauskrecht M, Hee Ryu O, Pelikan R, Valko M, et al. Identification of microbial and proteomic biomarkers in early childhood cCaries. Int J Dent. 2011;2011:196721.

60. Zhai X, Yu J, Lin C, Wang L, Zheng S. Combining proteomics, serum biomarkers and bioinformatics to discriminate between esophageal squamous cell carcinoma and pre-cancerous lesion. J Zhejiang Univ Sci B. 2012;13:964–71.

61. Magni P, Ferrazzi F, Sacchi L, Bellazzi R. TimeClust: a clustering tool for gene expression time series. Bioinformatics. 2008;24:430–2.

62. Conesa A, Nueda MJ, Ferrer A, Talón M. maSigPro: a method to identify significantly differential expression profiles in time-course microarray experiments. Bioinformatics. 2006;22:1096–102.

63. Tai Y. timecourse: statistical analysis for developmental microarray time course data. 2007.

64. Pedro Cardoso, Francois Rigal JCC. BAT. R Package.

65. Aryee M: betr: identify differentially expressed genes in microarray time-course data. R 2011.

66. Peng J. fpca: restricted MLE for functional principal components analysis. R Package.

67. Martini P, Sales G, Calura E, Cagnin S, Chiogna M, Romualdi C. timeClip: pathway analysis for time course data without replicates. BMC Bioinformatics. 2014;15 Suppl 5:S3.

68. Sangurdekar D. Rnits: R normalization and inference of time series data.

69. Cameletti M. STEM. R Package.

70. Schilling R, Costa IG, Schliep A. pGQL: a probabilistic graphical query language for gene expression time courses. BioData Min. 2011;4:9.

71. Sinha A, Markatou M. A platform for processing expression of short time series (PESTS). BMC Bioinformatics. 2011;12:13.

72. Tchagang AB, Phan S, Famili F, Shearer H, Fobert P, Huang Y, et al. Mining biological information from 3D short time-series gene expression data: the OPTricluster algorithm. BMC Bioinformatics. 2012;13:54.

73. Sivriver J, Habib N, Friedman N. An integrative clustering and modeling algorithm for dynamical gene expression data. Bioinformatics. 2011;27:i392–400.

74. Ashburner M, Ball CA, Blake JA, Botstein D, Butler H, Cherry JM, et al. Gene ontology: tool for the unification of biology. The Gene Ontology Consortium. Nat Genet. 2000;25:25–9.

75. Bassel GW, Glaab E, Marquez J, Holdsworth MJ, Bacardit J. Functional network construction in Arabidopsis using rule-based machine learning on large-scale data sets. Plant Cell. 2011;23:3101–16.

76. Kanehisa M, Goto S. KEGG: Kyoto Encyclopedia of Genes and Genomes. Nucleic Acids Res. 2000;28:27–30.

77. Subramanian A, Tamayo P, Mootha VK, Mukherjee S, Ebert BL, Gillette MA, et al. Gene set enrichment analysis: a knowledge-based approach for interpreting genome-wide expression profiles. Proc Natl Acad Sci U S A. 2005;102:15545–50.

78. IPA. [http://www.ingenuity.com/products/ipa]

79. Smoot ME, Ono K, Ruscheinski J, Wang P-L, Ideker T. Cytoscape 2.8: new features for data integration and network visualization. Bioinformatics. 2011;27:431–2.

80. Pathway Commons. A resource for biological pathway analysis. [http://www.pathwaycommons.org/about/]

81. HumanCyc. Encyclopedia of human genes and metabolism. [http://humancyc.org/]

82. PathVisio - pathway drawing and pathway analysis tool. [http://www.pathvisio.org/]

83. 3Omics. A web based systems biology visualization tool for integrating human transcriptomic, proteomic and metabolomic data. [http://3omics.cmdm.tw/]

84. Chang JT, Nevins JR. GATHER: a systems approach to interpreting genomic signatures. Bioinformatics. 2006;22:2926–33.

85. PANTHER - gene list analysis. [http://pantherdb.org/]

86. Wu X, Al Hasan M, Chen JY. Pathway and network analysis in proteomics. J Theor Biol. 2014;362:44–52.

87. Webber J, Stone TC, Katilius E, Smith BC, Gordon B, Mason MD, et al. Proteomics analysis of cancer exosomes using a novel modified aptamer-based array (SOMAscan™) platform. Mol Cell Proteomics. 2014;13:1050–64.

88. Pride. [http://www.ebi.ac.uk/pride/archive/]

89. Peptideatlas. [http://www.peptideatlas.org/]

90. Deutsch EW, Mendoza L, Shteynberg D, Farrah T, Lam H, Tasman N, et al. A guided tour of the Trans-Proteomic Pipeline. Proteomics. 2010;10:1150–9.

91. Welcome to MassIVE. [http://massive.ucsd.edu/ProteoSAFe/static/massive.jsp]

92. CCMS The Center for Computational Mass Spectrometry. [http://proteomics.ucsd.edu/]

93. Chorus - Home. [https://chorusproject.org/pages/index.html]

94. GPMdb. [http://omictools.com/gpmdb-s3019.html]

95. ProteomeXchange. [http://www.proteomexchange.org/]

96. Dennis G, Sherman BT, Hosack DA, Yang J, Gao W, Lane HC, et al. DAVID: Database for Annotation, Visualization, and Integrated Discovery. Genome Biol. 2003;4:P3.

97. Schaefer CF, Anthony K, Krupa S, Buchoff J, Day M, Hannay T, et al. PID: the Pathway Interaction Database. Nucleic Acids Res. 2009;37(Database issue):D674–9.

98. Shannon P, Markiel A, Ozier O, Baliga NS, Wang JT, Ramage D, et al. Cytoscape: a software environment for integrated models of biomolecular interaction networks. Genome Res. 2003;13:2498–504.

99. Chen JY, Mamidipalli S, Huan T: HAPPI: an online database of comprehensive human annotated and predicted protein interactions. BMC Genomics 2009, 10 (Suppl 1):S16.

100. Matthews L, Gopinath G, Gillespie M, Caudy M, Croft D, de Bono B, et al. Reactome knowledgebase of human biological pathways and processes. Nucleic Acids Res 2009,37(Database issue):D619–22.

101. Nishimura D: BioCarta. Biotech Softw Internet Rep 2001, 2:117–120.

102. Chowbina SR, Wu X, Zhang F, Li PM, Pandey R, Kasamsetty HN, et al. HPD: an online integrated human pathway database enabling systems biology studies. BMC Bioinformatics 2009, 10 (Suppl 1):S5.

103. Huang H, Wu X, Sonachalam M, Mandape SN, Pandey R, MacDorman KF, et al. PAGED: a pathway and gene-set enrichment database to enable molecular phenotype discoveries. BMC Bioinformatics. 2012, 13 (Suppl 1):S2.

104. Keshava Prasad TS, Goel R, Kandasamy K, Keerthikumar S, Kumar S, Mathivanan S, et al. Human Protein Reference Database–2009 update. Nucleic Acids Res. 2009;37(Database):D767–D772.

105. Kamburov A, Stelzl U, Lehrach H, Herwig R: The ConsensusPathDB interaction database: 2013 update. Nucleic Acids Res 2013, 41(Database issue):D793–800.

106. Maere S, Heymans K, Kuiper M. BiNGO: a Cytoscape plugin to assess overrepresentation of gene ontology categories in biological networks. Bioinformatics 2005, 21:3448–9.

9

Experience of a multidisciplinary task force with exome sequencing for Mendelian disorders

S. Fokstuen[1†], P. Makrythanasis[1,2†], E. Hammar[1†], M. Guipponi[1,2†], E. Ranza[1†], K. Varvagiannis[1,2†], F. A. Santoni[1,2†], M. Albarca-Aguilera[1], M. E. Poleggi[1], F. Couchepin[1], C. Brockmann[1], A. Mauron[4], S. A. Hurst[4], C. Moret[4], C. Gehrig[1], A. Vannier[1], J. Bevillard[2], T. Araud[1], S. Gimelli[1], E. Stathaki[1], A. Paoloni-Giacobino[1], A. Bottani[1], F. Sloan-Béna[1], L. D'Amato Sizonenko[1], M. Mostafavi[1], H. Hamamy[2], T. Nouspikel[1], J. L. Blouin[1] and S. E. Antonarakis[1,2,3*]

Abstract

Background: In order to optimally integrate the use of high-throughput sequencing (HTS) as a tool in clinical diagnostics of likely monogenic disorders, we have created a multidisciplinary "Genome Clinic Task Force" at the University Hospitals of Geneva, which is composed of clinical and molecular geneticists, bioinformaticians, technicians, bioethicists, and a coordinator.

Methods and results: We have implemented whole exome sequencing (WES) with subsequent targeted bioinformatics analysis of gene lists for specific disorders. Clinical cases of heterogeneous Mendelian disorders that could potentially benefit from HTS are presented and discussed during the sessions of the task force. Debate concerning the interpretation of identified variants and the content of the final report constitutes a major part of the task force's work. Furthermore, issues related to bioethics, genetic counseling, quality control, and reimbursement are also addressed.

Conclusions: This multidisciplinary task force has enabled us to create a platform for regular exchanges between all involved experts in order to deal with the multiple complex issues related to HTS in clinical practice and to continuously improve the diagnostic use of HTS. In addition, this task force was instrumental to formally approve the reimbursement of HTS for molecular diagnosis of Mendelian disorders in Switzerland.

Background

Since the technological and bioinformatics developments of high-throughput sequencing (HTS) and the use of exome sequencing for the discovery of new genes causative of Mendelian disorders [1, 2], this technology has been rapidly and widely integrated in the clinical setting [3] as it outperforms previously used methods in diagnostic yield, time, and cost-effectiveness [4]. However, the use of HTS technology in the clinical setting brings its own set of challenges (7), although many of them were already

encountered during the introduction of other genomic diagnostic methods such as array CGH. The main challenges of diagnostic HTS include pre- and post-HTS counseling with appropriate and adapted informed consent [5, 6], bioinformatics analysis setup and validation [7], variant interpretation and classification [8–10], specific policies concerning the identification and disclosure of variants not directly linked to the patient's phenotype [11], validation of HTS as a diagnostic test that conforms to quality control standards [12], data storage and accessibility, and reimbursement issues [13], as well as updates and follow-up strategies. In order to optimally integrate HTS into the clinical practice and to continuously improve this novel and rapidly evolving diagnostic approach, we have realized quite early in the process

---
* Correspondence: Stylianos.Antonarakis@unige.ch
†Equal contributors
[1]Service of Genetic Medicine, University Hospitals of Geneva, Geneva, Switzerland
[2]Department of Genetic Medicine and Development, University of Geneva, 1 rue Michel-Servet, 1211 Geneva, Switzerland

the need for a multidisciplinary approach. Accordingly, the Genome Clinic Task Force (GCTF) was established in 2012, with the specific objective to provide a platform for regular exchanges of all involved specialists in order to find solutions for the various types of problems and concerns that we may encounter by performing HTS in our clinic. Currently, this task force meets once per week and is composed of roughly 25 specialists and a coordinator, including clinical geneticists (consultants and trainees), molecular biologists, scientists, bioinformaticians, bioethicists, and technicians (Fig. 1).

In this review, we present the composition, practices, and workflow of the GCTF, the results obtained to date, the challenges we have encountered, the reimbursement directives that were officially introduced in Switzerland in January 2015 by the Swiss Federal Office of Public Health (SFOPH), and the lessons learned from this experience.

The Genome Clinic Task Force (GCTF) of the University Hospitals of Geneva

Figure 1 shows the organization of the GCTF working group as well as the tasks that the two sections (clinical and laboratory) have to fulfill. The head of our Genetics Institute, an MD, PhD, is the director of the task force. The coordinator is a trained PhD molecular biologist with experience in health policy and diagnostic issues. The role of the coordinator is to perform the preparatory work of each GCTF session, to formalize the procedures, to record the minutes of all GCTF sessions, and to handle relevant administrative tasks. The clinical section consists of the clinical geneticists of our service, who present patients to the task force and critically examine the indications of HTS for each patient, as well as providing their input regarding the clinical interpretation of identified variants. The HTS laboratory section is headed by a senior molecular biologist with appropriate qualifications for molecular diagnostic services and

subdivided in a sequencing, bioinformatics, and analysis groups. Finally, two bioethicists from the Institute of Bioethics of the University of Geneva are participating in the weekly meetings. Their participation helps to immediately address ethical issues that may arise during the discussions. The profession of the genetic counselor (as it is defined in the USA) is not formally recognized as such in Switzerland, and thus genetic counselors are not included in the task force.

Standard operating procedure

The different steps of the diagnostic workflow are shown in Fig. 2 and illustrated by an example. This standard operating procedure was among the initial objectives of the GCTF and is regularly reviewed according to the evolution of this diagnostic field. As minutes of all GCTF meetings are written, all discussions and decisions taken can be referred to and reevaluated according to new experiences, international recommendations, and practical considerations. As shown in the flowchart, every case that undergoes diagnostic HTS is discussed at least three times in the GCTF: a first time before the test is performed (step 2), a second time during the preliminary report (step 7), and a third time (step 8) during the presentation and debate of the final report. The following paragraphs provide more details on the operating procedure.

Pre-test considerations (steps 1, 2, and 3 of Fig. 2)

Clinical cases that have been seen for genetic counseling at the Genome Clinic (step 1) and who may benefit from HTS are presented by one of the clinical geneticists at the GCTF (step 2) in order to evaluate within the task force the clinical and reimbursement aspects of the case and to decide whether the patient is appropriate for a HTS approach. In general, we accept patients who suffer from a likely heterogeneous Mendelian disorder with at least one known clearly pathogenic gene or patients with a developmental delay of unknown origin. The cost of the HTS analysis has to be less than that of Sanger sequencing for the corresponding genes. Physicians from other medical specialties may also present their cases during the sessions. These presentations include a detailed family history, personal medical history, photos if available, genetic tests that have already been performed, and the list of genes proposed to be tested. So far, 246 patients have been presented to the GCTF and 240 cases were accepted. Two were redirected towards research projects due to the absence of clearly known pathogenic genes causing their respective phenotypes, while two other cases were rejected because their phenotypes were multifactorial with genetic predispositions identified through GWAS studies but without a known monogenic cause. One case was not accepted because a specific

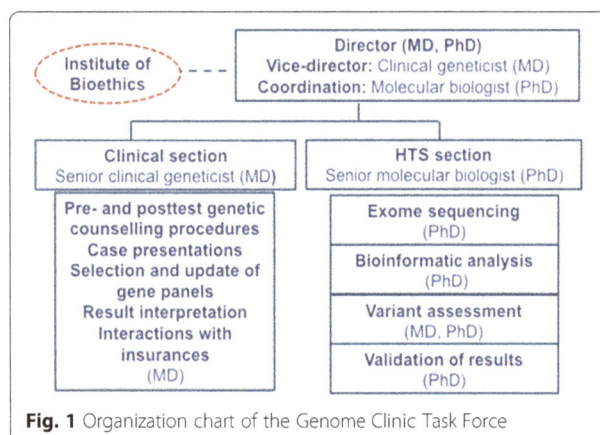

Fig. 1 Organization chart of the Genome Clinic Task Force

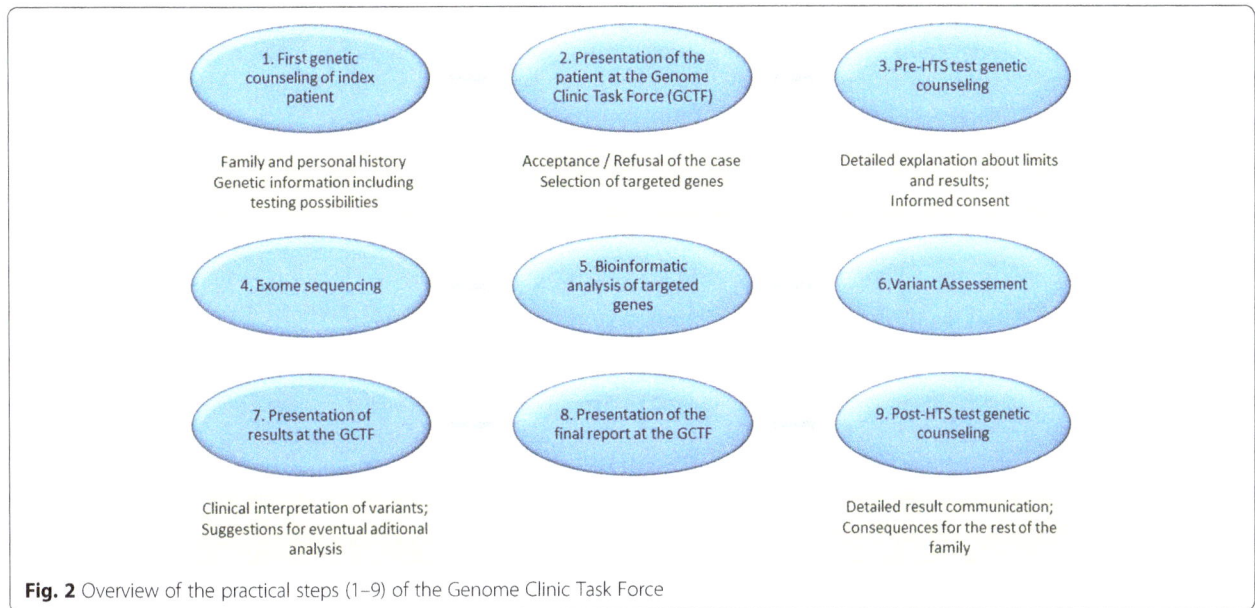

Fig. 2 Overview of the practical steps (1–9) of the Genome Clinic Task Force

standard genetic test was judged more appropriate than HTS, and another case was rejected because it consisted of a prenatal diagnosis based on ultrasound findings without a specific hypothesis for a Mendelian disorder.

Once a case is accepted for HTS, the GCTF members discuss and define the most appropriate targeted gene panel. Since we use exome sequencing followed by bioinformatics selection of genes of interest, the clinical geneticists continuously reevaluate and update the gene lists. Before a gene panel is created or reevaluated, a detailed research of the literature is performed for recent review papers and available proposed gene panels (academic and commercial). Frequently used gene lists, such as the intellectual disability panel, are reviewed biannually or sooner at the request of the referring physician, while rarely used lists are reevaluated each time they are needed. As illustrated by the example, we initially used rather restrictive gene panels in order to minimize the incidental findings. In cases where no pathogenic variant was identified, we had the tendency to add a second or even a third bioinformatics analysis using additional genes. This multi-step approach was eventually deemed time consuming. Thus, we have recently decided to directly include all potentially causative genes in the panels.

After the pre-test GCTF decision, the patients or the legal representative are seen for a pre-HTS genetic consultation by one of the clinical geneticists (step 3). The patients are informed about the procedure and the possible results, including incidental findings in case of large panels. If they agree to be tested, we discuss and explain the specific HTS informed consent form, which we have developed in collaboration with the Swiss Society of

Medical Genetics (ww.sgmg.ch) and with the input of the bioethicists. The patients can opt-in for the following categories of incidental findings, which are available only for the specific set of genes that will be analyzed:

1. Disorders for which no medical intervention (curative or preventive) is possible
2. Disorders for which medical intervention (curative or preventive) is possible
3. Carrier status for recessive disorders.

HTS and bioinformatics analysis (steps 4, 5, and 6 of Fig. 2)
One important decision was the choice of the HTS strategy [14]. The different options included (i) WES and analysis of all genes implicated in Mendelian disorders; (ii) WES and bioinformatics targeted analysis of gene lists; (iii) targeted gene panels only, and (iv) whole genome sequencing (WGS) and bioinformatics targeted analysis of gene lists. Based on the previous experience from research projects [15, 16], we have chosen to perform whole exome sequencing (WES) followed by targeted analysis of specific sets of genes. Trio exome sequencing (patient and parents) was not considered due to limitations imposed by the Swiss medical insurance reimbursement regulations.

Exome capture is performed using the SureSelect Human ALL Exon technology (Agilent Technologies, Santa Clara, CA, USA), and the sequencing is realized in an Illumina HiSeq 2000 instrument (usually 100 bp paired-end, Illumina, San Diego, CA, USA). Read mapping, variant calling, and variant annotation are performed using a locally developed bioinformatics tool that surveys the sequential progress of the data from BWA [14] for mapping (hg19) and Samtools [15] and Pindel [16] for SNV and indel calling. Targeted bioinformatics analysis of the selected genes of interest is performed through locally developed pipelines, which select only the variants from the specified genes of interest, masking the rest of the data. We have validated the sequencing and variant detection quality by sequencing DNA samples from the individual for which sets of high-quality variants are publicly available (Platinum Genomes, Illumina®) and obtained 99.6 % concordance for the SNVs and 97.8 % for the indels. The bioinformaticians from the HTS section follow the updates of the databases (Table 1). New versions are implemented 1 to 3 months after they become publicly available. Software updates (Table 1) follow a more

Table 1 Databases and tools routinely used for variant annotation and classification. Additional databases and tools are used as deemed necessary

Population, disease-specific, and sequence databases	
Population databases	
Exome Aggregation Consortium	http://exac.broadinstitute.org/
1000 Genomes	http://browser.1000genomes.org
dbSNP	http://www.ncbi.nlm.nih.gov/snp
Disease databases	
ClinVar	http://www.ncbi.nlm.nih.gov/clinvar
OMIM	http://www.omim.org
Human Gene Mutation Database	http://www.hgmd.org
Leiden Open Variation Database	http://www.lovd.nl
Sequence databases	
NCBI Genome Source	http://www.ncbi.nlm.nih.gov/genome
RefSeqGene	http://www.ncbi.nlm.nih.gov/refseq/rsg
In-silico predictive algorithms	
Missense prediction	
SIFT	http://sift.jcvi.org
MutationTaster	http://www.mutationtaster.org
PolyPhen-2	http://genetics.bwh.harvard.edu/pph2
Splice site prediction	
GeneSplicer	http://www.cbcb.umd.edu/software/ GeneSplicer/gene_spl.shtml
Human Splicing Finder	http://www.umd.be/HSF/
NetGene2	http://www.cbs.dtu.dk/services/NetGene2
NNSplice	http://www.fruitfly.org/seq_tools/ splice.html
Conservation scores	
GERP	http://mendel.stanford.edu/sidowlab/ downloads/gerp/
PhastCons	http://compgen.bscb.cornell.edu/phast/
PhyloP	http://compgen.bscb.cornell.edu/phast/

lengthy cycle and are annually reevaluated. Newer versions are implemented only when a significant improvement over previous results can be demonstrated.

Potential pathogenicity of the variants (step 6) is evaluated by two senior scientists with extensive experience in the analysis of exome data. All analyses are performed independently and the results are then merged for the presentation and debate at the GCTF (step 7). Data pathogenicity estimation adheres to published guidelines [9] that has allowed the standardization of the process and has also guided a more structured approach towards the available databases (Table 1) that are now used in order to support or reject specific criteria. We are planning to set up a specific training program in order to increase the number of people involved and thus increase the capacity to perform exome analysis for the patients.

Intermediate report and decisions on pathogenicity (step 7 of Fig. 2)

The intermediate report, produced for each patient, includes a technical and a variant interpretation section. The technical section documents all the HTS and bioinformatics analysis aspects including the specific filtering steps and quality metrics (e.g. genes, coverage, and number of identified variants). For the interpretation section we use the currently accepted 5 class variant classification system (8) and several tools for the classification (Table 1) [8, 9].

All class 3, 4, and 5 variants [9] are documented in the intermediate report with a summary of available literature and presented at the GCTF session (step 7). The classification performed by the analysis team is debated and occasionally the variants are reclassified after thorough evaluation of potential phenotype-genotype correlations. In cases where no clear pathogenic variant is found, it is discussed whether further analyses are warranted (e.g., MLPA for deletions/duplications, broader bioinformatics analysis that includes analysis of additional genes, Sanger sequencing of individual exons that are insufficiently covered) or if familial segregation analysis of a variant is justified. If necessary, we extend such segregation analysis up to second-degree relatives (cousins, nephews/nieces) but this depends mainly on disease status, demographic circumstances, and the interfamilial relationships.

Verification, final report, and post-test considerations (steps 8 and 9 in Fig. 2)

All identified variants (100 %) that are disclosed in the final report are currently confirmed by Sanger sequencing and the content of the final reports are discussed during the GCTF sessions (step 8). All class 4 and 5 variants identified in genes compatible with the phenotype are reported. The disclosure of class 4 and 5 variants considered as incidental findings is done according to the patient's pre-test decision. Class 3 variants are only disclosed if they are found in a gene causing a phenotype which is compatible with the clinical presentation of the patient. In these cases, a remark that the variants should be reevaluated in 1–2 years according to the evolving knowledge is added.

Once the final report has been validated by the GCTF, the report is signed by the senior molecular biologist, allowing the clinical geneticists to subsequently arrange genetic counseling for communication of the results (step 9).

Example illustrating the operating procedure of the GCTF

Dizygotic 12-month-old male twins were addressed for genetic counseling because of seizures since the age of 4 months associated with severe developmental delay. Family history was unremarkable. An array-CGH (resolution 180 KB) was performed and identified a 417 kb paternally inherited duplication at 6p12.2 that was considered non-pathogenic. Extensive paraclinical workup and brain imaging did not reveal the cause. The clinical geneticist in charge presented the situation at the weekly GCTF meeting. Given the lack of diagnosis and the severe presentation, we decided to perform WES with targeted bioinformatics analysis of 120 selected epilepsy genes. This initial analysis did not reveal any potential pathogenic variants. It was then decided to extend the analysis by including all known syndromic and non-syndromic epilepsy genes. The second panel consisted of 395 genes and revealed a novel missense variant NM_020473.3:c.481G>A: p.(Glu161Lys) in the gene *PIGA* on chromosome Xp22.2 (MIM 311770). This variant concerns a very well-conserved nucleotide (GERP: 5.89) and was predicted to be pathogenic by all three bioinformatics tools (SIFT: 0, PolyPhen/HumVar:0.931, Mutation Taster:0.999). Pathogenic mutations in *PIGA* cause paroxysmal nocturnal hemoglobinuria (MIM 300818) and multiple congenital anomalies-hypotonia-seizures syndrome 2 (MCAHS2, MIM 300868). The latter phenotypic description was concordant with the children's clinical presentation. The variant was transmitted from their unaffected mother. No additional family members were available for clinical testing. Based on the aforementioned evidence, the variant was reported as pathogenic. Our approach allowed us to expand the bioinformatics analysis to additional genes without the need for resequencing; however, the turnaround time was prolonged as we did not immediately include all potentially causative genes. Based on this experience, and on similar other situations, we decided to change our procedure and to analyze directly the largest possible gene panel.

Reimbursement of HTS for Mendelian disorders

Another aim of the GCTF was to initiate together with the Swiss Society of Medical Genetics the administrative process with the Swiss Federal Office of Public Health (SFOPH) in order to integrate HTS as a reimbursable genetic test in the Swiss health care system. Genetic tests in Switzerland are reimbursed according to a positive "list of analyses" (LA) [17], which specifies each disorder and testing method covered by the compulsory, albeit private, medical insurance scheme. The LA includes an additional nonspecific entry for orphan diseases, applicable to rare Mendelian disorders, not otherwise registered in the LA. In January 2015, after 30 months of continuous negotiations with the Swiss Federal Office of Public Health, HTS was officially introduced in the LA as a reimbursable genetic test for Mendelian disorders [17]. In addition, the Swiss Society of Medical Genetics (SSGM) [18] developed a document of "good practice" for the use of HTS in clinical setting [19], which was required by the SFOPH, covers pre- and post-HTS genetic counseling issues, an informed consent form adapted to HTS genetic testing, laboratory requirements and specifications, regulations for secure data storage and quality control, and disclosure of secondary findings and variants of unknown clinical significance, as well as recommendations for the reporting of results.

The costs of HTS are based on the sum of three distinct entries within the LA: laboratory costs of high-throughput sequencing, bioinformatics analysis, and additional confirmatory laboratory analyses such as Sanger sequencing and/or MLPA. More specifically:

1. High-throughput sequencing has a fixed price of 2300 CHF, irrespective of the sequencing technology used (WES or targeted panel approach).
2. Bioinformatics analysis costs vary according to the number of genes analyzed: 600 CHF for 1-10 genes, 1000 CHF for 11-100 genes and 1500 CHF for more than 100 genes.
3. Confirmation of variants using Sanger sequencing (215 CHF per variant): a maximum of two Sanger confirmations for 1–10 genes, four for 11–100 genes, and six for more than 100 genes. In all cases, a maximum of four multiplex ligation-dependent amplification (MLPA) analyses can also be added to the total cost (350 CHF per MPLA).

This modular setting enables flexibility for the diagnostic laboratories and allows each step to be performed and charged separately. In particular, it allows performing additional bioinformatics analyses without resequencing, which is arguably cost-effective.

Additional requirements have been set forth by the SFOPH for the reimbursement of HTS: diagnostic laboratories performing HTS must participate in quality assessment schemes (EQAs), according to the Swiss law [20], and become accredited for HTS by the Swiss Accreditation Service (SAS), before December 31, 2017. Furthermore, all the steps of HTS need to be performed within Switzerland. However, it is not necessary that they are all performed within the same institution. Finally, because of the complexity of pre- and post-HTS counseling, only board-certified medical geneticists [21] are allowed to prescribe HTS tests of more than 10 genes. Physicians from all other medical specialties can only prescribe HTS for less than 10 genes. In addition, the requirements for expert genetic counseling are well specified in the existing law for genetic analyses [22]. It is planned to regularly update and reevaluate the requirements from the SFOPH as well as the "good practice" document according to the new developments and international recommendations in the field.

Experience to date

Until now we have designed 51 different gene lists containing 2 to 1038 genes. On average, 160 (SD = 18) million reads are produced per sample. After removal of duplicate reads, 132 (SD = 22) million reads remain, 78 (SD = 8.5) million of which are on target (target = total coding sequence as defined by RefSeq). These reads represent an average coverage of the coding portion of the RefSeq genes of at least 20× for 94.73 % (±1.18 SD) and of at least 30× for 92.08 % (±1.66 SD). On average, 21,565 (±1,125 SD) variants (SNVs and small indels) are detected per individual.

So far, we enrolled 240 patients with our HTS approach. Thirty-two percent (77/240) displayed developmental delay with or without other anomalies; the remaining 68 % (163/240) presented with various heterogeneous Mendelian disorders such as short rib polydactyly syndrome, juvenile Parkinson disease, connective tissue disorders, Cornelia de Lange syndrome, microcephalic primordial dwarfism (MPD), Kallmann syndrome, arthrogryposis, Gitelman syndrome, various inherited cardiac diseases, Charcot-Marie-Tooth disease, Kabuki syndrome, hereditary spastic paraparesis, or likely monogenic epilepsy. We completed the final report for 139 of these patients (47 with developmental delay and 92 with other Mendelian disorders).

Pathogenic variants (class 4 or 5) were detected in 28 % (13/47) of the patients with developmental delay (Fig. 3) and in 46 % (42/92) of the patients with other Mendelian phenotypes (Fig. 4), which gives an average detection rate of class 4 or 5 variants of 40 % (55/139) (Table 2). Variants of unknown clinical significance (VUS, class 3) were found in 23 % (11/47) of the patients with developmental delay (Fig. 3) and in 10 % (9/92) of the patients with other Mendelian phenotypes (Fig. 4).

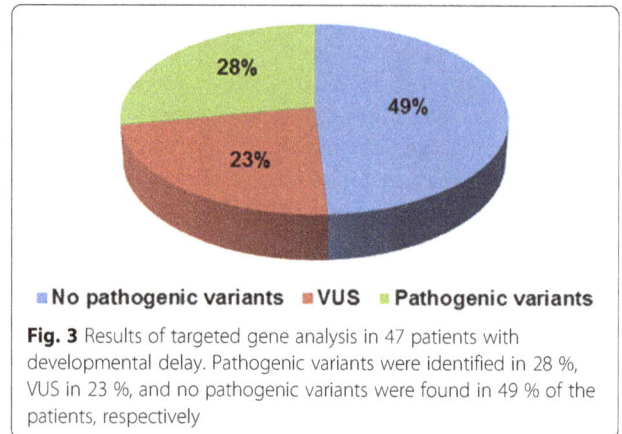

Fig. 3 Results of targeted gene analysis in 47 patients with developmental delay. Pathogenic variants were identified in 28 %, VUS in 23 %, and no pathogenic variants were found in 49 % of the patients, respectively

So far, we reported 8 out of 20 identified VUS in the final report.

Sanger sequencing of not well-covered exons in genes that were considered to be highly compatible with the phenotype was performed in two cases (2/139, 1.4 %) and in one of them the causative variant was identified. In six other cases (6/139, 4.3 %), bioinformatics analysis of further added genes to the originally determined gene panel resulted in the identification of the causative variants.

Lessons learned

HTS has provided exciting new diagnostic opportunities in the investigation of genetically heterogeneous Mendelian disorders and has expanded the capacity to test simultaneously a large number of candidate genes in a timely fashion and for a reasonable cost. The advantages of our approach include the following: (i) the use of one common diagnostic test for all patients suffering from Mendelian disorders with known pathogenic genes, (ii) it is amenable to a customized and flexible bioinformatics analysis, (iii) it allows us to invite all patients with a

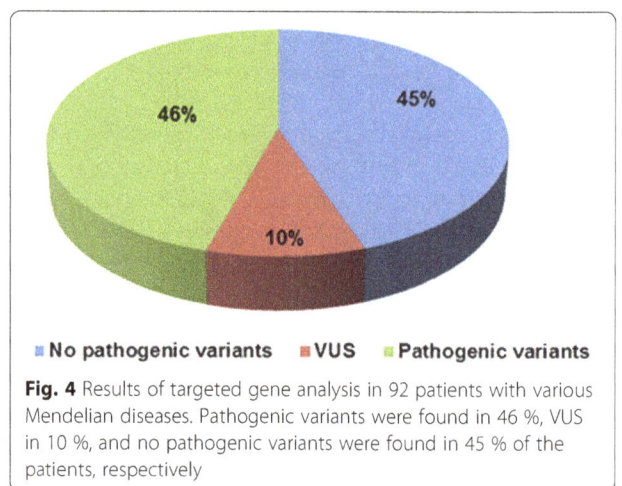

Fig. 4 Results of targeted gene analysis in 92 patients with various Mendelian diseases. Pathogenic variants were found in 46 %, VUS in 10 %, and no pathogenic variants were found in 45 % of the patients, respectively

Table 2 Causative variants identified in the resolved cases

	Phenotype	Gene panel	Identified pathogenic variant(s)
1	Short rib polydactyly	Short rib polydactyly panel (10 genes)	NM_001377.2(DYNC2H1_v001):c.1953G>A:p.(=)
			NM_001377.2(DYNC2H1_v001):c.4625 C>T:p.(Ala1542Val)
2	Severe ID	ID (536 genes)	NM_000489.4(ATRX_v001):c.6122G>A:p.(Ser2041Asn)
3	Intellectual disability, microcephaly	ID (536 genes)	NM_004380.2(CREBBP_v001):c.4665A>C:p.(Glu1555Asp)
4	Cornelia de Lange syndrome	Cornelia de Lange panel (5 genes)	NM_015384.4(NIPBL_v001):c.5483G>A:p.(Arg1828Gln)
5	Intellectual disability	ID (536 genes)	NM_004187.3(KDM5C_v001):c.769_770del :p.(Leu257Alafs*5)
6	Glomerulopathy	Glomerulopathy and Alport panel (61 genes)	NM_000495.4(COL4A5_v001):c.2288G>A:p.(Gly763Glu)
7	Intellectual disability, psychotic symptoms	ID (536 genes)	NM_033517.1(SHANK3_v001):c.3637dup:p.(His1213Profs*83)
8	Microcephalic primordial dwarfism	MPD panel (18 genes)	NM_002312.3(LIG4_v001):c.2321T>C:p.(Leu774Pro)
			NM_002312.3(LIG4_v001):c.2440C>T c.2440 C>T p.(Arg814*)
9	Kallmann syndrome	Kallmann panel (21 genes)	NM_015850.3(FGFR1_v001):c.1444del:p.(Leu482Trpfs*25)
10	Dyskinesia, dystonia, myoclonia	Dystonia panel (8 genes)	NM_003919.2(SGCE_v001):c.783dup :p.(Phe262Ilefs*8)
11	Cardiac arrest	Cardiomyopathy panel (66 genes)	NM_001035.2(RYR2_v001):c.14711G>A:p.(Gly4904Asp)
12	Periodic fever syndrome	Periodic fever panel (4 genes)	NM_004895.4(NLRP3_v001):c.1049C>T:p.(Thr350Met)
13	Intellectual disability, microcephaly, strabismus	ID (536 genes)	NM_021140.3(KDM6A_v001):c.3598C>T :p.(Leu1200Phe)
14	Hereditary spastic paraplegia	Hereditary spastic paraplegia panel (45 genes)	NM_014846.3(KIAA0196_v001):c.1857G>C:p.(Leu619Phe)
15	Epileptic encephalopathy	Epilepsy panel (395 genes)	NM_020473.3(PIGA_v001):c.481G>A:p.(Glu161Lys)
16	Gitelman syndrome	Gitelman syndrome panel (2 genes)	NM_000339.2(SLC12A3_v001):c.1924C>G:p.(Arg642Gly)
17	Autism, Intellectual disability, trigonocephaly	ID (536 genes)	NM_001111125.2(IQSEC2_v001):c.2477T>C:p.(Met826Thr)
18	Aortic dissection	Aneurysm panel (20 genes)	NM_000138.4(FBN1_v001):c.6616G>A:p.(Asp2206Asn)
19	Epileptic encephalopathy	Epileptic encephalopathy (141 genes)	NM_004518.4(KCNQ2_v001):c.821C>T :p.(Thr274Met)
20	Kabuki syndrome	Kabuki panel (2 genes)	NM_003482.3(KMT2D_v001):c.12661C>T:p.(Gln4221*)
21	Hereditary Spastic paraparesis	Spastic paraparesis panel (11 genes)	NM_199436.1(SPAST_v001):c.1015C>T :p. (Leu339Phe)
22	Ohdo syndrome	KAT6B gene	NM_001256468.1(KAT6B_v001):c.4652_4661dup: p.(Gln1554Hisfs*41)
23	Neurofibramotosis type 1	NF panel (2 genes)	NM_000267.3 (NF1_v001):c1381C>T: p.(Arg461*)
24	Inclusion body myositis	Inclusion body myosotis panel (10 genes)	NM_001927.3 (DES_v001):c.1155G>T:p.(Asp399Tyr)
25	Noonan syndrome	Noonan and rasopathy syndrome (12 genes)	NM_002834.3 (PTPN11_v001):c.797G>C:p.(Glu139Asp)
26	Periodic fever	Personalized periodic fever panel (207 genes)	NM_000243.2 (MEFV_v001):c.2084A>G:p.(Lys695Arg)
27	Charcot Marie Tooth type 2	CMT2 panel (23 genes)	NM_001005373.3 (LRSAM1_v001):c.2069T>C:p.(Cys690Arg)
28	Hypoglycemia on congenital hyperinsulinemia	Congenital hyperinsulinemia panel (10 genes)	NM_000525.3 (KCNJ11_v001):c.400T>C:p.(Leu147Pro)
			NM_000525.3 (KCNJ11_v001):c.154C>T:p.(Gln52*)
29	Cardiomyopathy	Cardiomyopathy panel (66 genes)	NM_001018008.1 (TPM1_v001):c.304G>A:p.(Glu102Lys)
30	Intellectual disability, epilepsy	Intellectual disability panel (537 genes)	NM_000834.3 (GRIN2B_v001):c.1598G>A:p.(Gly533Asp)
31	X-linked intellectual disability	Intellectual disability panel (990 genes)	NM_003916.4 (AP1S2_v001):c.1-3C>A
32	Lissencephaly	Lissencephaly panel (12 genes)	NM_000403.3 (PAFAH1B1_v001):c.162dupA:p.(Trp55Metfs*6)
33	Vascular leukoencephalopathy	Vascular leukoencephalopathy panel (7 genes)	NM_002775.4 (HTRA1_v001):c.854C>T:p.(Pro285Leu)
34	Cardiomyopathy	Cardiomyopathy panel (66 genes)	NM_000256.3 (MYBPC3_v001):c.3324-3325del:p. (Lys1108Asnfs*41)
35	Cardiomyopathy	Cardiomyopathy panel (66 genes)	NM_000256.3 (MYBPC3_v001):c.3697C>T:p.(Gln1233*)
36	Cardiomyopathy and connective tissue disorder	Cardiomyopathy and connective tissue disorder panel (166 genes)	NM_0004415.2 (DSP_v001):c.4003C>T:p.(Gln1335*)

Table 2 Causative variants identified in the resolved cases *(Continued)*

37	Intellectual disability	Intellectual disability panel (990 genes)	NM_002834.3 (PTPN11_v001):c.794G>A:p.(Arg265Gln)
38	Cystinuria	Cystinuria panel (2 genes)	NM_001243036 (SLC7A9_v001):c.1225-4678_1324del
39	Noonan syndrome	Noonan panel (12 genes)	NM_002834.3 (PTPN11_v001):c.923A>G:p.(Asn308Ser)
40	Intellectual disability, microcephaly	Personalized panel (2 genes: *DYRK1A* and *DDX3X*)	NM_00139.3 (DYRK1A_v001):c.1491delC:p.(Ala498Profs*94)
41	Neonatal encephalopathy	Encephalopathy panel (225 genes)	NM_001909.4 (CTSD_v001):c.686_688del:p.(Phe229del)
42	Intellectual disability, cryptorchidism	Intellectual disability panel (990 genes)	NM_001243234.1 (TCF4_v001):c.656dupT:p.(Leu219Phefs*9)
43	Intellectual disability, obesity	Intellectual disability panel (990 genes)	NM_032531.3 (KIRREL3_v001):c.2019G>A:p.(Met673Ile)
44	Epilepsy, vertigo, episodic ataxia	Epilepsy (396 genes)	NM_0010540143.1 (SCN2A_v001):c.2960G>T:p.(Ser987Ile)
45	Intellectual disability	Intellectual disability panel (990 genes)	NM_015559.2 (SETBP1_v001):c.2016-2017insT:p.(Lys673*)
46	Kabuki syndrome	Kabuki panel (2 genes)	NM_003482.3 (KMT2D_v001):c.2994delT:p(Met999*)
47	Long QT syndrome	Arythmia panel (47 genes)	NM_000238.3 (KCNH2_v001):c.1786C>G:p(Pro596Ala)
48	Rubinstein-Taybi syndrome	Rubinstein-Taybi syndrome panel (2 genes).	NM_004380.2 (CREBBP_v001). Variant found by MLPA
49	Aneurysm and dyslipidemia	Aneurysm and dyslipidemia panel (50 genes)	NM_000041.3 (APOE_v001):c.461G>T:p.(Arg154Leu)
50	Marfan syndrome	Marfan syndrome panel (8 genes)	NM_000138.4 (FBN1_v001):c.7339G>A:p.(Glu2447Lys)
51	Ehlers-Danlos syndrome	Ehlers-Danlos panel (4 genes)	NM_000093.4 (COL5A1_v001):c.2203dupC:p.(Gln735Profs*25)
52	Epileptic encephalopathy and intellectual disability	Intellectual disability and epilepsy panel (1038 genes)	NM_001127648.1 (GABRA1_v001):c.641G>A:p.(Arg214His)
53	Intellectual and communication disability	Whole exome	NM_001197104.1 (MLL/KMT2A_v001):c.2633G>A: p.(Arg878Gln)
54	Catecholaminergic polymorphic ventricular tachycardia, arrhythmia	Cardiomyopathy panel (66 genes)	NM_001018008.1 (TPM1_v001):c.304G>A:p.(Glu102Lys)
55	Dilated non compaction cardiomyopathy	Arythmia and cardiomyopathy panel (97 genes)	NM_003319.4 (TTN_v001):c.49905dup:p.(Pro16636Thrfs*9)

negative result to contact us again within an interval of 1–2 years, in order to reevaluate the results and possibly expand the analysis according to new discoveries without the need to re-sequence, (iv) it minimizes the probability of incidental findings not related to the patient's disease.

However, WES followed by targeted bioinformatics also has drawbacks. Firstly, from the sequencing depth and coverage point of view, the capture procedure of the whole exome is less efficient when compared to well-established targeted gene panel enrichment. Nevertheless, with the use of the most recently marketed whole exome capture reagents, this is gradually becoming less of an issue [23], although there is a need to continuously monitor the coverage. This drawback may be partially resolved by filling up the sequencing gaps using traditional methods such as targeted Sanger sequencing of poorly covered exons. This complementary procedure was recently introduced in the proposed draft of the Eurogentest guidelines for next-generation sequencing [24].

The second difficulty concerns the selection of genes of interest, which is not yet standardized, and thus remains idiosyncratic and to some extent subjective.

Although our approach allows a very flexible and liberal selection of genes, a human omission or a change in the gene name increases the risk of false negative results. Furthermore, the possibility exists to include, particularly in large gene lists such as the one for intellectual disability, some actionable genes with broad phenotypic spectrum of manifestations or not directly linked to the investigated phenotype. Our experience has shown that it is crucial to regularly reevaluate the gene lists, ideally by two independent individuals using criteria adapted for diagnostic testing, such as the level of evidence for pathogenicity and its correlation with a known phenotype. Yet, despite rigorous monitoring of the literature, the risk of wrongful inclusion or exclusion of genes remains. We strongly encourage the establishment of international norms and criteria for selecting gene panels, in order to render the diagnostic possibilities universal. For example, achievements such as the release of the Eurogentest guidelines are a welcome development [25].

We do not routinely perform trio sequencing (father, mother, affected offspring) despite the fact that the trio approach seems to have a slightly higher diagnostic yield

[26]. The main reasons for this decision are financial, as the health insurances only reimburse the HTS costs of the proband but not that of the parents.

The most demanding challenge in clinical HTS arguably remains the issue of variant interpretation. Since the number of variants identified is roughly proportional to the number of analyzed genes, the task of variant interpretation rises accordingly. In our group, each variant's pathogenicity is first assessed by the analysis team according to international guidelines [9] and then presented, discussed, and evaluated for concordance with the phenotype of the patient during the GCTF sessions. The input of expert physicians familiar with the patient's phenotype is invaluable as well as the interpretation workup and knowhow of the laboratory team. The emerging databases for the pathogenicity of variants are also extremely important, and sharing of the interpretation of variants among diagnostic centers is crucial. Clinical knowledge and experience help to (mostly) exclude variants as non-relevant to the phenotype or to consider them as likely pathogenic [9, 10]. This complementary exchange is in our opinion indispensable for adequate variant interpretation, especially in cases of large gene lists with several potentially pathogenic variants identified. Accordingly, in a few cases, the consensus of GCTF was to consider the identified variant as likely pathogenic (class 4), despite VUS classification (class 3) by the analysis team according to the guidelines [9].

In order to further facilitate and improve variant interpretation, the need for international sharing of variants and phenotypes is of paramount importance and cannot be overemphasized. We have put in place a semi-automatic submission of variants of classes 3/4/5 in ClinVar. Furthermore, false positive "pathogenic" variants that have been misclassified in the past need to be updated in relevant databases, so that false diagnoses will not be perpetuated.

In parallel to the technical advances, ethical aspects have to be constantly considered at the present stage of HTS genetic testing. One important issue concerns the pre- and post-HTS genetic counseling challenges, including informed consent. The expert participation of ethicists within the GCTF was of considerable value for the development of a specific informed consent form for HTS application in accordance with the Swiss law on genetic testing [22] and also for the continuous reevaluation of our procedures. The informed consent form respects the rights to know and not to know, especially concerning secondary findings in actionable genes and carrier status for recessive disorders. Our experience to date has shown that the majority of patients and parents want to know about treatable diseases or diseases for which effective preventive measures exist, but decide not to know about non-treatable disorders and carrier status.

Additionally, almost all have agreed that their DNA and sequencing data could be stored for prospective future research projects. All these aspects need to be systematically studied on large cohorts in order to provide statistically meaningful conclusions.

A further key effort of the GCTF together with the Swiss Society of Medical Genetics was regarding the reimbursement of HTS as a diagnostic test by the insurance companies. It necessitated 2.5 years of continuous negotiations with the SFOPH until the proposal for formal reimbursement was accepted by the federal health authorities. To our knowledge, Switzerland is the first European country for which a specific formal policy for reimbursing of HTS for Mendelian disorders has been introduced. In a few European countries, reimbursement is achieved through general genetic testing policies; while in most other countries, HTS is still being funded by research projects or by non-reimbursable payments from the consumers. We hope that reimbursement policies will be developed in other countries in order to achieve a widespread acceptance and use of HTS for the diagnosis of genetic disorders.

In conclusion, the multidisciplinary GCTF has allowed us to implement a number of local procedures and criteria necessary to ensure high standard clinical services within the new field of diagnostic HTS, as well as to achieve in collaboration with the Swiss Society of Medical Genetics the formal federal decision for HTS reimbursement for monogenic disorders.

Acknowledgements

We thank the patients and the various referring clinicians, especially Joël Fluss and Christian Korff from the pediatric neurology University Hospitals of Geneva for their contributions, and Mike Morris and Mathias Lidgren for helpful discussions. We also thank the department of Genetic and Laboratory Medicine of the University Hospitals of Geneva for the financial and administrative support and specifically Denis Hochstrasser, Christine Guillaume, Sophie Christen, and David Breda; the Faculty of Medicine of the University of Geneva for the genomics infrastructure that allowed the development of the diagnostic Genome Clinic, and the board of the Swiss Society of Medical Genetics for their contribution in the submission of the application to the Swiss Federal Office of Public Health.

Authors' contributions

All the authors are past and present members of the Genome Clinic Task Force of the University Hospitals of Geneva. All authors read and approved the final manuscript.

Competing interests

The authors declare that they have no competing interests.

Author details

[1]Service of Genetic Medicine, University Hospitals of Geneva, Geneva, Switzerland. [2]Department of Genetic Medicine and Development, University of Geneva, 1 rue Michel-Servet, 1211 Geneva, Switzerland. [3]iGE3, Institute of Genetics and Genomics of Geneva, Geneva, Switzerland. [4]Institute for Ethics, History, and the Humanities, Geneva University Medical School, Geneva, Switzerland.

References

1. Marx A, Hoenger A, Mandelkow E. Structures of kinesin motor proteins. Cell Motil Cytoskeleton. 2009;66(11):958–66.

2. Ng SB et al. Exome sequencing identifies the cause of a Mendelian disorder. Nat Genet. 2010;42(1):30–5.

3. Makrythanasis P, Antonarakis SE. High-throughput sequencing and rare genetic diseases. Mol Syndromol. 2012;3(5):197–203.

4. Metzker ML. Sequencing technologies—the next generation. Nat Rev Genet. 2010;11(1):31–46.

5. Biesecker LG. Opportunities and challenges for the integration of massively parallel genomic sequencing into clinical practice: lessons from the ClinSeq project. Genet Med. 2012;14(4):393–8.

6. Ayuso C et al. Informed consent for whole-genome sequencing studies in the clinical setting. Proposed recommendations on essential content and process. Eur J Hum Genet. 2013;21(10):1054–9.

7. Rehm HL et al. ACMG clinical laboratory standards for next-generation sequencing. Genet Med. 2013;15(9):733–47.

8. Plon SE et al. Sequence variant classification and reporting: recommendations for improving the interpretation of cancer susceptibility genetic test results. Hum Mutat. 2008;29(11):1282–91.

9. Richards S, et al. Standards and guidelines for the interpretation of sequence variants: a joint consensus recommendation of the American College of Medical Genetics and Genomics and the Association for Molecular Pathology. Genet Med. 2015;17(5):405–24.

10. MacArthur DG et al. Guidelines for investigating causality of sequence variants in human disease. Nature. 2014;508(7497):469–76.

11. Green RC et al. ACMG recommendations for reporting of incidental findings in clinical exome and genome sequencing. Genet Med. 2013;15(7):565–74.

12. Linderman MD et al. Analytical validation of whole exome and whole genome sequencing for clinical applications. BMC Med Genomics. 2014;7:20.

13. Deverka PA, Kaufman D, McGuire AL. Overcoming the reimbursement barriers for clinical sequencing. JAMA. 2014;312(18):1857–8.

14. Li H, Durbin R. Fast and accurate short read alignment with Burrows-Wheeler transform. Bioinformatics. 2009;25(14):1754–60.

15. Li H et al. The Sequence Alignment/Map format and SAMtools. Bioinformatics. 2009;25(16):2078–9.

16. Ye K et al. Pindel: a pattern growth approach to detect break points of large deletions and medium sized insertions from paired-end short reads. Bioinformatics. 2009;25(21):2865–71.

17. Office_Fédérale_de_la_Santé_Publique_(OFSP). Liste des Analyses (LA). 2016 [cited 2016 27 Februay]; Available from: www.bag.admin.ch/la.

18. *Swiss Society of Medical Genetics*. 2016 [cited 2016 February 27]; Available from: http://www.sgmg.ch/.

19. Swiss_Society_of_Medical_Genetics. Bonnes Pratiques Pour les applications cliniques du Séquençage à haut débit (SHD). 2014 [cited 2016 Februyry 27]; Available from: http://www.bag.admin.ch/themen/krankenversicherung/02874/02875/11440/index.html?lang=fr&download=NHzLpZig7t, lnp6l0NTU042l2Z6ln1ae2lZn4Z2qZpnO2Yuq2Z6gpJCMdYR7gWym162dpYbUzd,Gpd6emK2Oz9aGodetmqaN19Xl2ldvoaCUZ,s-.

20. Office_Fédérale_de_la_Santé_Publique_(OFSP). Obligations des laboratoires de diagnostic génétique médical. 2011 [cited 2016 February 27]; Available from: http://www.cscq.ch/SiteCSCQ/FichierPDF_FR/BAG_Lab-B_10.2011_F.pdf.

21. Fédération_des_Médecins_Suisses_(FMH). Génétique Médicale. 2016 [cited 2016 February 27]; Available from: http://www.fmh.ch/fr/formation-isfm/domaines-specialises/titres-formation-postgraduee/genetique-medicale.html.

22. Gouvernment_Suisse_Droit_Féderale. Loi fédérale sur l'analyse génétique humaine (LAGH). 2014 [cited 2016 February 27]; Available from: https://www.admin.ch/opc/fr/classified-compilation/20011087/index.html.

23. Sun Y, et al. Next generation diagnostics: gene panel, exome or whole genome? Hum Mutat. 2015;36(6):648–55.

24. Matthijs G et al. Guidelines for diagnostic next-generation sequencing. Eur J Hum Genet. 2016;24(1):2–5.

25. Eurogentest. Clinical Utility Gene Cards, NGS panel database. [cited 2016 February 27]; Available from: http://www.eurogentest.org/index.php?id=668.

26. Lee H et al. Clinical exome sequencing for genetic identification of rare Mendelian disorders. JAMA. 2014;312(18):1880–7.

10

Exome sequencing discloses *KALRN* homozygous variant as likely cause of intellectual disability and short stature in a consanguineous pedigree

Periklis Makrythanasis[1,2], Michel Guipponi[2], Federico A. Santoni[1], Maha Zaki[3], Mahmoud Y. Issa[3], Muhammad Ansar[1], Hanan Hamamy[1*] and Stylianos E. Antonarakis[1,2,4*]

Abstract

Background: The recent availability of whole-exome sequencing has opened new possibilities for the evaluation of individuals with genetically undiagnosed intellectual disability.

Results: We report two affected siblings, offspring of first-cousin parents, with intellectual disability, hypotonia, short stature, growth hormone deficiency, and delayed bone age. All members of the nuclear family were genotyped, and exome sequencing was performed in one of the affected individuals. We used an in-house algorithm (CATCH v1.1) that combines homozygosity mapping with exome sequencing results and provides a list of candidate variants. One identified novel homozygous missense variant in *KALRN* (NM_003947.4:c.3644C>A: p.(Thr1215Lys)) was predicted to be pathogenic by all pathogenicity prediction software used (SIFT, PolyPhen, Mutation Taster). *KALRN* encodes the protein kalirin, which is a GTP-exchange factor protein with a reported role in cytoskeletal remodeling and dendritic spine formation in neurons. It is known that mice with ablation of *Kalrn* exhibit age-dependent functional deficits and behavioral phenotypes.

Conclusion: Exome sequencing provided initial evidence linking *KALRN* to monogenic intellectual disability in man, and we propose that *KALRN* is the causative gene for the autosomal recessive phenotype in this family.

Keywords: Exome sequencing, Intellectual disability, Short stature, Consanguineous, KALRN

Background

Consanguinity is practiced in a large number of human populations with rates reaching 20–50 % in several countries in North Africa and the Middle East [1, 2]. Offspring of consanguineous marriages are at a higher risk of having congenital anomalies caused by pathogenic variants in genes following autosomal recessive inheritance [3–5]. In a study in our laboratory, we employed whole-exome sequencing and genotype analysis to screen members of consanguineous families with likely recessive disorders. Our hypothesis was that because of the homozygosity of the causative defect, the

diagnostic strategy would be successful in identifying the molecular basis of the disorder in at least a proportion of the participating patients. Exome sequencing identified the causative variant in up to 34 % of the progeny of consanguineous parents affected by undiagnosed autosomal recessive disorders sequenced in our laboratory [6], as well as a number of novel candidate genes for autosomal recessive disorders [7–9].

Early-onset intellectual disability originating before the age of 18 years with an IQ below 70 is estimated to affect 1–3 % of western populations but could be more common in highly consanguineous populations [10] due to the role of autosomal recessive variants. Autosomal recessive intellectual disability (ARID) is extremely heterogeneous, and causative genes may reach thousands with the vast majority still unknown [10]. With the

* Correspondence: hananhamamy@yahoo.com; Hanan.AlHamami@unige.ch; Stylianos.Antonarakis@unige.ch
[1]Department of Genetic Medicine and Development, University of Geneva, 1 Rue Michel-Servet, 1211 Geneva, Switzerland
Full list of author information is available at the end of the article

introduction of NGS technologies, new ID genes are being identified including over 300 genes for ARID [11]. Many of these were identified in studies on consanguineous families. Among 136 consanguineous Iranian families with various forms of ID, pathogenic variants were identified in 78 families including homozygous mutations in 23 genes previously implicated in ID and in 50 candidate genes for ARID [12]. Examples from recent studies on consanguineous families with ID revealing pathogenic variants include variants in TNIK gene [13], SLC6A17 gene [14], and c12orf4 gene [15].

To add to the repository of ARID candidate genes, we report two affected siblings, offspring of a consanguineous marriage, with intellectual disability (ID), hypotonia, short stature, growth hormone deficiency, and delayed bone age. Exome sequencing and homozygosity mapping identified only one strong candidate variant in *KALRN* (NM_003947.4:c.3644C>A: p.(Thr1215Lys)). Kalirin-7, a major isoform of kalirin, is known to regulate spine density in hippocampal and cortical neurons [16], growth hormone functional secretion [17], and bone homeostasis [18].

Methods

Genotyping, exome sequencing, and variant analysis were performed as previously described [6]. All the family members were genotyped with a dense genotype array in order to define runs of homozygosity (ROH) in all the family members. DNA from the proband was used for exome sequencing, and after calling variants,

the ROH were used in order to identify the variants that were in an ROH in the affected siblings, which was not shared by the unaffected sibling and the parents. In more detail, DNA samples from affected family members, their unaffected sibling, and their parents were genotyped using the HumanOmniExpress Bead Chip by Illumina Inc.® (720K SNPs). ROH for every individual were identified using PLINK. We defined as ROH the regions with 50 consecutive SNPs irrespective of the total size of the genomic region, allowing for one mismatch. The ROH region was the one demarcated by the first heterozygous SNPs flanking each established homozygous region. The exome of one affected individual (IV:1) was captured using the SureSelect Human All Exonsv5 reagents by Agilent Inc.®. Sequencing was performed in an Illumina HiSeq 2000 sequencer. Results were analyzed with BWA [19], SAMtools [19], Pindel [20] and ANNOVAR [21], and the exonic variants in combination with the ROH were used by CATCH v1.1 [22] that provided the final list of variants

Results

Clinical report

Figure 1 shows the pedigree of the consanguineous family originating from Egypt. The parents of two affected offspring are paternal first cousins. The eldest affected is a girl (IV:1), first seen at the genetic clinic at the age of 13 years with intellectual disability, short stature, and delayed puberty. Her height was 128 cm (−5.1SD), weight was 25 kg (−2.3SD), and head circumference (HC)

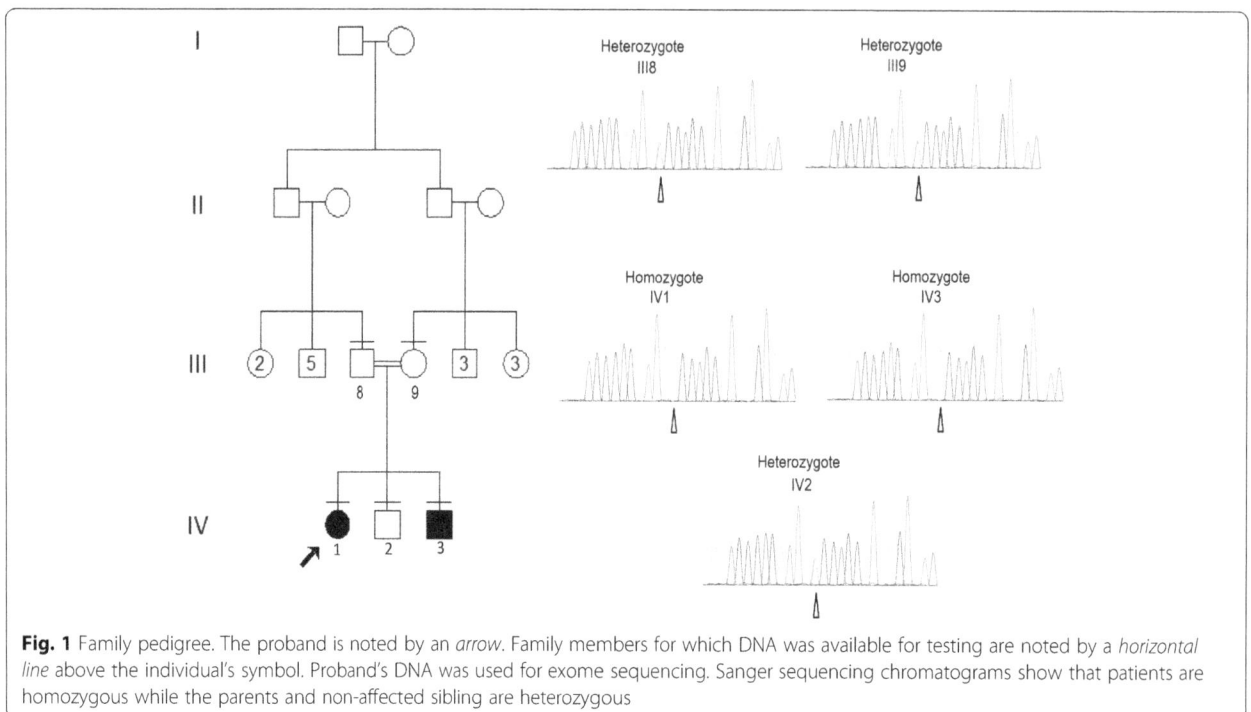

Fig. 1 Family pedigree. The proband is noted by an *arrow*. Family members for which DNA was available for testing are noted by a *horizontal line* above the individual's symbol. Proband's DNA was used for exome sequencing. Sanger sequencing chromatograms show that patients are homozygous while the parents and non-affected sibling are heterozygous

was 51 cm (−2SD). Birth history was uncomplicated. She had hypotonia and motor delay; sat unsupported at 2 years and walked at 3 years. Speech development was also delayed. At 13 years of age, she shows hypotonia and intellectual disability (IQ = 58, Stanford-Binet scale). Her dysmorphic facial features (Fig. 2) include sparse eyebrows and eyelashes, pigmented sclerae, high forehead, prominent nose, low-set ears, and abnormal palmar creases. At 13 years, she showed no signs of puberty with axillary hair at A1, pubic hair at P1, breast development at B1, and absence of menarche. Investigations revealed a deficiency of growth hormone as measured by ITT and clonidinetesting. Radiologic examination revealed delayed bone age. Brain MRI, EEG, hearing and ophthalmologic testing were without abnormalities. Banded karyotype was 46, XX. The proband was reexamined at the age of 15 6/12 years showing a height of 135 cm (−3.8SD), weight of 32 kg (−2.3SD), and HC of 51.2 cm (−2.2 SD). Menarche was reached at the age of 15 years.

The affected brother (IV:3) (Fig. 1) was first seen at the genetic clinic at the age of 5 years. He was delivered by cesarean section and noticed to have unilateral ptosis at birth. He had hypotonia; sitting unsupported at 2 years and walking at 3.5 years. Speech development was also delayed with first words at 2.5 years. Examination revealed a height of 85 cm (−4.8SD), weight of 13 kg (−3SD), and HC of 50 cm (−1.2SD). He has intellectual disability with similar facies to his affected sister (Fig. 2) including sparse eyebrows and eyelashes, pigmented sclerae, unilateral ptosis, high forehead, prominent nose, long philtrum, low-set ears, and abnormal palmar creases. Investigations have shown similar results to his sister with growth hormone deficiency, delayed bone age, and no abnormalities in brain MRI, EEG, hearing, and ophthalmologic testing. Karyotype was 46, XY. He was reexamined at the age of 7 years showing a height of 94 cm (−4.8SD), weight of 14 kg (−2.9SD) and HC of 52 cm (within the mean) with intellectual disability (IQ = 68, Stanford-Binet scale).

Genetic analysis

In both parents, 56 ROH were detected with a total size of 44 and 45.8 Mb and average size of 785 and 817 kb, respectively. In the children, the equivalent values were 58, 65, and 70 ROH, with a total size of 365, 312, and 114 Mb and average size of 6.3, 4.8, and 1.6 Mb, respectively. The combination of the ROH provided the position of the target areas in which the candidate genes would be searched (Fig. 3).

After exome sequencing, 139,353,457 unique reads were on target, resulting in coverage of at least eight times for 98.1 % of the protein coding fraction of the genome (RefSeq, version 58). Twenty-two thousend seven hundred sixty exonic variants of high quality were

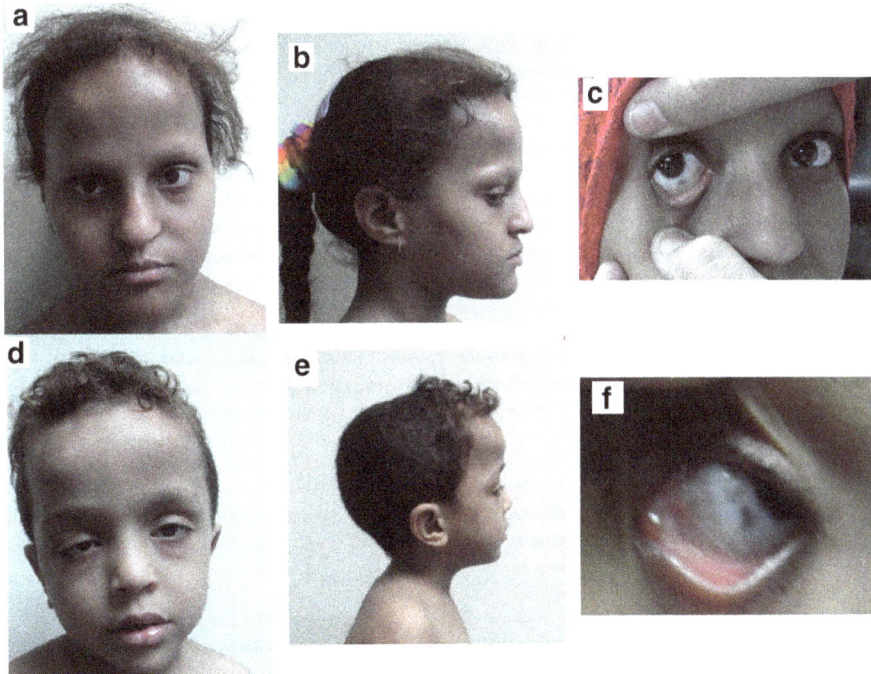

Fig. 2 Phenotypes of proband IV1 and brother IV3. **a**, **b** Proband's facial features with high forehead, prominent nose, and sparse eyebrows and eyelashes. **c** Scleral pigmentation in proband. **d**, **e** Facial features with high forehead, prominent nose sparse eyebrows and eyelashes, and unilateral ptosis in brother IV3. **f** Scleral pigmentation in brother IV3

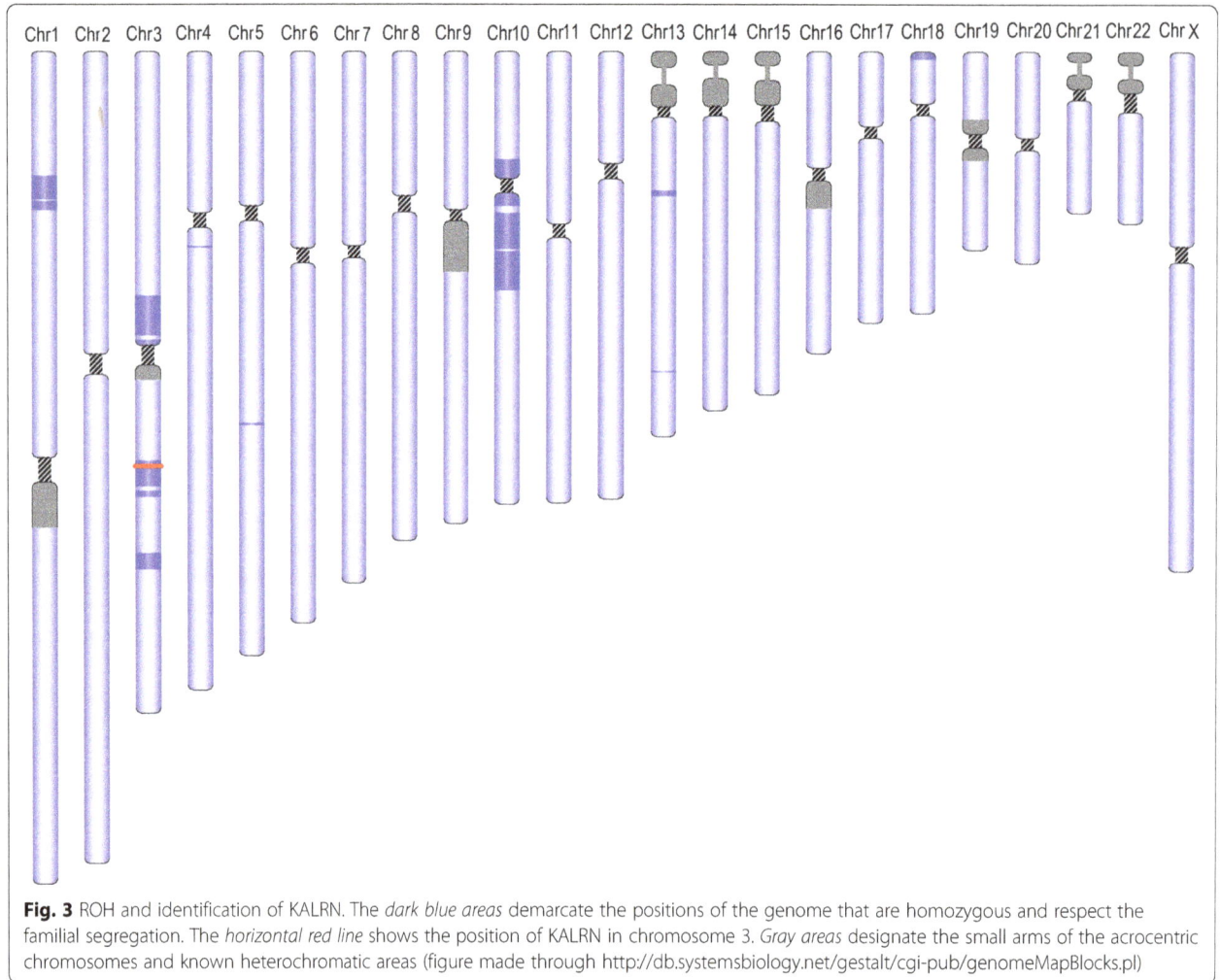

Fig. 3 ROH and identification of KALRN. The *dark blue areas* demarcate the positions of the genome that are homozygous and respect the familial segregation. The *horizontal red line* shows the position of KALRN in chromosome 3. *Gray areas* designate the small arms of the acrocentric chromosomes and known heterochromatic areas (figure made through http://db.systemsbiology.net/gestalt/cgi-pub/genomeMapBlocks.pl)

detected, 11,373 of which were synonymous (50.1 %), 10,044 (44.25 %) were missense, 74 (0.33 %) were nonsense, and 653 were indels (2.87 %).

Three variants passed the frequency criterion (MAF <0.01) and respected the segregation in the family. Of these, only one was predicted to be pathogenic by at least two pathogenicity prediction programs, NM_003947.4:c.3644C>A:p.(Thr1215Lys) in *KALRN*. The values predicted by SIFT [23], PolyPhen2 [24], and Mutation Taster [25] were 0.02, 0.8, and 0.996, respectively, and the region is very well conserved throughout the species (Fig. 4a). The variant was found in one of the previously defined target genomic areas of 6.6 Mbp (Fig. 3). Sanger sequencing confirmed the homozygosity in the affected individuals and the heterozygosity of the identified variant in both parents and the non-affected sibling. The variant is found in the ninth and last spectrin repeat of kalirin [26]. The spectrin repeats of kal7 are important for its role in spine morphogenesis [27, 28] (Fig. 4b).

The other two variants which were not predicted to be pathogenic were identified in *CFAP57* (previously named *WDR65*) (NM_001195831.2:c.3299C>T:p.(Pro1100Leu)) and in the last exon in *DPH2* (NM_001384.4:c.1429 C>T:p.(Arg477*)). *CFAP57* was previously correlated with a Van der Woode syndrome (VDW) variant [29] but was later reclassified as variant of unknown clinical significance (VUS). Scarce data is known for *DPH2* gene. In mouse, the function of the gene is mostly related to aging and to the skeleton, with expression in liver and biliary tissue. This data led us to consider this variant as unlikely to cause the phenotype in this family [30, 31].

Discussion

Kalirin is a prime candidate for a role in ID since alterations in signaling pathways involving the Rho family of small GTPases, key regulators of the actin, and microtubule cytoskeletons contribute to both syndromic and non-syndromic intellectual disability disorders [32]. In mice, loss of one or both copies of *Kalrn* leads to

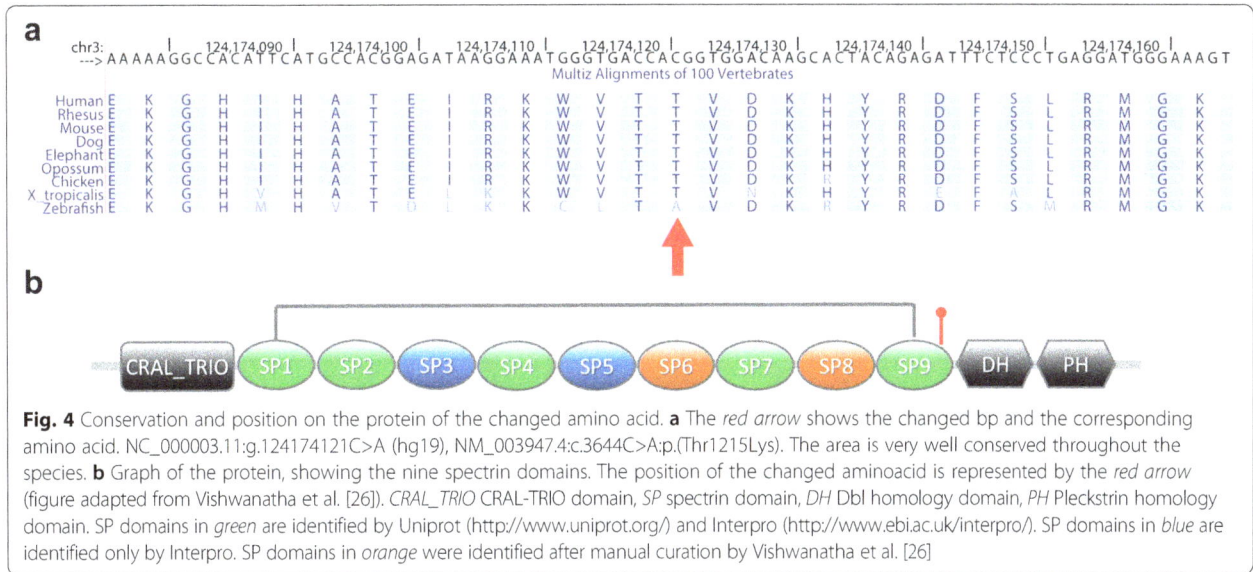

Fig. 4 Conservation and position on the protein of the changed amino acid. **a** The *red arrow* shows the changed bp and the corresponding amino acid. NC_000003.11:g.124174121C>A (hg19), NM_003947.4:c.3644C>A:p.(Thr1215Lys). The area is very well conserved throughout the species. **b** Graph of the protein, showing the nine spectrin domains. The position of the changed aminoacid is represented by the *red arrow* (figure adapted from Vishwanatha et al. [26]). *CRAL_TRIO* CRAL-TRIO domain, *SP* spectrin domain, *DH* Dbl homology domain, *PH* Pleckstrin homology domain. SP domains in *green* are identified by Uniprot (http://www.uniprot.org/) and Interpro (http://www.ebi.ac.uk/interpro/). SP domains in *blue* are identified only by Interpro. SP domains in *orange* were identified after manual curation by Vishwanatha et al. [26]

reduction in neuron spine densities in certain brain regions [33]. Heterozygote and knockout mice showed variable impairments in cognitive functions related to working memory, social recognition, and social approach, demonstrating the role of kalirin in the regulation of cortical ultrastructure and spine structural plasticity [34]. Elimination of Kalrn expression in POMC cells reduces anxiety-like behavior in mice [35]. Kalirin-7 in rodents (equivalent to human NM_003947.4) is the most abundant kalirin isoform in the adult rodent brain and is localized at the postsynaptic side of excitatory synapses [36, 37]. Knockout and overexpression studies revealed important roles for Kal7 in dendritic spine formation and synaptic function in rodents [16, 33, 34, 36].

The processes that regulate the morphological development of dendrites and dendritic spines have a significant impact on the establishment and function of synapses and on neuronal circuits [16]. The major functional group of ID-related proteins corresponds to proteins enriched at synaptic compartments [38], and investigations in children and adolescents with unclassified ID confirm the reduced density and spine dysgenesis in apical dendrites of the prefrontal cortex [39]. Abnormal neural cell spine morphology was the only anatomical alteration reported in cases of non-syndromic ID [33]. A rare coding variant, kalirin-7-D1338N, was identified in a schizophrenia patient and his sibling with major depressive disorder where both subjects carrying the polymorphism displayed reduced cortical volume in the superior temporal sulcus (STS), a region implicated in schizophrenia [40]. These data suggest that single amino acid changes in proteins involved in dendritic spine function can have significant effects on the structure and function of the cerebral cortex [40]. Other

reports have indicated that dysfunctions in Rho-GEF signaling pathways are associated with various ID syndromes [41] and with Alzheimer disease [42]. Moreover, a correlation has been reported between the levels of Kalirin expression and the pathology of dendritic spines in some psychiatric and neurological disorders such as Huntington's disease, Alzheimer's disease, ischemic stroke, schizophrenia, depression, and cocaine addiction [43, 44].

In the tested family, there were clinical features other than ID, which could be correlated with Kalirin dysfunction. Both patients suffer from short stature and growth hormone deficiency. Mice with ablated spectrin domains of Kalirin had deficient growth rate and dysfunctional secretion of growth hormone [17]. Hypotonia was also present in our patients and may be related to the pre- and postsynaptic deficits in the neuromuscular junction reported in mice [17].

Kalirin has also been implicated in bone homeostasis where its deletion was reported to affect directly osteoclast and osteoblast activity. It may also play a role in paracrine and/or endocrine signaling events that control skeletal bone remodeling and the maintenance of bone mass [18]. This could be consistent with the observed delayed bone age in the patients.

The other two variants identified in *CFAP57* and in the last exon in *DPH2* were not predicted to be pathogenic. The patients in the studied family do not have any of the features of VWS including the pits and/or sinuses of the lower lip, and cleft lip and/or cleft palate, which leads us to consider *CFAP57* variant as a VUS in respect to the patient's phenotype. *DPH2*, one of several enzymes involved in the synthesis of diphthamide linked to diphtheria toxin (OMIM 603456), was also considered as unlikely to cause the phenotype in this family.

Conclusions

In this report, the clinical evaluation combined with the power and efficiency of genomic analysis defined a new candidate gene, *KALRN*, as the possible underlying cause of the syndromic autosomal recessive ID in the studied family. The role of *KALRN* in causing cognitive impairment in human and the full phenotypic spectrum will be established when other families with ID show pathogenic variants in *KALRN* and the exact molecular pathophysiology is understood.

Abbreviations

ID, intellectual disability; OMIM, Online Mendelian Inheritance in Man (http://omim.org/); ROH, run of homozygosity

Acknowledgements

We are grateful to the members of the family enrolled in this study.

Funding

This study was supported by the grants from the Gebert ruf Stiftung foundation (to SEA), von Meissner foundation, and Bodossaki foundation (to PM).

Authors' contributions

HH, PM, and SEA designed the study and wrote the manuscript. PM performed the exome analyses. HH coordinated the patient collection. MZ and MI examined the patients, described the phenotypic characteristics, and contributed the DNA samples. MG performed the exome sequencing. FAS conceived the used algorithms and MA coordinated Sanger sequencing. All authors contributed to the manuscript and approved the final version.

Competing interests

The authors declare that they have no competing interests.

Consent for publication

Parents of the affected children gave written consent for publication of photos.

Author details

[1]Department of Genetic Medicine and Development, University of Geneva, 1 Rue Michel-Servet, 1211 Geneva, Switzerland. [2]Service of Genetic Medicine, University Hospitals of Geneva, Geneva, Switzerland. [3]Department of Clinical Genetics, National Research Centre, Cairo, Egypt. [4]iGE3, Institute of Genetics and Genomics of Geneva, Geneva, Switzerland.

References

1. Al-Gazali L, Hamamy H, Al-Arrayad S. Genetic disorders in the Arab world. BMJ. 2006;333(7573):831–4.
2. Hamamy H, et al. Consanguineous marriages, pearls and perils: Geneva International Consanguinity Workshop Report. Genet Med. 2011;13(9):841–7.
3. Sheridan E, et al. Risk factors for congenital anomaly in a multiethnic birth cohort: an analysis of the Born in Bradford study. Lancet. 2013;382(9901):1350–9.
4. Stoll C, et al. Parental consanguinity as a cause for increased incidence of births defects in a study of 238,942 consecutive births. Ann Genet. 1999;42(3):133–9.
5. Stoltenberg C, et al. Consanguinity and recurrence risk of birth defects: a population-based study. Am J Med Genet. 1999;82(5):423–8.
6. Makrythanasis P, et al. Diagnostic exome sequencing to elucidate the genetic basis of likely recessive disorders in consanguineous families. Hum Mutat. 2014;35(10):1203–10.
7. Hamamy H, et al. Recessive thrombocytopenia likely due to a homozygous pathogenic variant in the FYB gene: case report. BMC Med Genet. 2014;15:135.
8. Makrythanasis P, et al. A novel homozygous mutation in FGFR3 causes tall stature, severe lateral tibial deviation, scoliosis, hearing impairment, camptodactyly, and arachnodactyly. Hum Mutat. 2014;35(8):959–63.
9. Makrythanasis P, et al. Pathogenic variants in PIGG cause intellectual disability with seizures and hypotonia. Am J Hum Genet. 2016;98(4):615–26.
10. Musante L, Ropers HH. Genetics of recessive cognitive disorders. Trends Genet. 2014;30(1):32–9.
11. Vissers LE, Gilissen C, Veltman JA. Genetic studies in intellectual disability and related disorders. Nat Rev Genet. 2016;17(1):9–18.
12. Najmabadi H, et al. Deep sequencing reveals 50 novel genes for recessive cognitive disorders. Nature. 2011;478(7367):57–63.
13. Anazi S, et al. A null mutation in TNIK defines a novel locus for intellectual disability. Hum Genet. 2016;135(7):773–8.
14. Iqbal Z, et al. Homozygous SLC6A17 mutations cause autosomal-recessive intellectual disability with progressive tremor, speech impairment, and behavioral problems. Am J Hum Genet. 2015;96(3):386–96.
15. Philips AK, et al. Identification of C12orf4 as a gene for autosomal recessive intellectual disability. Clin Genet. 2016. Jun 17. doi:10.1111/cge.12821. [Epub ahead of print]
16. Cahill ME, et al. Kalirin regulates cortical spine morphogenesis and disease-related behavioral phenotypes. Proc Natl Acad Sci U S A. 2009;106(31):13058–63.
17. Mandela P, et al. Kalrn plays key roles within and outside of the nervous system. BMC Neurosci. 2012;13:136.
18. Huang S, et al. The Rho-GEF kalirin regulates bone mass and the function of osteoblasts and osteoclasts. Bone. 2014;60:235–45.
19. Li H, et al. The sequence alignment/map format and SAMtools. Bioinformatics. 2009;25(16):2078–9.
20. Ye K, et al. Pindel: a pattern growth approach to detect break points of large deletions and medium sized insertions from paired-end short reads. Bioinformatics. 2009;25(21):2865–71.
21. Wang K, Li M, Hakonarson H. ANNOVAR: functional annotation of genetic variants from high-throughput sequencing data. Nucleic Acids Res. 2010;38(16), e164.
22. Santoni FA, Makrythanasis P, Antonarakis SE. CATCHing putative causative variants in consanguineous families. BMC Bioinformatics. 2015;16(1):310.
23. Ng PC, Henikoff S. Predicting deleterious amino acid substitutions. Genome Res. 2001;11(5):863–74.
24. Adzhubei IA, et al. A method and server for predicting damaging missense mutations. Nat Methods. 2010;7(4):248–9.
25. Schwarz JM, et al. MutationTaster2: mutation prediction for the deep-sequencing age. Nat Methods. 2014;11(4):361–2.
26. Vishwanatha KS, et al. Structural organization of the nine spectrin repeats of kalirin. Biochemistry. 2012;51(28):5663–73.
27. Schiller MR, et al. Autonomous functions for the Sec14p/spectrin-repeat region of kalirin. Exp Cell Res. 2008;314(14):2674–91.
28. Ma XM, et al. Nonenzymatic domains of kalirin7 contribute to spine morphogenesis through interactions with phosphoinositides and Abl. Mol Biol Cell. 2014;25(9):1458–71.
29. Rorick NK, et al. Genomic strategy identifies a missense mutation in WD-repeat domain 65 (WDR65) in an individual with Van der Woude syndrome. Am J Med Genet A. 2011;155A(6):1314–21.
30. Eppig JT, et al. The Mouse Genome Database (MGD): facilitating mouse as a model for human biology and disease. Nucleic Acids Res. 2015;43(Database issue):D726–36.
31. Smith CM, et al. The mouse Gene Expression Database (GXD): 2014 update. Nucleic Acids Res. 2014;42(Database issue):D818–24.
32. Newey SE, et al. Rho GTPases, dendritic structure, and mental retardation. J Neurobiol. 2005;64(1):58–74.
33. Vanleeuwen JE, Penzes P. Long-term perturbation of spine plasticity results in distinct impairments of cognitive function. J Neurochem. 2012;123(5):781–9.
34. Xie Z, Cahill ME, Penzes P. Kalirin loss results in cortical morphological alterations. Mol Cell Neurosci. 2010;43(1):81–9.

35. Mandela P, et al. Elimination of Kalrn expression in POMC cells reduces anxiety-like behavior and contextual fear learning. Horm Behav. 2014;66(2):430–8.

36. Ma XM, et al. Kalirin-7, an important component of excitatory synapses, is regulated by estradiol in hippocampal neurons. Hippocampus. 2011;21(6):661–77.

37. Miller MB, et al. Neuronal Rho GEFs in synaptic physiology and behavior. Neuroscientist. 2013;19(3):255–73.

38. Pavlowsky A, Chelly J, Billuart P. Emerging major synaptic signaling pathways involved in intellectual disability. Mol Psychiatry. 2012;17(7):682–93.

39. Kaufmann WE, Moser HW. Dendritic anomalies in disorders associated with mental retardation. Cereb Cortex. 2000;10(10):981–91.

40. Russell TA, et al. A sequence variant in human KALRN impairs protein function and coincides with reduced cortical thickness. Nat Commun. 2014;5:4858.

41. Ramakers GJ. Rho proteins, mental retardation and the cellular basis of cognition. Trends Neurosci. 2002;25(4):191–9.

42. Youn H, et al. Kalirin is under-expressed in Alzheimer's disease hippocampus. J Alzheimers Dis. 2007;11(3):385–97.

43. Mandela P, Ma XM. Kalirin, a key player in synapse formation, is implicated in human diseases. Neural Plast. 2012;2012:728161.

44. Remmers C, Sweet RA, Penzes P. Abnormal kalirin signaling in neuropsychiatric disorders. Brain Res Bull. 2014;103:29–38.

11

MicroRNAs in acute kidney injury

Pei-Chun Fan[1,2], Chia-Chun Chen[3], Yung-Chang Chen[1], Yu-Sun Chang[3] and Pao-Hsien Chu[4,5,6,7*] (iD)

Abstract

Acute kidney injury (AKI) is an important clinical issue that is associated with significant morbidity and mortality. Despite research advances over the past decades, the complex pathophysiology of AKI is not fully understood. The regulatory mechanisms underlying post-AKI repair and fibrosis have not been clarified either. Furthermore, there is no definitively effective treatment for AKI. MicroRNAs (miRNAs) are endogenous single-stranded noncoding RNAs of 19~23 nucleotides that have been shown to be crucial to the post-transcriptional regulation of various cellular biological functions, including proliferation, differentiation, metabolism, and apoptosis. In addition to being fundamental to normal development and physiology, miRNAs also play important roles in various human diseases. In AKI, some miRNAs appear to act pathogenically by promoting inflammation, apoptosis, and fibrosis, while others may act protectively by exerting anti-inflammatory, anti-apoptotic, anti-fibrotic, and pro-angiogenic effects. Thus, miRNAs have not only emerged as novel biomarkers for AKI; they also hold promise to be potential therapeutic targets.

Keywords: MicroRNAs, Acute kidney injury, Renal fibrosis

Abbreviations: AA, aristolochic acid; ADIPOR2, adiponectin receptor 2; AGO, argonaute; AKI, acute kidney injury; ATF3, activating transcription factor 3; B, blood; BCL-2, B cell lymphoma 2; BUMPT-306 cell, Boston University mouse proximal tubule cell clone 306; $CdCl_2$, cadmium chloride; CRL-2753 cell, rat mesangial cell line; CKD, chronic kidney disease; CXCR4, chemokine receptor type 4; DGCR8, Di-George syndrome critical region gene 8 or Pasha; DM, diabetes mellitus; EMT, epithelial-to-mesenchymal transition; ER, endoplasmic reticulum; ERK-2, extracellular signal-regulated kinase 2; FGF-2, fibroblast growth factor 2; Foxo3, forkhead box O3; FSGS, focal segmental glomerulosclerosis; H2A. X, H2A histone family member X; HEK cell, human embryonic kidney cell; HepG2 cell, human hepatocellular liver carcinoma cell line; HIF-1α, hypoxia-inducible factor 1 alpha, HK-2 cell, human kidney 2 cell; HO-1, heme oxygenase-1; HPTEC, human proximal tubular epithelial cell; HUVEC, human umbilical vein endothelial cell; ICU, intensive care units; IGF1R, insulin-like growth factor 1 receptor; IL, interleukin; IKKb, inhibitor of NF-kB kinases b; IRAK-1, interleukin-1 receptor-associated kinase 1; IRI, ischemia-reperfusion injury; $K_2Cr_2O_7$, potassium dichromate; LC3-II, light chain 3-II; MCP-1, monocyte chemoattractant protein-1; MDM2, murine double-minute 2; miRNA, microRNA; mRNA, messenger RNA; NF-kB, nuclear factor-kappaB; NRK-52E cell, rat renal proximal tubular cell line; PDCD4, programmed cell death protein 4; Pparα, peroxisome proliferator activated receptor alpha; PTC, proximal tubular cell; PTEN, phosphatase and tensin homolog; Rab-11a, Ras-related proteins in brain 11 a; RISC, RNA-induced silencing complex; ROS, reactive oxygen species; S1PR1, sphingosine-1-phosphate receptor 1; SHRSP, stroke-prone spontaneously hypertensive rat; STZ, streptozocin; T, tissue; TEC, tubular epithelial cell; TEnC, tubular endothelial cell; TEpC, tubular epithelial cell; TNF, tumor necrosis factor; TGF-β, transforming growth factor beta; TRAF-6, TNF receptor-associated factor 6; Treg, regulator T cell; U, urine; UTR, untranslated region; UUO, unilateral ureteral obstruction; VEGF, vascular endothelial growth factor; VEGFR2, vascular endothelial growth factor receptor 2; ZEB1/ZEB2, zinc finger E-box-binding homeobox

* Correspondence: taipei.chu@gmail.com
[4]Division of Cardiology, Department of Internal Medicine, Chang Gung Memorial Hospital, College of Medicine, Chang Gung University, Taipei, Taiwan
[5]Healthcare Center, Chang Gung Memorial Hospital, College of Medicine, Chang Gung University, Taipei, Taiwan
Full list of author information is available at the end of the article

Background

Acute kidney injury

Acute kidney injury (AKI) is a complex syndrome that occurs in a variety of settings with clinical manifestations ranging from a minimal elevation in serum creatinine to anuric renal failure. AKI conveys significant morbidity and mortality, is a major risk factor of chronic kidney disease, and is thus associated with huge health and socioeconomic burdens [1, 2]. Despite research advances in the past decades, however, the complex pathophysiology of AKI is not fully understood. The regulatory mechanisms underlying post-AKI repair and fibrosis remain to be clarified. Furthermore, there is no definitively effective treatment for AKI.

MicroRNA biogenesis and function

MicroRNAs (miRNAs) are endogenous single-stranded noncoding mRNAs of 19~23 nucleotides. They were first discovered in *Caenorhabditis elegans* by Ambros's group in 1993 [3] and show surprisingly high conservation across species. The evidence accumulated over the past two decades shows that miRNAs play a critical role in the post-transcriptional regulation of almost all biological cell functions, including proliferation, differentiation, metabolism, and apoptosis [4]. miRNAs, which are expressed in a tissue-specific manner, are fundamental to normal development and physiology [4] and are involved in the pathologic pathways of many disease models.

To date, more than 2000 miRNAs have been discovered in the human genome. The miRNA-encoded genes are found as either independent genes having their own promoters, or as sequences in the introns of protein-coding genes [5]. RNA polymerase II transcribes an miRNA gene into a primary transcript (called a pri-miRNA) of several kilobases that can encode either an individual miRNA or a polycistronic cluster of two or more miRNAs. The RNase III enzyme, DROSHA, and its cofactor DGCR8 (Di-George syndrome critical region gene 8 or Pasha), cleave a pri-miRNA at its stem-loop structure, generating an approximately 70-nucleotide intermediate called the pre-miRNA. Exportin-5 exports the pre-miRNA from the nucleus to the cytoplasm, and the RNase III enzyme, DICER, further cleaves it to yield a single-stranded mature miRNA. To perform its function, an miRNA is incorporated along with the argonaute (AGO) protein to form an effector complex called the RNA-induced silencing complex (RISC). RISC binds to the 3′-untranslated region (UTR) of a target messenger RNA (mRNA), leading to the repression of either protein translation or mRNA degradation. Unlike small interfering RNAs in plants, miRNAs do not require complete complementarity to bind their targets. Instead, the evidence suggests that the "seed sequence" (nucleotides 2 through 8 of the miRNA) is the most important region for the ability of an miRNA to bind and regulate its target gene(s). Once bound, miRNAs induce repression by blocking the initiation or elongation of translation or de-adenylating the mRNA transcripts. Because miRNAs do not require complete complementarity to repress gene expression, a given miRNA can regulate multiple mRNA transcripts and a given mRNA transcript can be repressed by multiple miRNAs. It is estimated that miRNAs regulate more than half of the protein-coding genes in human [6]. Moreover, miRNAs have been implicated in various human diseases [7, 8], including kidney diseases, such as polycystic kidney disease [9], renal cell carcinoma [10], diabetic nephropathy [11], lupus nephritis, [12] and renal allograft rejection [13]. In the past few years, researchers have begun to address the relevance of miRNAs to AKI.

miRNAs in acute kidney injury

The miRNAs that have been implicated in AKI are summarized in Tables 1 and 2, and those with potential pathological or protective roles are summarized in Table 3. The first evidence of miRNAs having pathological roles in AKI was reported by Wei et al. who developed a *Dicer*-knockout mouse model, in which *Dicer* was specifically deleted from proximal tubular cells. These mice exhibit a global down-regulation of microRNAs in the renal cortex. They have normal renal function and histology under control conditions but show resistance to the AKI that follows bilateral renal ischemia-reperfusion (IRI). Under the latter conditions, *Dicer*-null mice show significantly better renal function, less tissue damage, less tubular apoptosis, and better survival than their wild-type counterparts [14].

miR-10a is renal tubule-specific miRNA that is released from kidney tissues upon injury. In rodent models of renal IRI and streptozocin (STZ)-induced diabetic nephropathy, the levels of miR-10a are increased decreased in urine and kidney tissue, respectively [15, 16]. miR-10a is thought to exert protective actions during injury by targeting IL-12/IL-23p40 and the pro-apoptotic protein BIM [17]. In humans, decreased plasma levels of miR-10a have been shown to predict AKI in critical patients of intensive care units (ICUs) [18].

The members of the miR-17 family have been shown to be induced by pro-inflammatory cytokines, and their tissue expressions are increased in rodent models of renal IRI [19, 20].

miR-21 appears to play a dual role; on the one hand, it protects against injury by inhibiting apoptosis and inflammation; on the other hand, it may amplify the injury response and promote fibrosis. Studies have shown that miR-21 inhibits apoptosis by down-regulating programmed cell death protein 4 (PDCD4), down-regulating

Table 1 miRNAs implicated in acute kidney injury

miRNA	Samples	Species	Model	Expression	Reference
k12-3	In vitro	HK-2 cells	Oxidative stress	Down then up	[51]
let-7a	T, B	Rat, human	Contrast nephropathy, contrast nephropathy[a]	Down	[52]
let-7a-1-3p	T, U	Rat	Cisplatin nephropathy	Up (urine), down (tissue)	[33]
let-7a-2*	In vitro	HK-2 cells, primary PTCs	AA nephropathy	Up	[38]
let-7b	B	Human	ICU AKI[a]	Down	[46]
let-7d	U	Rat	Gentamicin nephropathy	Down	[53]
let-7e	T, in vitro	Mouse, HK-2 cells	IRI	Up, down	[23, 54]
let-7f	B, T	Human, mouse	ICU AKI[a], IRI	Down	[34, 46]
let-7g	T, U	Mouse, rat	Cisplatin nephropathy	Up, down	[33, 35]
miR-7	T, in vitro	Mouse, HK-2 cells	IRI, oxidative stress	Up	[14, 51]
miR-7a-1-3p	T, U	Rat	Cisplatin nephropathy	Up (urine), down (tissue)	[33]
miR-10a	T, U, B	Mouse, human, rat	IRI, DM-CKD (STZ), FSGS[a], ICU AKI[a]	Up, down	[15, 16, 18]
miR-10b*	T	Mouse	Cisplatin nephropathy	Down	[35]
miR-15	U	Rat	Cisplatin nephropathy	Up	[55]
miR-15b-5p	T, in vitro	Mouse, HK-2 cells	IRI	Down	[54]
miR-16	B, U	Human, rat	ICU AKI[a], cisplatin nephropathy	Up, down	[46, 55]
miR-17-3p	T	Mouse	IRI	Up	[14, 49]
miR-17-5p	T, U	Mouse, rat	IRI, cisplatin nephropathy	Up, down	[19, 20, 33]
miR-18a	T, B, U, in vitro	Mouse, rat, human, HPTECs	IRI, gentamicin nephropathy, folic acid, CdCl$_2$, arsenic trioxide, AA, K$_2$Cr$_2$O$_7$, cisplatin, UUO, allograft rejection[a], renal fibrosis[a]	Up, down	[14, 34, 47, 56, 57]
miR-19a	T	Mouse	IRI	Up	[34]
miR-20a	T, in vitro, U	Mouse, TECs, rat, HK-2 cells	Cisplatin nephropathy, IRI	Up, down	[37, 54, 55]
miR-20b-5p	T, U, in vitro	Rat, mouse, HK-2 cells	Cisplatin nephropathy, IRI	Up (urine), down (tissue)	[33, 54]
miR-21	B, U, T, in vitro	Human, rat, mouse, TEC, CRL-2753 cells, NRK52E cells, HK-2 cells	IRI, TGF-β, anti-Thy 1.1, UUO, SHRSP, gentamicin nephropathy, folic acid, CdCl$_2$, arsenic trioxide, AA, K$_2$Cr$_2$O$_7$, allograft rejection[a], renal fibrosis[a], AKI[a]	Up, down	[19–30, 34, 37, 45, 56–58]
miR-24	B, T, in vitro	Human, rat, CRL-2753 cells, NRK52E cells, HK-2 cells, HUVECs	ICU AKI[a], transplantation[a], UUO, IRI	Up, down	[31, 45, 46]
miR-24-2	T	Mouse	IRI	Up	[34]
miR-25-3p	T, U	Rat	Cisplatin nephropathy	Up (urine), down (tissue)	[33]
miR-26a	In vitro, T	HK-2 cells, mouse	IRI, oxidative stress, cisplatin nephropathy	Down	[32, 35, 51]
miR-26b	T, in vitro, U, B	Rat, CRL-2753 cells, NRK52E cells, human	UUO, cisplatin nephropathy, ICU AKI[a]	Down (tissue, blood), up (urine)	[18, 33, 45]
miR-27a-3p	B	Human	ICU AKI[a]	Down	[18]
miR-29a	T, in vitro, B	HK-2 cells, human	Oxidative stress, ICU AKI[a]	Up, down	[18, 51, 59]
miR-29b	T, in vitro	Rat, HK-2 cells	Oxidative stress	Up	[51, 59]
miR-29c	T	Mouse	IRI	Up	[34, 59]

Table 1 miRNAs implicated in acute kidney injury *(Continued)*

miR-30a-5p	T, U, in vitro, B	Rat, mouse, HK-2 cells, human	Cisplatin nephropathy, IRI, contrast-induced nephropathy, contrast-induced nephropathy[a]	Up (urine, blood, tissue), down (tissue)	[33, 52, 54]
miR-30c	T, in vitro, B	Rat, CRL-2753 cells, NRK52E cells, mouse, human	TGF-β, UUO, SHRSP, contrast-induced nephropathy, contrast-induced nephropathy[a]	Up, down	[34, 45, 52]
miR-30c-1	T	Mouse	IRI	Up	[34]
miR-30c-2*	In vitro	HK-2 cells	Oxidative stress	Down	[51]
miR-30d	T, U, B	Mouse, human	IRI, DM-CKD (STZ), FSGS[a]	Up, down, unchanged	[16]
miR-30d*	B	Human	ICU AKI[a]	Down	[46]
miR-30e	T, B	Mouse, rat, human	Cisplatin nephropathy, contrast-induced nephropathy, contrast-induced nephropathy[a]	Up, down	[35, 52]
miR-34a	T, in vitro	Mouse, BUMPT-306 cells, NRK-52E cells, RTECs	Cisplatin nephropathy, IRI	Up	[35, 60, 61]
miR-34b	T	Mouse	IRI	Up	[47]
miR-92a	T	Mouse	IRI	Up	[34]
miR-92b*	B	Human	ICU AKI[a]	Up	[46]
miR-93-3p	B	Human	ICU AKI[a], AKI post-cardiac surgery[a]	Down	[18]
miR-93-5p	T, U	Rat	Cisplatin nephropathy	Up (urine), down (tissue)	[33]
miR-99b	In vitro, T	HK-2 cells, mouse	ER stress, IRI	Down	[51, 54]
miR-101-3p	B	Human	ICU AKI[a]	Down	[18]
miR-101a	T, in vitro	Mouse, HK-2 cells	UUO	Down	[25]
miR-106a-5p	T, in vitro	Mouse, HK-2 cells, primary PTCs, rat	IRI, AA nephropathy	Up, down	[19, 20, 38]
miR-122	T	Mouse	Cisplatin nephropathy, IRI	Down	[35, 49]
miR-123	T	Mouse	IRI	Up	[49]
miR-125a-5p	T, in vitro	Mouse, HK-2 cells	IRI	Down	[54]
miR-125b	T, in vitro	Mouse, HepG2 cells, HEK293 cells, NRK52E cells	Cisplatin nephropathy		[62]
miR-126-3p	B	Human	ICU AKI[a]	Down	[18]
miR-126-5p	T, in vitro	Mouse, rat, TEnCs, TEpCs	IRI	Up	[34, 43, 44, 63]
miR-127-3p	T, in vitro, B	Rat, mouse, NRK-52E cells, HK-2 cells, human	IRI, ICU AKI[a], AKI post-cardiac surgery[a]	Up, down	[14, 18, 36, 49]
miR-129-3p	T	Mouse	IRI	Up	[34]
miR-129-5p	In vitro	HK-2 cells, primary PTCs	AA nephropathy	Down	[38]
miR-130b-3p	T, U	Rat	Cisplatin nephropathy	Up (urine), down (tissue)	[33]
miR-132	T, in vitro	Mouse, human, HPTECs	IRI, folic acid, $CdCl_2$, arsenic trioxide, AA, $K_2Cr_2O_7$, cisplatin, UUO, allograft rejection[a], renal fibrosis[a]	Up	[14, 57]
miR-133a	In vitro	HK-2 cells	ER stress	Down	[51]
miR-134	T	Mouse	IRI	Up	[47]
miR-135b	T	Mouse	IRI	Down	[14, 49]
miR-140-3p	T, U	Rat	Cisplatin nephropathy	Up (urine), down (tissue)	[33]
miR-141	T	Mouse	IRI	Up	[34]
miR-142-3p	T, in vitro	Mouse, HK-2 cells	UUO	Up	[25]

Table 1 miRNAs implicated in acute kidney injury *(Continued)*

miR-142-5p	T, in vitro	Mouse, HK-2 cells	UUO	Up	[25]
miR-145	T, in vitro	Rat, mouse, CRL-2753 cells, NRK52E cells, CD133+ renal medullary progenitor cells	TGF-β, SHRSP salt challenge	Up, down	[39, 45, 64]
miR-146a	T, in vitro, B	Mouse, TECs, human	IRI, ICU AKI[a]	Down (blood), up (tissue)	[18, 37]
miR-146b-5p	T, in vitro	Mouse, human, HPTECs	IRI, folic acid, CdCl$_2$, arsenic trioxide, AA, K$_2$Cr$_2$O$_7$, cisplatin, UUO, allograft rejection[a], renal fibrosis[a]	Up	[57]
miR-149	T	Mouse	IRI	Down	[34]
miR-150	T, in vitro	Mouse, immortalized mouse cardiac endothelial cell lines	IRI, AMI using LAD ligation	Down	[65]
miR-155	B, U, T, in vitro	Rat, human, mouse, HK-2 cells	IRI, gentamicin nephropathy, Cisplatin nephropathy, AKI[a]	Up, down	[54, 56, 66]
miR-181a*	In vitro	HK-2 cells	ER stress	Up	[51]
miR-181a-2*	In vitro	HK-2 cells	ER stress	Down	[51]
miR-181d	T	Mouse	IRI	Down	[34]
miR-182	T	Mouse	IRI	Up	[47]
miR-183-5p	T, U	Rat	Cisplatin nephropathy	Up (urine), down (tissue)	[33]
miR-187	T, in vitro	Mouse, TECs	IRI	Down	[37]
miR-188-5p	T	Mouse	IRI	Up	[34]
miR-191a-5p	T, U	Rat	Cisplatin nephropathy	Up (urine), down (tissue)	[33]
miR-192	T, in vitro, B, U	Mouse, rat, CRL-2753 cells, NRK52E cells, TECs, HK-2 cells, primary PTCs	IRI, UUO, SHRSP, AA nephropathy, cisplatin nephropathy, contact freezing, Dahl salt-sensitive rat with high salt administration	Up, down	[15, 33, 45, 55]
miR-193	T, in vitro, U	Mouse, HK-2 cells, Rat	UUO, cisplatin nephropathy	Down (tissue), up (urine)	[25, 33, 35, 55]
miR-194	T, in vitro, B, U	Mouse, rat, TECs, HK-2 cells, primary PTCs	IRI, AA nephropathy, contact freezing, Dahl salt-sensitive rat with high salt administration	Up, down	[15, 37–39]
miR-197	T	Mouse	IRI	Down	[34]
miR-199a-3p	T, in vitro	Mouse, TECs	IRI	Up	[37]
miR-200a	T, B, U	Human, Rat, mouse	Contact freezing, Dahl salt-sensitive rat with high salt administration, contrast-induced nephropathy, contrast-induced nephropathy[a]	Up, down	[39, 52]
miR-200b	T, in vitro, B, U	Rat, CRL-2753 cells, NRK52E cells, human	TGF-β, UUO, contact freezing, early CKD (Dahl salt-sensitive rat with high salt administration)	Up, down	[34, 39, 45]
miR-200c	T, in vitro, U, B	Rat, CRL-2753 cells, NRK52E cells, human	TGF-β, contact freezing, early CKD (Dahl salt-sensitive rat with high salt administration), ICU and transplant AKI[a]	Up, down	[29, 39, 45]
miR-202	In vitro	HK-2 cells	ER stress	Down	[51]
miR-203	U	Rat	Gentamicin nephropathy	Down	[53]
miR-205	In vitro	HK-2 cells, primary PTCs	Oxidative stress, ER stress, AA nephropathy	Down	[38, 51]
miR-207	T	Mouse	IRI	Up, down	[14, 34]
miR-210	B, T, in vitro, U	Human, mouse, HUVEC-12 cells, HK-2 cells, primary PTCs, rat	IRI, Oxidative stress, AA nephropathy, cisplatin nephropathy, ICU AKI[a]	Up, down	[18, 34, 38, 46, 47, 51, 55]

Table 1 miRNAs implicated in acute kidney injury *(Continued)*

miR-211	T	Mouse	IRI	Down	[34]
miR-212	T	Mouse, human, HPTECs	IRI, folic acid, CdCl$_2$, arsenic trioxide, AA, K$_2$Cr$_2$O$_7$, cisplatin, UUO, allograft rejection[a], renal fibrosis[a]	Up, down	[34, 57]
miR-214	T, in vitro	Mouse, rat, HK-2 cells, TECs, CRL-2753 cells, NRK52E cells, human	TGF-β, anti-Thy 1.1, UUO, SHRSP, IRI, diabetic nephropathy[a]	Up	[23–25, 37, 45, 47]
miR-215	In vitro	HK-2 cell	ER stress	Down	[51]
miR-218	T, in vitro	Mouse, HK-2 cells	UUO	Down	[25]
miR-218-1	T	Mouse	IRI	Up	[34]
miR-218a-5p	T, U	Rat	Cisplatin nephropathy	Up (urine),down (tissue)	[33]
miR-221*	In vitro	HK-2 cells	Oxidative stress	Up	[51]
miR-223	T, in vitro	Mouse, HK-2 cells	UUO	Up	[25]
miR-290-3p	T	Mouse	IRI	Up	[34]
miR-296	T, in vitro	Rat, mouse, TEnCs, TEpCs	IRI	Up, down	[14, 43]
miR-302b	T	Mouse	IRI	Up	[34]
miR-302c	T	Mouse	IRI	Up	[34]
miR-320	B, T, U	Human, mouse, rat	IRI, cisplatin nephropathy, gentamicin nephropathy, contrast-induced nephropathy, ICU AKI[a], contrast-induced nephropathy[a]	Up, down	[23, 33, 34, 46, 52, 53]
miR-322	T	Mouse	IRI	Down	[14]
miR-324-3p	T	Mouse	IRI	Down	[14]
miR-326	T	Mouse	IRI	Down	[34]
miR-328	T	Mouse	IRI	Down	[34]
miR-328a-3p	T, U	Rat	Cisplatin nephropathy	Up (urine),down (tissue)	[33]
miR-329	T, in vitro	Rat, CRL-2753 cells, NRK52E cells	UUO	Down	[45]
miR-335	T, U	Rat	Cisplatin nephropathy	Up (urine),down (tissue)	[33]
miR-340-5p	T, U	Rat	Cisplatin nephropathy	Up (urine),down (tissue)	[33]
miR-346	T	Mouse	IRI	Down	[34]
miR-362-5p	T	Mouse	IRI	Up	[14, 34]
miR-365*	In vitro	HK-2 cells, primary PTCs	AA nephropathy	Down	[38]
miR-378a-5p	T, U	Rat	Cisplatin nephropathy	Up (urine),down (tissue)	[33]
miR-379	T	Mouse	IRI	Down	[14, 49]
miR-382	In vitro	HK-2 cells, primary PTCs	AA nephropathy	Up	[38]
miR-423	U	Human	ICU and transplant AKI[a]	Up	[29]
miR-449	In vitro	NRK-52E cells	Cisplatin nephropathy	Up	[67]
miR-450a-3p	T, in vitro	Mouse, HK-2 cells, primary PTCs	IRI, AA nephropathy	Up, down	[34, 38]
miR-451	T	Mouse	IRI	Up	[34]
miR-455-3p	T	Mouse	IRI	Down	[14]
miR-466a-5p	T	Mouse	IRI	Up	[34]
miR-466b-5p	T	Mouse	IRI	Up	[34]

Table 1 miRNAs implicated in acute kidney injury *(Continued)*

miR-466c-5p	T	Mouse	IRI	Down	[34]
miR-466f-3p	T	Mouse	IRI	Down	[34]
miR-466g	T	Mouse	IRI	Down	[34]
miR-466i	T	Mouse	IRI	Down	[34]
miR-467	T	Mouse	IRI	Up	[14]
miR-467a	T	Mouse	IRI	Down	[34]
miR-467b	T	Mouse	IRI	Down	[34]
miR-467e	T	Mouse	IRI	Down	[34]
miR-467f	T	Mouse	IRI	Down	[34]
miR-467g	T	Mouse	IRI	Down	[34]
miR-468	T	Mouse	IRI	Down	[34]
miR-483	T	Mouse	IRI	Up, down?	[34]
miR-484	T	Mouse	IRI	Down	[34]
miR-486	T	Mouse	IRI	Up	[14]
miR-487b	T	Mouse	IRI	Down	[14]
miR-489	T	Mouse	IRI	Up	[14]
miR-491	T	Mouse	IRI	Down	[14]
miR-494	T, U, B	Mouse, human	IRI, ICU AKI[a]	Up, unchanged	[48]
miR-495	T	Mouse	IRI	Up	[14]
miR-503	In vitro	HK-2 cells	ER stress	Down	[51]
miR-532-3p	T, U	Mouse, rat	IRI, Cisplatin nephropathy	Up, down	[33, 34]
miR-542-3p	In vitro	HK-2 cells, primary PTCs	AA nephropathy	Up	[38]
miR-547-3p	T	Mouse	IRI	Down	[34]
miR-574-5p	In vitro	HK-2 cells, primary PTCs	AA nephropathy	Down	[38]
miR-617	B	Human	ICU AKI[a]	Up	[46]
miR-620	B	Human	ICU AKI[a]	Down	[46]
miR-625*	In vitro	HK-2 cells, primary PTCs	AA nephropathy	Down	[38]
miR-630	In vitro	HK-2 cells	Oxidative stress	Up	[51]
miR-638	B	Human	ICU AKI[a]	Up	[46]
miR-663b	B	Human	ICU AKI[a]	Up	[46]
miR-668	T	Mouse	IRI	Up	[14]
miR-669a	T	Mouse	IRI	Down	[34]
miR-669f	T	Mouse	IRI	Down	[34]
miR-669h-3p	T	Mouse	IRI	Down	[34]
miR-671-3p	In vitro	HK-2 cells, primary PTCs	AA nephropathy	Down	[38]
miR-671-5p	T	Mouse	IRI	Up	[34]
miR-674	T	Mouse	IRI	Down	[34]
miR-680	T	Mouse	IRI	Up	[34]
miR-684	T	Mouse	IRI	Up	[34]
miR-685	T	Mouse	IRI	Up	[14, 34, 49]
miR-687	T, in vitro	Mouse, BUMPT-306 cells, HEK cells	IRI	Up	[14, 49]
miR-689	T	Mouse	IRI	Up	[34]
miR-694	T	Mouse	IRI	Up	[14]

Table 1 miRNAs implicated in acute kidney injury *(Continued)*

miR-705	T	Mouse	IRI	Up	[34]
miR-708	T	Mouse	IRI	Up	[34]
miR-714	T, B	Mouse	IRI	Up	[68]
miR-718	T	Mouse	IRI	Down	[34]
miR-721	T	Mouse	IRI	Up	[34]
miR-744-5p	T, U	Rat	Cisplatin nephropathy	Up (urine),down (tissue)	[33]
miR-805	T, in vitro	Mouse, TECs	IRI	Down	[34, 37]
miR-875-5p	T	Mouse	IRI	Down	[34]
miR-876-5p	T	Mouse	IRI	Up	[34]
miR-877	T	Mouse	IRI	Up, down?	[34]
miR-877*	T, B	Mouse	IRI	Up	[68]
miR-1187	T	Mouse	IRI	Down	[34]
miR-1188	T, B	Mouse	IRI	Up	[68]
miR-1196	T	Mouse	IRI	Down	[34]
miR-1198	T	Mouse	IRI	Down	[34]
miR-1224	T, B	Mouse	IRI	Up	[68]
miR-1244	B	Human	ICU AKI[a]	Down	[46]
miR-1249	In vitro	HK-2 cells, primary PTCs	AA nephropathy	Up	[38]
miR-1839-5p	T, U	Rat	Cisplatin nephropathy	Up (urine),down (tissue)	[33]
miR-1892	T	Mouse	IRI	Up	[34]
miR-1894-3p	T	Mouse	IRI	Up	[34]
miR-1897-3p	T, B	Mouse	IRI	Up	[68]
miR-4521	In vitro	HK-2 cells, primary PTCs	AA nephropathy	Down	[38]
miR-4640	U	Human	ICU and transplant AKI[a]	Down	[29]
miR-4716-5p	In vitro	HK-2 cells, primary PTCs	AA nephropathy	Up	[38]
miR-4730	In vitro	HK-2 cells, primary PTCs	AA nephropathy	Up	[38]
miR-4747-3p	In vitro	HK-2 cells, primary PTCs	AA nephropathy	Up	[38]

[a]Human studies

phosphatase and tensin homolog (PTEN), activating the AKT pathway, up-regulating B cell lymphoma 2 (BCL-2), and decreasing the levels of active caspase-3 and caspase-8 proteins [21, 22]. Up-regulation of miR-21 also inhibits inflammation by decreasing nuclear factor-kappaB (NF-kB), tumor necrosis factor (TNF), interleukin 6 (IL-6), and IL-18, and by increasing IL-10 [21]. Experimental up-regulation of miR-21 provides morphologic and functional renoprotection in animal models of AKI [21–23]. miR-21 is induced by transforming growth factor beta (TGF-β)/Smad, hypoxia inducible factor 1 alpha (HIF-1α), TNF, and fibroblast growth factor 2 (FGF-2) [24, 25], and this miRNA promotes fibrosis by targeting peroxisome proliferator-activated receptor alpha (Pparα) and altering lipid metabolism [26]. miR-21 also targets Mpv17l, a mitochondria inhibitor of reactive oxygen species (ROS) [26]. miR-21 inhibits autophagy by targeting Ras-related proteins in brain

11 a (Rab-11a), decreasing light chain 3-II (LC3-II), decreasing beclin-1, and increasing p62 [27]. In vivo blockade of miR-21 reduces renal fibrosis and macrophage infiltration in animal models. Moreover, increased urinary and plasma levels of miR-21 have been observed in various clinical AKI settings [26, 28, 29]. For example, urine and plasma miR-21 levels were shown to correlate with AKI severity and hospital mortality and to predict the need for postoperative renal replacement therapy [28]. Interestingly, one study found decreased, but not increased, expression of miR-21 in AKI patients. Lower baseline plasma levels of miR-21 have been demonstrated to predict cardiac surgery-associated AKI [30].

miR-24 is up-regulated in mouse kidney after IRI and in patients after kidney transplantation. This miRNA enhances apoptosis by down-regulating sphingosine-1-phosphate receptor 1 (S1PR1), H2A histone family

Table 2 miRNAs implicated in human studies related to kidney injury

miRNA	Kidney injury	Expression		Reference
		Up	Down	
hsa-let-7b	AKI in ICU		Blood	[46]
hsa-let-7f	AKI in ICU		Blood	[46]
hsa-miR-10a	Focal segmental sclerosis	Urine		[16]
	AKI in ICU		Blood	[18]
hsa-miR-16	AKI in ICU		Blood	[46]
hsa-miR-21	AKI, chronic renal allograft dysfunction, renal allograft rejection, renal fibrosis	Tissue, blood, urine		[24, 26, 28, 29, 56]
	AKI after cardiac surgery		Blood	[30]
hsa-miR-24	AKI in ICU		Blood	[46]
	Transplanted renal graft with prolonged cold ischemia time	Tissue		[31]
hsa-miR-26b	AKI in ICU		Blood	[18]
hsa-miR-27a-3p	AKI in ICU		Blood	[18]
hsa-miR-29a	AKI in ICU		Blood	[18]
hsa-miR-30a-5p	Contrast-induced nephropathy	Blood		[52]
hsa-miR-30c	Contrast-induced nephropathy	Blood		[52]
hsa-miR-30d	Focal segmental sclerosis	Urine		[16]
hsa-miR-30d*	AKI in ICU		Blood	[46]
hsa-miR-30e	Contrast-induced nephropathy	Blood		[52]
hsa-miR-92b*	AKI in ICU	Blood		[46]
hsa-miR-93-3p	AKI in ICU, AKI post-cardiac surgery		Blood	[18]
hsa-miR-101-3p	AKI in ICU		Blood	[18]
hsa-miR-126-3p	AKI in ICU		Blood	[18]
hsa-miR-127-3p	AKI in ICU, AKI post-cardiac surgery		Blood	[18]
hsa-miR-146a	AKI in ICU		Blood	[18]
hsa-miR-155	AKI	Urine		[56]
hsa-miR-200c	AKI in ICU, AKI in renal transplant	Urine		[29]
hsa-miR-210	AKI in ICU	Blood		[46]
	AKI in ICU		Blood	[18]
hsa-miR-214	Diabetes related chronic kidney disease stage 4	Tissue		[24]
hsa-miR-320	AKI in ICU		Blood	[46]
hsa-miR-423	AKI in ICU, AKI in renal transplant	Urine		[29]
hsa-miR-494	AKI in ICU	Urine		[48]
hsa-miR-617	AKI in ICU	Blood		[46]
hsa-miR-620	AKI in ICU		Blood	[46]
hsa-miR-638	AKI in ICU	Blood		[46]
hsa-miR-663b	AKI in ICU	Blood		[46]
hsa-miR-1244	AKI in ICU		Blood	[46]
hsa-miR-4640	AKI in ICU, AKI in renal transplant		Urine	[29]

member X (H2A.X), and heme oxygenase-1 (HO-1). Inhibition of miR-24 was shown to prevent renal injury in animal models [31].

miR-26a represses IL-6 expression to promote the expansion of regulator T cells (Tregs). The tissue levels of miR-26a is down-regulated in animal models of AKI, and experimental overexpression attenuates renal IRI and improves renal recovery [32]. miR-26b is down-regulated in the tissue and blood, yet up-regulated in the urine [18, 33]. Decreased blood levels of miR-26a and miR-

Table 3 Functional roles of miRNAs in acute kidney injury

Protective		Pathogenic	Kidney enriched, released from injured kidney tissues

Anti-inflammation
  miR-10a
  miR-21
  miR-26a
  miR-126
  miR-146a
  miR-199a
  miR-296
Anti-apoptosis
  miR-10a
  miR-21
  miR-122
  miR-126
  miR-199a
  miR-296
  miR-494
Anti-fibrosis
  miR-29a
  miR-200b
  miR-200c
	Pro-angiogenesis
	  miR-126
	  miR-214
	  miR-296
	Enhancing tubular proliferation
	  miR-126
	  miR-296
	Cytoskeleton, cell-matrix, cell-cell
	  adhesion, cell trafficking
	  miR-127a
		Pro-inflammation
		  miR-21
		  miR-210
		  miR-494
		Pro-apoptosis
		  miR-24
		  miR-192
		  miR-494
		  miR-687
		Pro-fibrosis
		  miR-21
		  miR-192
		  miR-214
			miR-10a
			miR-30c
			miR-30d
			miR-200 family

27a predict AKI in the ICU. Decreased blood levels of miR-26a and miR-27a prior to cardiac surgery also predict AKI later on [18].

miR-29a is highly expressed in the kidney, where it acts against fibrosis by suppressing collagen expression in tubular cells. Decreased serum levels of miR-29a have been shown to predict AKI in ICU patients, and correlate with AKI severity [18].

miR-30c, which is essential for normal kidney homoeostasis, targets several genes important for kidney structure and function. miR-30c is up-regulated in the tissue, blood, and urine obtained from animal models of contrast nephropathy and IRI [34].

miR-30d, which is released to the urine from kidney tissues following injury, down-regulates the apoptotic proteins, caspase 3 and p53, and may provide protective effects during IRI [16].

miR-101-3p is highly expressed in the kidney, and decreased serum levels of this miRNA have been shown to predict AKI in the ICU [18].

miR-122 is down-regulated in the mice kidneys of mice subjected to cisplatin-induced AKI [35]. It exerts anti-apoptotic effects by down-regulating forkhead box O3 (Foxo3).

miR-127a, which is induced by HIF-1α, participates in protecting the cytoskeleton protection (by preventing actin depolmerization), maintaining cell-matrix and cell-cell adhesion maintenance (by preventing focal adhesion complexes disassembly and tight junctions disorganization), and promoting intracellular trafficking (by targeting kinesin family member 3B) [36]. Decreased blood levels of miR-127a were shown to predict AKI in the ICU. Decreased blood levels of miR-127a prior

to cardiac surgery were found be predict AKI later on [18].

miR-146a is down- and up-regulated in the blood and kidney, respectively, during AKI. Decreased blood levels have been shown to predict AKI in the ICU and correlate with the severity of AKI [18]. It is induced by NF-kB and exerts anti-inflammatory effect by down-regulating TNF receptor-associated factor 6 (TRAF-6) and interleukin-1 receptor-associated kinase 1 (IRAK-1) [37].

miR-192 is enriched in kidneys and the small intestine. It is induced by TGF-β during the stress response. It promotes fibrosis by down-regulating SIP1. It also down-regulates E3 ubiquitin ligase and murine double-minute 2 (MDM2) and results in de-repression of p53 and G2/M arrest [38]. miR-194 is also enriched in kidneys and small intestine. It is induced during the stress response, and its levels in tissue, blood, and urine levels are increased during AKI [15, 38, 39].

miR-199a exerts anti-inflammatory effect by down-regulating inhibitor of NF-kB kinases b (IKKb) [40], exhibits anti-proliferatory effect by down-regulating the proto-oncogene MET [41], and confers anti-apoptosis effect by down-regulating extracellular signal–regulated kinase 2 (ERK-2) and HIF-1α [41, 42]. Therefore, it may help limit kidney injury.

miR-126 and miR-296 have been identified in microvesicles from endothelial progenitor cells and are thought to exert renoprotective effects via their abilities to decrease apoptosis and leukocyte infiltration, while promotes angiogenesis and tubular cell proliferation [43]. Hematopoietic overexpression of miR-126 enhances stromal cell-derived factor 1/chemokine receptor type 4 (CXCR4) -dependent vasculogenic progenitor cell mobilization and promotes vascular integrity and supports renal recovery after IRI [44]. Decreased serum levels of miR-126 have been shown to predict AKI in ICU patients, and correlate with the severity of AKI [18].

Members of the miR-200 family are highly expressed in tubular structures such as renal tubules, lungs, the small intestine, and various exocrine glands. miR-200b and miR-200c have been proposed to be anti-fibrotic. They down-regulate TGFβR1 and zinc finger E-box-binding homeobox (ZEB1/ZEB2), which are transcriptional repressors of E-cadherin, and thereby prevent the epithelial-to-mesenchymal transition (EMT) induced by TGF-β [45].

miR-210 is induced by HIF1-α and released by renal endothelial cell. It regulates angiogenesis by down-regulating ephrin-A3 and up-regulating vascular endothelial growth factor (VEGF) and vascular endothelial growth factor receptor 2 (VEGFR2). It also regulates mitochondria ROS production. Increased blood levels of miR-210 was shown to predict post-AKI mortality in critically ill patients [46]. In another study, decreased blood levels of miR-210 were shown to predict AKI in the ICU and correlate with the severity of AKI [18].

miR-214 is induced by TGF-β and promotes fibrosis; it has been shown to down-regulate PTEN, up-regulate the AKT pathway and inhibit apoptosis of monocytes and macrophages. miR-214 is up-regulated in various models of AKI and renal fibrosis [24, 45, 47] as well as in monocytes of animal with chronic kidney disease. Experimental antagonism of miR-214 has been shown to ameliorate renal fibrosis [24].

miR-494 is up-regulated early in AKI, with increased urine levels detected in rodent models of renal IRI and patients with AKI. It has been reported to promote apoptosis and inflammation by down-regulating activating transcription factor 3 (ATF3) and increasing IL-6, monocyte chemoattractant protein-1 (MCP-1), p-selectin [48]. Pathway analysis has suggested that it also targets adiponectin receptor 2 (ADIPOR2), BCL-2 facilitator, and insulin-like growth factor 1 receptor (IGF1R), which would increase inflammation and lead to more damage. However, miR-494 also targets pro-apoptotic proteins in the AKT pathway, and to exert protective effects. The mechanism responsible for regulating the balance between these anti- and pro- apoptotic effects requires further study.

Finally, miR-687 is induced by HIF-1, and enhances apoptosis by down-regulating PTEN. Animal studies have shown that miR-687 blockade preserves PTEN expression and attenuates cell cycle activation and decreases apoptosis, resulting in protection against kidney injury [49].

Conclusions

Many miRNAs have been implicated in the AKI. Some of them contribute to the pathogenesis by regulating apoptosis and inflammation, to amplifying or reduce acute injury responses, while others regulate fibrosis and angiogenesis, to participate in renal recovery or the progression to fibrosis. The biological and pathological functions of many miRNAs in AKI are still not fully understood in AKI. Some studies have yielded inconsistent data regarding the expression pattern of miRNAs across different samples, species, disease models, and time points. These discrepancies warrant investigations.

In addition to their tissue expressions, miRNAs may be detected in various extracellular human body fluids, such as serum, urine, saliva, and cerebral spinal fluid. miRNAs are contained in exosomes and may remained stable over prolonged periods. They may be specifically up-regulated or down-regulated in response to injury signals and/or released into body fluids from resident tissues. Certain miRNAs have been investigated for their potential to serve as novel biomarkers for the early detection or prognostication of AKI. Given the complex pathophysiology and the dynamic nature of AKI, an miRNA panel may be more feasible rather than a single miRNA. Further validation studies are needed to evaluate the clinical utility of such a panel.

Some miRNAs may be potential therapeutic targets for AKI. Recently, an miRNA inhibitor has been proven to successfully suppress the replication of hepatitis C virus in a clinical trial [50]. Systemic or local administration of specific miRNAs mimics or antagonists in vivo could offer a strategy for preventing or ameliorating AKI or barring its progression to chronic kidney disease.

In the post-genome era, miRNAs are promising rising stars in translational medicine as they offer the potential to guide the individualized diagnosis and treatment of human diseases including AKI.

Acknowledgements
Not applicable.

Funding
This work was supported by the Chang Gung Memorial Hospital Research Program grant CMRPG3D1452, CMRPG 3F0561, CMRPG1B0581 and CIRPG3B0042.; Ministry of Science and Technology 104-2314-B-182A-131 and 105-2314-B-182A-121.

Authors' contributions
The manuscript was written by PCF. All authors critically revised the manuscript. All authors read and approved the final manuscript.

Competing interests
The authors declare that they have no competing interests.

Consent for publication
Not applicable.

Author details

[1]Kidney Research Center, Department of Nephrology, Chang Gung Memorial Hospital, Linkou Medical Center, Taoyuan, Taiwan. [2]Graduate Institute of Clinical Medical Sciences, Chang Gung University, Taoyuan, Taiwan. [3]Molecular Medicine Research Center, Chang Gung University, Taoyuan, Taiwan. [4]Division of Cardiology, Department of Internal Medicine, Chang Gung Memorial Hospital, College of Medicine, Chang Gung University, Taipei, Taiwan. [5]Healthcare Center, Chang Gung Memorial Hospital, College of Medicine, Chang Gung University, Taipei, Taiwan. [6]Heart Failure Center, Chang Gung Memorial Hospital, College of Medicine, Chang Gung University, Taipei, Taiwan. [7]Department of Cardiology, Chang Gung Memorial Hospital, College of Medicine, Chang Gung University, 199 Tung Hwa North Road, Taipei 105, Taiwan.

References

1. Schrier RW, Wang W, Poole B, Mitra A. Acute renal failure: definitions, diagnosis, pathogenesis, and therapy. J Clin Invest. 2004;114(1):5–14. doi:10.1172/jci22353.

2. Bellomo R, Ronco C, Kellum JA, Mehta RL, Palevsky P. Acute renal failure - definition, outcome measures, animal models, fluid therapy and information technology needs: the Second International Consensus Conference of the Acute Dialysis Quality Initiative (ADQI) Group. Crit Care (London, England). 2004;8(4):R204–12. doi:10.1186/cc2872.

3. Lee RC, Feinbaum RL, Ambros V. The C. elegans heterochronic gene lin-4 encodes small RNAs with antisense complementarity to lin-14. Cell. 1993; 75(5):843–54.

4. Krol J, Loedige I, Filipowicz W. The widespread regulation of microRNA biogenesis, function and decay. Nat Rev Genet. 2010;11(9):597–610. doi:10.1038/nrg2843.

5. Bhatt K, Mi QS, Dong Z. microRNAs in kidneys: biogenesis, regulation, and pathophysiological roles. Am J Physiol Renal Physiol. 2011;300(3):F602–10. doi:10.1152/ajprenal.00727.2010.

6. Bartel DP. MicroRNAs: target recognition and regulatory functions. Cell. 2009;136(2):215–33. doi:10.1016/j.cell.2009.01.002.

7. Huang JB, et al. MiR-196a2 rs11614913 T > C polymorphism is associated with an increased risk of Tetralogy of Fallot in a Chinese population. Acta Cardiol Sin. 2015;31(1):18–23.

8. Hsu A, Chen SJ, Chang YS, Chen HC, Chu PH. Systemic approach to identify serum microRNAs as potential biomarkers for acute myocardial infarction. Biomed Res Int. 2014;2014:418628. doi:10.1155/2014/418628.

9. Pandey P, et al. Microarray-based approach identifies microRNAs and their target functional patterns in polycystic kidney disease. BMC Genomics. 2008; 9:624. doi:10.1186/1471-2164-9-624.

10. Nakada C, et al. Genome-wide microRNA expression profiling in renal cell carcinoma: significant down-regulation of miR-141 and miR-200c. J Pathol. 2008;216(4):418–27. doi:10.1002/path.2437.

11. Kato M, Natarajan R. MicroRNA circuits in transforming growth factor-beta actions and diabetic nephropathy. Semin Nephrol. 2012;32(3):253–60. doi:10.1016/j.semnephrol.2012.04.004.

12. Dai Y, et al. Comprehensive analysis of microRNA expression patterns in renal biopsies of lupus nephritis patients. Rheumatol Int. 2009;29(7):749–54. doi:10.1007/s00296-008-0758-6.

13. Sui W, et al. Microarray analysis of MicroRNA expression in acute rejection after renal transplantation. Transpl Immunol. 2008;19(1):81–5. doi:10.1016/j.trim.2008.01.007.

14. Wei Q, et al. Targeted deletion of Dicer from proximal tubules protects against renal ischemia-reperfusion injury. J Am Soc Nephrol. 2010;21(5):756–61. doi:10.1681/asn.2009070718.

15. Wang JF, et al. Screening plasma miRNAs as biomarkers for renal ischemia-reperfusion injury in rats. Med Sci Monit. 2014;20:283–9. doi:10.12659/msm.889937.

16. Wang N, et al. Urinary microRNA-10a and microRNA-30d serve as novel, sensitive and specific biomarkers for kidney injury. PLoS One. 2012;7(12): e51140. doi:10.1371/journal.pone.0051140.

17. Ho J, et al. The pro-apoptotic protein Bim is a microRNA target in kidney progenitors. J Am Soc Nephrol. 2011;22(6):1053–63. doi:10.1681/asn.2010080841.

18. Aguado-Fraile E, et al. A pilot study identifying a set of microRNAs as precise diagnostic biomarkers of acute kidney injury. PLoS One. 2015;10(6): e0127175. doi:10.1371/journal.pone.0127175.

19. Kaucsar T, et al. Activation of the miR-17 family and miR-21 during murine kidney ischemia-reperfusion injury. Nucleic Acid Ther. 2013;23(5):344–54. doi:10.1089/nat.2013.0438.

20. Ma L, et al. Changes of miRNA-17-5p, miRNA-21 and miRNA-106a level during rat kidney ischemia-reperfusion injury. Zhonghua yi xue za zhi. 2015;95(19):1488–92.

21. Hu H, Jiang W, Xi X, Zou C, Ye Z. MicroRNA-21 attenuates renal ischemia reperfusion injury via targeting caspase signaling in mice. Am J Nephrol. 2014;40(3):215–23. doi:10.1159/000368202.

22. Jia P, et al. Xenon protects against septic acute kidney injury via miR-21 target signaling pathway. Crit Care Med. 2015;43(7):e250–9. doi:10.1097/ccm.0000000000001001.

23. Xu X, et al. Delayed ischemic preconditioning contributes to renal protection by upregulation of miR-21. Kidney Int. 2012;82(11):1167–75. doi: 10.1038/ki.2012.241.

24. Denby L, et al. MicroRNA-214 antagonism protects against renal fibrosis. J Am Soc Nephrol. 2014;25(1):65–80. doi:10.1681/asn.2013010072.

25. Zarjou A, Yang S, Abraham E, Agarwal A, Liu G. Identification of a microRNA signature in renal fibrosis: role of miR-21. Am J Physiol Renal Physiol. 2011; 301(4):F793–801. doi:10.1152/ajprenal.00273.2011.

26. Chau BN, et al. MicroRNA-21 promotes fibrosis of the kidney by silencing metabolic pathways. Sci Transl Med. 2012;4(121):121ra18. doi:10.1126/scitranslmed.3003205.

27. Liu X, et al. MiR-21 inhibits autophagy by targeting Rab11a in renal ischemia/ reperfusion. Exp Cell Res. 2015;338(1):64–9. doi:10.1016/j.yexcr.2015.08.010.

28. Du J, et al. MicroRNA-21 and risk of severe acute kidney injury and poor outcomes after adult cardiac surgery. PLoS One. 2013;8(5):e63390. doi:10.1371/journal.pone.0063390.

29. Ramachandran K, et al. Human miRNome profiling identifies microRNAs differentially present in the urine after kidney injury. Clin Chem. 2013;59(12):1742–52. doi:10.1373/clinchem.2013.210245.

30. Gaede L, et al. Plasma microRNA-21 for the early prediction of acute kidney injury in patients undergoing major cardiac surgery. Nephrol Dial Transplant. 2016. doi:10.1093/ndt/gfw007

31. Lorenzen JM, et al. MicroRNA-24 antagonism prevents renal ischemia reperfusion injury. J Am Soc Nephrol. 2014;25(12):2717–29. doi:10.1681/asn.2013121329.

32. Liang S, Wang W, Gou X. MicroRNA 26a modulates regulatory T cells expansion and attenuates renal ischemia-reperfusion injury. Mol Immunol. 2015;65(2):321–7. doi:10.1016/j.molimm.2015.02.003.

33. Kanki M, et al. Identification of urinary miRNA biomarkers for detecting cisplatin-induced proximal tubular injury in rats. Toxicology. 2014;324:158–68. doi:10.1016/j.tox.2014.05.004.

34. Liu F, et al. Upregulation of microRNA-210 regulates renal angiogenesis mediated by activation of VEGF signaling pathway under ischemia/perfusion injury in vivo and in vitro. Kidney Blood Press Res. 2012;35(3):182–91. doi:10.1159/000331054.

35. Lee CG, et al. Discovery of an integrative network of microRNAs and transcriptomics changes for acute kidney injury. Kidney Int. 2014;86(5): 943–53. doi:10.1038/ki.2014.117.

36. Aguado-Fraile E, et al. miR-127 protects proximal tubule cells against ischemia/ reperfusion: identification of kinesin family member 3B as miR-127 target. PLoS One. 2012;7(9):e44305. doi:10.1371/journal.pone.0044305.

37. Godwin JG, et al. Identification of a microRNA signature of renal ischemia reperfusion injury. Proc Natl Acad Sci U S A. 2010;107(32):14339–44. doi:10.1073/pnas.0912701107.

38. Jenkins RH, et al. miR-192 induces G2/M growth arrest in aristolochic acid nephropathy. Am J Pathol. 2014;184(4):996–1009. doi:10.1016/j.ajpath.2013.12.028.

39. Kito N, Endo K, Ikesue M, Weng H, Iwai N. miRNA profiles of tubular cells: diagnosis of kidney injury. Biomed Res Int. 2015;2015:465479. doi:10.1155/2015/465479.

40. Chen R, et al. Regulation of IKKbeta by miR-199a affects NF-kappaB activity in ovarian cancer cells. Oncogene. 2008;27(34):4712–23. doi:10.1038/onc.2008.112.

41. Kim S, et al. MicroRNA miR-199a* regulates the MET proto-oncogene and the downstream extracellular signal-regulated kinase 2 (ERK2). J Biol Chem. 2008;283(26):18158–66. doi:10.1074/jbc.M800186200.

42. Rane S, et al. Downregulation of miR-199a derepresses hypoxia-inducible factor-1 alpha and Sirtuin 1 and recapitulates hypoxia preconditioning in cardiac myocytes. Circ Res. 2009;104(7):879–86. doi:10.1161/circresaha.108.193102.

43. Cantaluppi V, et al. Microvesicles derived from endothelial progenitor cells protect the kidney from ischemia-reperfusion injury by microRNA-dependent reprogramming of resident renal cells. Kidney Int. 2012;82(4):412–27. doi:10.1038/ki.2012.105.

44. Bijkerk R, et al. Hematopoietic microRNA-126 protects against renal

ischemia/reperfusion injury by promoting vascular integrity. J Am Soc Nephrol. 2014;25(8):1710–22. doi:10.1681/asn.2013060640.

45. Denby L, et al. miR-21 and miR-214 are consistently modulated during renal injury in rodent models. Am J Pathol. 2011;179(2):661–72. doi:10.1016/j.ajpath.2011.04.021.

46. Lorenzen JM, et al. Circulating miR-210 predicts survival in critically ill patients with acute kidney injury. Clin J Am Soc Nephrol. 2011;6(7):1540–6. doi:10.2215/cjn.00430111.

47. Cui R, Xu J, Chen X, Zhu W. Global miRNA expression is temporally correlated with acute kidney injury in mice. PeerJ. 2016;4:e1729. doi:10.7717/peerj.1729.

48. Lan YF, et al. MicroRNA-494 reduces ATF3 expression and promotes AKI. J Am Soc Nephrol. 2012;23(12):2012–23. doi:10.1681/asn.2012050438.

49. Bhatt K, et al. MicroRNA-687 induced by hypoxia-inducible factor-1 targets phosphatase and tensin homolog in renal ischemia-reperfusion injury. J Am Soc Nephrol. 2015;26(7):1588–96. doi:10.1681/asn.2014050463.

50. Janssen HL, et al. Treatment of HCV infection by targeting microRNA. N Engl J Med. 2013;368(18):1685–94. doi:10.1056/NEJMoa1209026.

51. Muratsu-Ikeda S, et al. Downregulation of miR-205 modulates cell susceptibility to oxidative and endoplasmic reticulum stresses in renal tubular cells. PLoS One. 2012;7(7):e41462. doi:10.1371/journal.pone.0041462.

52. Gutierrez-Escolano A, Santacruz-Vazquez E, Gomez-Perez F. Dysregulated microRNAs involved in contrast-induced acute kidney injury in rat and human. Ren Fail. 2015;37(9):1498–506. doi:10.3109/0886022x.2015.1077322.

53. Nassirpour R, et al. Identification of tubular injury microRNA biomarkers in urine: comparison of next-generation sequencing and qPCR-based profiling platforms. BMC Genomics. 2014;15:485. doi:10.1186/1471-2164-15-485.

54. Wang IK, et al. MiR-20a-5p mediates hypoxia-induced autophagy by targeting ATG16L1 in ischemic kidney injury. Life Sci. 2015;136:133–41. doi:10.1016/j.lfs.2015.07.002.

55. Pavkovic M, Riefke B, Ellinger-Ziegelbauer H. Urinary microRNA profiling for identification of biomarkers after cisplatin-induced kidney injury. Toxicology. 2014;324:147–57. doi:10.1016/j.tox.2014.05.005.

56. Saikumar J, et al. Expression, circulation, and excretion profile of microRNA-21, -155, and -18a following acute kidney injury. Toxicol Sci. 2012;129(2):256–67. doi:10.1093/toxsci/kfs210.

57. Pellegrini KL, et al. Application of small RNA sequencing to identify microRNAs in acute kidney injury and fibrosis. Toxicol Appl Pharmacol. 2015. doi:10.1016/j.taap.2015.12.002.

58. Jia P, et al. miR-21 contributes to xenon-conferred amelioration of renal ischemia-reperfusion injury in mice. Anesthesiology. 2013;119(3):621–30. doi:10.1097/ALN.0b013e318298e5f1.

59. Shen B, et al. Revealing the underlying mechanism of ischemia reperfusion injury using bioinformatics approach. Kidney Blood Press Res. 2013;38(1):99–108. doi:10.1159/000355759.

60. Bhatt K, et al. MicroRNA-34a is induced via p53 during cisplatin nephrotoxicity and contributes to cell survival. Mol Med (Cambridge, Mass). 2010;16(9-10):409–16. doi:10.2119/molmed.2010.00002.

61. Liu XJ, et al. MicroRNA-34a suppresses autophagy in tubular epithelial cells in acute kidney injury. Am J Nephrol. 2015;42(2):168–75. doi:10.1159/000439185.

62. Joo MS, Lee CG, Koo JH, Kim SG. miR-125b transcriptionally increased by Nrf2 inhibits AhR repressor, which protects kidney from cisplatin-induced injury. Cell Death Dis. 2013;4:e899. doi:10.1038/cddis.2013.427.

63. Bijkerk R, et al. Silencing of miRNA-126 in kidney ischemia reperfusion is associated with elevated SDF-1 levels and mobilization of Sca-1+/Lin-progenitor cells. MicroRNA (Shariqah, United Arab Emirates). 2014;3(3):144–9.

64. Bussolati B, et al. Hypoxia modulates the undifferentiated phenotype of human renal inner medullary CD133+ progenitors through Oct4/miR-145 balance. Am J Physiol Renal Physiol. 2012;302(1):F116–28. doi:10.1152/ajprenal.00184.2011.

65. Ranganathan P, et al. MicroRNA-150 deletion in mice protects kidney from myocardial infarction-induced acute kidney injury. Am J Physiol Renal Physiol. 2015;309(6):F551–8. doi:10.1152/ajprenal.00076.2015.

66. Pellegrini KL, et al. MicroRNA-155 deficient mice experience heightened kidney toxicity when dosed with cisplatin. Toxicol Sci. 2014;141(2):484–92. doi:10.1093/toxsci/kfu143.

67. Qin W, Xie W, Yang X, Xia N, Yang K. Inhibiting microRNA-449 attenuates cisplatin-induced injury in NRK-52E cells possibly via regulating the SIRT1/P53/BAX pathway. Med Sci Monit. 2016;22:818–23.

68. Bellinger MA, et al. Concordant changes of plasma and kidney microRNA in the early stages of acute kidney injury: time course in a mouse model of bilateral renal ischemia-reperfusion. PLoS One. 2014;9(4):e93297. doi:10.1371/journal.pone.0093297.

12

Variation of global DNA methylation levels with age and in autistic children

Shui-Ying Tsang[1†], Tanveer Ahmad[1†], Flora W. K. Mat[1], Cunyou Zhao[1,2], Shifu Xiao[3*], Kun Xia[4*] and Hong Xue[1*]

Abstract

Background: The change in epigenetic signatures, in particular DNA methylation, has been proposed as risk markers for various age-related diseases. However, the course of variation in methylation levels with age, the difference in methylation between genders, and methylation-disease association at the whole genome level is unclear. In the present study, genome-wide methylation levels in DNA extracted from peripheral blood for 2116 healthy Chinese in the 2–97 age range and 280 autistic trios were examined using the fluorescence polarization-based genome-wide DNA methylation quantification method developed by us.

Results: Genome-wide or global DNA methylation levels proceeded through multiple phases of variation with age, consisting of a steady increase from age 2 to 25 ($r = 0.382$) and another rise from age 41 to 55 to reach a peak level of ~80 % ($r = 0.265$), followed by a sharp decrease to ~40 % in the mid-1970s (age 56 to 75; $r = -0.395$) and leveling off thereafter. Significant gender effect in methylation levels was observed only for the 41–55 age group in which methylation in females was significantly higher than in males ($p = 0.010$). In addition, global methylation level was significantly higher in autistic children than in age-matched healthy children ($p < 0.001$).

Conclusions: The multiphasic nature of changes in global methylation levels with age was delineated, and investigation into the factors underlying this profile will be essential to a proper understanding of the aging process. Furthermore, this first report of global hypermethylation in autistic children also illustrates the importance of age-matched controls in characterization of disease-associated variations in DNA methylation.

Keywords: Aging, Autism spectrum disorder, CpG methylation, Developmental epigenetics, Genome-wide methylation quantification

Background

Genetic changes can alter the genomic DNA sequence through point mutations, insertions, deletions, copy number variations, and chromosomal rearrangements while epigenetic modifications can modulate phenotype and gene expressions. DNA methylation is the most common epigenetic modification that plays an essential role in the regulation of tissue-specific gene expression, cellular differentiation, chromosome stabilization, genomic imprinting, and suppression of transposable element mobility [1, 2]. DNA methylation through DNA methyltransferases convert cytosine to 5-methycytocine, with the majority of the conversions occurring at CpG islands found in gene promoter regions. Aberrant DNA methylation patterns have long been associated with various human diseases including cancers, cardiovascular diseases, psychotic disorders, and autism [3–6].

Changes in epigenetics signatures, and in particular DNA methylation, have been reported to occur in normal physiological development and aging, and alterations in DNA methylation associated with the signaling and regulation of transcription have been demonstrated in some genes [7, 8]. Aging is the gradual deterioration of various body functions and represents an important risk factor for various age-related diseases such as cancer, neurodegenerative disorders, cardiovascular diseases, and type 2 diabetes mellitus [9]. Several studies have

* Correspondence: xiaoshifu@msn.com; xiakuncsu@gmail.com; hxue@ust.hk
†Equal contributors
3Department of Geriatric Psychiatry, Shanghai Mental Health Center, Shanghai Jiaotong University School of Medicine, Shanghai 200030, China
4The State Key Laboratory of Medical Genetics, Central South University, Changsha, Hunan 410078, China
1Division of Life Science, Applied Genomics Centre and Centre for Statistical Science, Hong Kong University of Science and Technology, Clear Water Bay, Hong Kong, China
Full list of author information is available at the end of the article

examined DNA methylation changes in old age as disease risk factor, focusing mostly on CpG islands and promoter regions in specific gene [10, 11]. However, the characterization of lifelong age-related changes in DNA methylation at the whole genome level has remained a largely unexplored area.

During the past decades, various HPLC-based, sequencing-based, (e.g., bisulfite-sequencing and methylated DNA immunoprecipitation) and microarray-based methods have been introduced to quantitate genomic DNA methylation [12]. Although these methods enable high-resolution and detailed methylation profiles of individual genes, they are time-consuming and incapable of measuring whole genome methylation levels accurately. Recently, a number of methods have been developed to render possible the measurement of whole genome methylation levels, including the LUminometric Methylation Assay (LUMA) method [13], the ELISA-based approach [14], and the fluorescence polarization DNA methylation (FPDM) method developed by us [15].

The objective of the present study was to analyze whole genome DNA methylation in the normal population in order to establish the quantitative relationship between global DNA methylation levels and age using the simple and accurate FPDM method, as well as delineate any gender differences. The DNA methylation-aging curve obtained for the normal population will provide a useful reference to facilitate an improved understanding of the regulation of DNA methylation in aging. Moreover, autism-associated changes in genome-wide methylation were investigated, which also served to demonstrate the importance of using age-matched controls in methylation-disease association studies.

Methods
Study population
The main study population in this study were enrolled from Beijing, Shanghai, Changsha, and Hong Kong and consisted of 2116 healthy Chinese subjects including 1108 (52.36 %) males and 1008 (47.64 %) females. The subjects spanned a wide age range from 2 to 97 years. The 280 autistic children (age 2-13 years) and their parents (n = 552; age 24–62 years) were recruited at Central South University in Changsha. Samples from the parents but not those from the children were included in the main study set for age-methylation analysis.

Genomic DNA extraction
Leukocytes were isolated from 5-ml peripheral blood samples. DNA was prepared by phenol extraction and chloroform extraction followed by isopropanol precipitation, washed with ethanol, and air-dried. Tris-EDTA buffer pH 8.0 was used to dissolve the final genomic DNA product.

Whole genome DNA methylation analysis by FPDM
To determine genome-wide or "global" DNA methylation by fluorescence polarization DNA methylation measurement, ~100 ng DNA sample was first subjected to separate restriction-enzyme digestions by HpaII and MspI as described [15]; the methylation-sensitive HpaII cut only un-methylated 5′CCGG-3′ sites, whereas the methylation-insensitive MspI cut both methylated and un-methylated 5′-CCGG-3′ sites. After completion of the restriction reactions, both digests were subjected to a one-label-extension reaction through incubation with fluorescent TAMRA-dCTP (5-propargylamino-dCTP-5/6-carboxytetra-methylrhodamine, Jena Bioscience) and Taq DNA polymerase. Measurements of fluorescence polarization on the two digests following the extension reaction yielded the percentile global methylation in the DNA sample. The global DNA methylation level in each instance was thus expressed in terms of the global percentage of CpG sites in genomic DNA that were methylated based on the average of triplicate measurements.

Statistical data analysis
Statistical analysis of data was performed using SPSS 19.0. Percentile methylation of each DNA sample represented the average of three independent measurements. To assess the relationship between global DNA methylation and age, methylation levels of samples in every 5-year age range were first analyzed for correlation with age using Pearson's correlation test. Based on these results, the samples were further grouped into five age ranges to represent different phases of methylation change with age and again analyzed using Pearson's correlation to yield an overall correlation coefficient for each age range. Differences in methylation levels between males and females were analyzed for all samples using independent sample t test as well as for each group of samples in the five age ranges using multivariable linear regression. Independent sample t test was used to analyze the methylation difference between autistic children and parents and between autistic children and age-matched healthy children. A p value <0.05 was regarded as statistically significant.

Results
The methylation data is given in Additional file 1: Table S1. Although the global DNA methylation data determined using the FPDM method displayed large standard deviations, when the subjects were divided into 5-year age groups and analyzed for within-group correlations with age, positive correlations were detected in the 16–20 and 51–55 groups, and a negative correlation was detected in the 56–60 group (Table 1). Based on the within-group correlations and the methylation-age plot (Fig. 1), multiple phases of change in methylation levels with age were discerned including a steady increase from year 2 to year 25 and

Table 1 Global DNA methylation levels of different age groups

Age range	Number	Global DNA methylation (%, mean ± SD)	Pearson's r[a]	p value[b]
1–5	160	47.52 ± 17.51	0.040	0.614
6–10	93	49.41 ± 17.03	0.053	0.613
11–15	68	54.73 ± 16.69	0.055	0.658
16–20	207	60.10 ± 17.78	0.205	*0.003*
21–25	100	66.92 ± 17.78	0.029	0.775
26–30	188	54.11 ± 16.10	0.008	0.916
31–35	321	66.68 ± 17.95	−0.059	0.292
36–40	202	68.24 ± 15.96	0.127	0.073
41–45	112	71.09 ± 16.08	0.150	0.113
46–50	129	77.17 ± 14.13	0.138	0.120
51–55	92	79.22 ± 10.95	0.213	*0.041*
56–60	68	69.84 ± 20.02	−0.294	*0.015*
61–65	120	53.20 ± 22.13	−0.028	0.765
66–70	71	51.54 ± 24.41	−0.061	0.612
71–75	54	39.13 ± 17.18	−0.133	0.339
76–80	75	54.73 ± 16.69	−0.064	0.584
81–85	36	40.95 ± 15.27	−0.023	0.896
>85	20	39.78 ± 12.08	−0.073	0.758

[a]Pearson's correlation coefficients pertaining to within-group analysis
[b]p values less than 0.05 are shown in italic font

another rise from year 40 onward to reach a peak level at year 55, followed by a sharp decrease up to year 75 and leveling off thereafter. Quantitatively, the increase from age 2 up to the age of 25 was significant to $p < 0.001$ with Pearson's correlation coefficient $r = 0.382$. From 26 to 40 years of age, there was no significant change in

methylation $(r = 0.028; p = 0.459)$. However, the DNA methylation levels again significantly increased with age between 41 to 55 years $(r = 0.265; p < 0.001)$. From 56 to 75 years of age, there was a significant decrease in global DNA methylation $(r = -0.395; p < 0.001)$ showing an inverse relationship between methylation and age, and no significant change in methylation levels was observed between 75 and 97 years of age $(r = -0.061; p = 0.486)$. The correlation data for these different age groups are given in Additional file 2: Table S2.

With respect to the genders (Table 2), a statistically significant gender effect was observed in the 41 to 55 age group (beta = 0.136; $p = 0.010$), where the average methylation levels in males $(73.49 ± 15.26)$ was higher than that in females $(77.52 ± 13.44)$. There was no significant gender effect in any of the other age groups or when all age groups were combined.

The global methylation levels for autistic children $(n = 280;$ mean age = 4.7) were compared to both those of their healthy parents $(n = 552;$ mean age = 33.8) and those of age-matched healthy children $(n = 236;$ mean age = 5.3). No significant difference $(p = 0.872)$ was observed between patients $(65.18 ± 16.69)$ and parents $(66.01 ± 19.98)$ (Additional file 3: Table S3); but the difference between patients and age-matched controls $(54.35 ± 21.37)$ was highly significant $(p < 0.001)$, with increased methylation in the patients (Fig. 2). There was no significant difference in methylation between autistic children and either their fathers or mothers separately (Additional file 3: Table S3).

Discussion

The purpose of the present study has been to analyze, using the FPDM method, the global DNA methylation

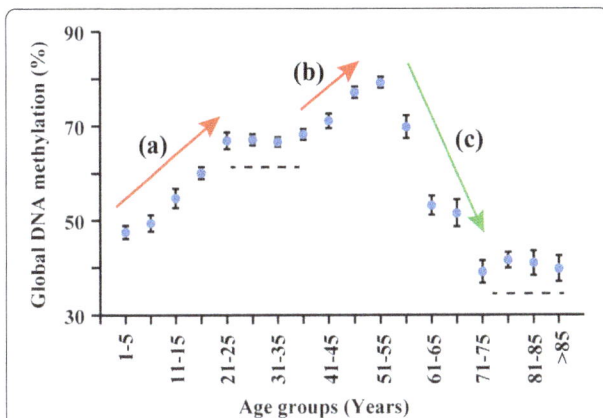

Fig. 1 Variation of global DNA methylation with age. Mean global DNA methylation of 5-year age groups are shown. *Error bars* indicate the standard errors of the mean, and the three *arrows* represent (*a*) increasing trend at age range 1–25 years, (*b*) increasing trend at age range 40–55 years, and (*c*) decreasing trend at age range 55–75 years. The two *dashed lines* indicate the two age ranges (25–40 and over 75 years) where there were no significant changes in methylation

Table 2 Global DNA methylation levels in different male and female age groups

Age groups	Gender	Number	Global DNA methylation (%, mean ± SD)	p value[a]
2–25	Male	414	54.99 ± 18.70	0.370
	Female	214	57.42 ± 18.92	
26–40	Male	356	67.17 ± 16.56	0.816
	Female	355	67.31 ± 17.29	
41–55	Male	151	73.49 ± 15.26	*0.010*
	Female	182	77.52 ± 13.44	
56–75	Male	127	56.42 ± 23.34	0.152
	Female	186	52.37 ± 23.50	
75–97	Male	60	40.72 ± 14.14	0.711
	Female	71	41.46 ± 14.20	
Overall	Male	1108	60.81 ± 19.86	0.060[b]
	Female	1008	62.47 ± 20.81	

$p < 0.05$ is shown in italic font
[a]p values using gender as a variable in multivariable linear regression analyses
[b]p value for between-gender comparison using Student's t test

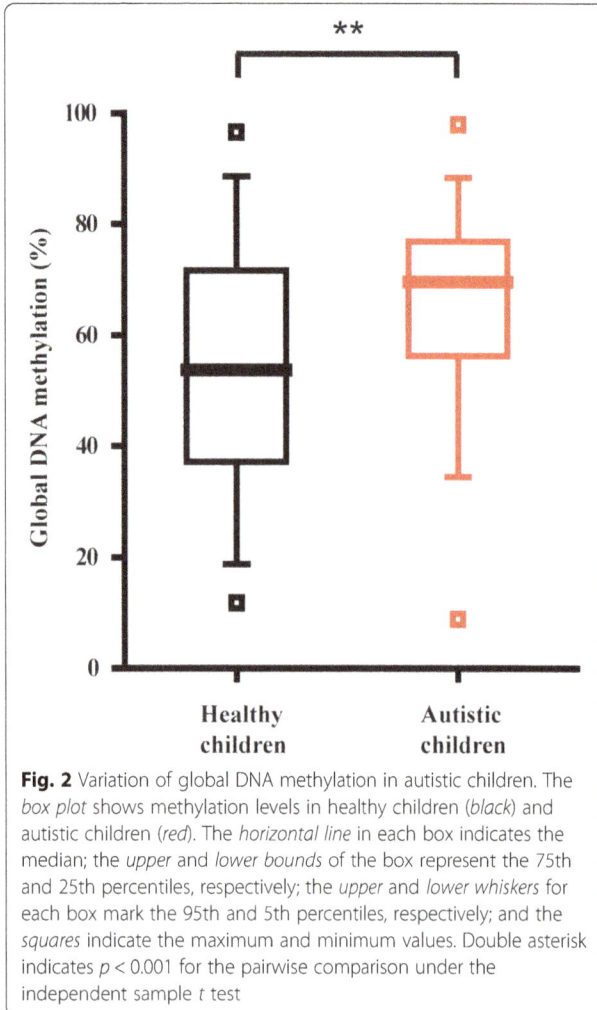

Fig. 2 Variation of global DNA methylation in autistic children. The *box plot* shows methylation levels in healthy children (*black*) and autistic children (*red*). The *horizontal line* in each box indicates the median; the *upper* and *lower bounds* of the box represent the 75th and 25th percentiles, respectively; the *upper* and *lower whiskers* for each box mark the 95th and 5th percentiles, respectively; and the *squares* indicate the maximum and minimum values. Double asterisk indicates $p < 0.001$ for the pairwise comparison under the independent sample t test

levels in leukocytes as a function of age in order to establish a continuous methylation-age curve for the population that could serve as a basis for describing phenotypic changes associated with aging and as an age-dependent standard for the detection of any significant deviation caused by disease. In providing a systematic characterization of the dependence of global DNA methylation on age, the study revealed that global DNA methylation as a genomic parameter of age was distinctly multiphasic in character. The global DNA methylation-age curve displayed evident increases over the adolescent age group of 10–20 and late middle-age age group of 40–50 and a sharp decrease over the age group of 55–70 to reach a hypomethylation level from age 71 onwards as a hallmark of old age. Previous studies on age-associated global DNA methylation have reported age-dependent decrease in methylation based on adult to old age populations. In addition, there are also reports of age-associated increase in methylation [7] and methylation increases in the early years of life [16]. Moreover, while many loci such as intergenic CpGs outside of CpG islands

display decreased methylation in later life, other loci such as promoter-associated CpG islands show increased methylation with age throughout the lifespan [8]. Therefore, in general global methylation decreases in old age, it has not been established that this decrease occurs continuously throughout life. Indeed, a general increase in DNA methylation with age before adulthood, followed by stabilization and an eventual decrease in old age, has emerged from studies on different age ranges using different methods [8]. The lifelong profile obtained in the present study is consistent with this general description, with the more comprehensive time curve revealing a second period of methylation increase in late middle-age prior to the methylation decline in old age.

Significant difference between male and female subjects was observed only in the late middle-age age group, suggesting that gender-related factors may contribute to this second period of methylation increase. The profile of global methylation variation between genders is somewhat unclear with previous reports of significant difference based on methylation levels in LINE-1 repeat elements for a 45–75 age group [17] but no difference for a 19–80 age group [18] and no significant difference based on the LUMA method for CCGG sites for a group with mean age of 24 [19]. The different results suggest that gender differences in "global" methylation levels are dependent on age as well as the subset of methylation sites examined in the quantification method.

DNA hypomethylation at old age has been reported in studies focused on a relatively limited number of gene loci and narrow age ranges, suggesting the possible association between DNA methylation and age-related diseases [9, 20, 21]. Indeed, the loss of global DNA methylation is one of the first epigenetic abnormalities in cancers [6], and advanced age represents a potent risk factor for human epithelial cancers with cancer incidence increasing sharply from age 60 onward, especially in males [22], which is in accord with the sharp decrease in DNA methylation over this age range shown in Fig. 1. Likewise, evidence of age-associated loss of DNA methylation in brain tissue suggests the significant role of DNA hypomethylation at old age in the pathogenesis of Alzheimer's disease [23]. The onset of Alzheimer's disease at age 65 is also in accord with the sharp decrease in global DNA methylation between the ages 55–70 shown in Fig. 1.

Unlike neurodegenerative and aging-related disorders, neurodevelopmental disorders such as autism and Down syndrome affect subject groups on the opposite end of the age spectrum. Autism is highly heritable and affects information processing in the brain, leading to symptoms that include impairments in social interaction and communication, restricted interest, and repetitive behavior [24]. These characteristic symptoms become apparent in early child-hood, typically before the age of three. Although the eti-ology of autism is mainly ascribed to genetic variations including single nucleotide polymorphisms and copy num-ber variations [25], epigenetic mechanisms have been in-voked to affect the environmental influences [4, 26]. As such, DNA methylation has been associated with dysregu-lation of biological pathways in autistic brains, with both hypomethylated and hypermethylated genomic regions be-ing identified [5]. Recently, in peripheral blood analysis, the *OXTR* promoter was shown to be hypomethylated in autism cases [27]. Here, we have demonstrated that the global methylation in autistic children was increased com-pared to healthy children (Fig. 2) with respect to the over-all effect across all CCGG sites recognized by the HpaII/ MspI enzymes, encompassing both hypermethylated and hypomethylated sites as well as unchanged sites. The over-all increase suggests that hypermethylated regions were more extensive than hypomethylated regions in the autis-tic genome. Moreover, in comparison with the time profile for methylation, the higher methylation level is that ex-pected of young to middle-aged adults and this could be interpreted to suggest an abnormally advanced methylome in autistic children. This is reflected in that no significant difference in methylation was found between autistic chil-dren and their parents. From another point of view, since a general increase in methylation takes place from young to middle age (Fig. 1), the comparison between children and parents is confounded by the age factor, and the result demonstrates the importance of using age-matched con-trols in analyzing methylation differences. Notably, envir-onmental and nutritional factors may also affect methylation, and application of the FPDM method will fa-cilitate the in-depth analysis of the quantitative effects of such external factors.

In conclusion, global DNA methylation measurements in the present study on leukocyte DNA have shown a multiphasic variation with age that leads to depression of methylation at old age to half its level at middle age, thereby providing strong evidence for DNA hypomethyla-tion at old age as a potent risk marker for various age-related disorders such as cancers, cardiovascular and neu-rodegenerative disorders, and type 2 diabetes. In contrast, hypermethylation was observed for autism, a neuro-developmental condition. These measurements, read-ily performed with the FPDM method, provide a simple and quantitative approach to investigate the multiple genetic and environment factors that deter-mine global DNA methylation. A delineation of glo-bal methylation changes will complement studies on gene-specific methylation changes to yield an increas-ingly comprehensive understanding of the regulation of DNA methylation and the roles of DNA methylation in age-related diseases.

Abbreviation
FPDM: Fluorescence polarization DNA methylation

Acknowledgements
We are grateful to the Hong Kong Red Cross and to Professor Yunlong Tan for their contributions to blood sample collection.

Funding
This study was supported by a research grant to HX from the University Grants Committee, Hong Kong SAR, for Knowledge Transfer (UGCOSE PCF.010.12/13). This work was also supported by the National Basic Research Program of China (2012CB517902). The funding bodies had no other roles in the study.

Authors' contributions
HX, SYT, and CZ conceived and designed the experiments; TA and FM performed the experiments; TA and SYT analyzed the data; SX and KX contributed to the clinical samples; TA, SYT, and HX wrote the manuscript. All authors read and approved the final manuscript.

Competing interests
The authors declare that they have no competing interests.

Consent for publication
Not applicable.

Author details
[1]Division of Life Science, Applied Genomics Centre and Centre for Statistical Science, Hong Kong University of Science and Technology, Clear Water Bay, Hong Kong, China. [2]Department of Medical Genetics, School of Basic Medical Science, Southern Medical University, Guangzhou, Guangdong 510515, China. [3]Department of Geriatric Psychiatry, Shanghai Mental Health Center, Shanghai Jiaotong University School of Medicine, Shanghai 200030, China. [4]The State Key Laboratory of Medical Genetics, Central South University, Changsha, Hunan 410078, China.

References
1. Gibney ER, Nolan CM. Epigenetics and gene expression. Heredity (Edinb). 2010;105:4–13.
2. Smith ZD, Meissner A. DNA methylation: roles in mammalian development. Nat Rev Genet. 2013;14:204–20.
3. Hamidi T, Singh AK, Chen T. Genetic alterations of DNA methylation machinery in human diseases. Epigenomics. 2015;7:247–65.
4. Ladd-Acosta C, Hansen KD, Briem E, Fallin MD, Kaufmann WE, Feinberg AP. Common DNA methylation alterations in multiple brain regions in autism. Mol Psychiatry. 2014;19:862–71.
5. Nardone S, Sams DS, Reuveni E, Getselter D, Oron O, Karpuj M, et al. DNA methylation analysis of the autistic brain reveals multiple dysregulated biological pathways. Transl Psychiatry. 2014;4:e433.
6. Petronis A. Epigenetics as a unifying principle in the aetiology of complex traits and diseases. Nature. 2010;465:721–7.
7. Bell JT, Tsai PC, Yang TP, Pidsley R, Nisbet J, Glass D, et al. Epigenome-wide scans identify differentially methylated regions for age and age-related phenotypes in a healthy ageing population. PLoS Genet. 2012;8:e1002629.
8. Jones MJ, Goodman SJ, Kobor MS. DNA methylation and healthy human aging. Aging Cell. 2015;14:924–32.
9. Lopez-Otin C, Blasco MA, Partridge L, Serrano M, Kroemer G. The hallmarks of aging. Cell. 2013;153:1194–217.
10. Finkel T, Serrano M, Blasco MA. The common biology of cancer and ageing. Nature. 2007;448:767–74.
11. Maegawa S, Gough SM, Watanabe-Okochi N, Lu Y, Zhang N, Castoro RJ, et al. Age-related epigenetic drift in the pathogenesis of MDS and AML. Genome Res. 2014;24:580–91.
12. Beck S, Rakyan VK. The methylome: approaches for global DNA methylation profiling. Trends Genet. 2008;24:231–7.
13. Karimi M, Johansson S, Stach D, Corcoran M, Grander D, Schalling M, et al. LUMA (LUminometric Methylation Assay)—a high throughput method to the analysis of genomic DNA methylation. Exp Cell Res. 2006;312:1989–95.

14. Gay MS, Li Y, Xiong F, Lin T, Zhang L. Dexamethasone treatment of
     newborn rats decreases cardiomyocyte endowment in the developing
     heart through epigenetic modifications. PLoS One. 2015;10:e0125033.
15. Zhao C, Xue H. A simple method for high-throughput quantification of
     genome-wide DNA methylation by fluorescence polarization. Epigenetics.
     2012;7:335–9.
16. Numata S, Ye T, Hyde TM, Guitart-Navarro X, Tao R, Wininger M, et al. DNA
     methylation signatures in development and aging of the human prefrontal
     cortex. Am J Hum Genet. 2012;90:260–72.
17. Zhang FF, Cardarelli R, Carroll J, Fulda KG, Kaur M, Gonzalez K, et al.
     Significant differences in global genomic DNA methylation by gender and
     race/ethnicity in peripheral blood. Epigenetics. 2011;6:623–9.
18. Zhu ZZ, Hou L, Bollati V, Tarantini L, Marinelli B, Cantone L, et al. Predictors
     of global methylation levels in blood DNA of healthy subjects: a combined
     analysis. Int J Epidemiol. 2012;41:126–39.
19. El-Maarri O, Becker T, Junen J, Manzoor SS, Diaz-Lacava A, Schwaab R, et al.
     Gender specific differences in levels of DNA methylation at selected loci
     from human total blood: a tendency toward higher methylation levels in
     males. Hum Genet. 2007;122:505–14.
20. Pogribny IP, Beland FA. DNA hypomethylation in the origin and
     pathogenesis of human diseases. Cell Mol Life Sci. 2009;66:2249–61.
21. Pogribny IP, Vanyushin BF. Age-related genomic hypomethylation. In: Tollefsbol
     TO, editor. Epigenetics of ageing. New York: Springer; 2010. p. 11–27.
22. DePinho RA. The age of cancer. Nature. 2000;408:248–54.
23. Wang SC, Oelze B, Schumacher A. Age-specific epigenetic drift in late-onset
     Alzheimer's disease. PLoS One. 2008;3:e2698.
24. Huguet G, Ey E, Bourgeron T. The genetic landscapes of autism spectrum
     disorders. Annu Rev Genomics Hum Genet. 2013;14:191–213.
25. State MW, Levitt P. The conundrums of understanding genetic risks for
     autism spectrum disorders. Nat Neurosci. 2011;14:1499–506.
26. Wong CC, Meaburn EL, Ronald A, Price TS, Jeffries AR, Schalkwyk LC, et al.
     Methylomic analysis of monozygotic twins discordant for autism spectrum
     disorder and related behavioural traits. Mol Psychiatry. 2014;19:495–503.
27. Elagoz Yuksel M, Yuceturk B, Karatas OF, Ozen M, Dogangun B. The altered
     promoter methylation of oxytocin receptor gene in autism.
     J Neurogenet 2016:1–5. [Epub ahead of print].

13

Navigating the dynamic landscape of long noncoding RNA and protein-coding gene annotations in GENCODE

Saakshi Jalali[1,2], Shrey Gandhi[1] and Vinod Scaria[1,2*]

Abstract

Background: Our understanding of the transcriptional potential of the genome and its functional consequences has undergone a significant change in the last decade. This has been largely contributed by the improvements in technology which could annotate and in many cases functionally characterize a number of novel gene loci in the human genome. Keeping pace with advancements in this dynamic environment and being able to systematically annotate a compendium of genes and transcripts is indeed a formidable task. Of the many databases which attempted to systematically annotate the genome, GENCODE has emerged as one of the largest and popular compendium for human genome annotations.

Results: The analysis of various versions of GENCODE revealed that there was a constant upgradation of transcripts for both protein-coding and long noncoding RNA (lncRNAs) leading to conflicting annotations. The GENCODE version 24 accounts for 4.18 % of the human genome to be transcribed which is an increase of 1.58 % from its first version. Out of 2,51,614 transcripts annotated across GENCODE versions, only 21.7 % had consistency. We also examined GENCODE consortia categorized transcripts into 70 biotypes out of which only 17 remained stable throughout.

Conclusions: In this report, we try to review the impact on the dynamicity with respect to gene annotations, specifically (lncRNA) annotations in GENCODE over the years. Our analysis suggests a significant dynamism in gene annotations, reflective of the evolution and consensus in nomenclature of genes. While a progressive change in annotations and timely release of the updates make the resource reliable in the community, the dynamicity with each release poses unique challenges to its users. Taking cues from other experiments with bio-curation, we propose potential avenues and methods to mend the gap.

Keywords: GENCODE, Long noncoding RNAs, Transcripts, Annotations

Introduction

The last decade has seen a tremendous improvement in our ability to understand the human genome and its transcriptional output at a much higher resolution than previously possible. This has largely been possible due to the availability of technologies which have enabled the annotation of transcripts at much higher depths and resolution. A number of systematic efforts to annotate the transcriptome in the human are also worth

mentioning. The earliest and most comprehensive approaches have been the H-invitational database consortium which aimed at assembling complementary DNA (cDNA) sequence information on the human genome through a global collaborative effort. This was followed by approaches including tiling arrays to characterize the transcriptional potential of the genome. Further, recent developments in deep sequencing approaches have greatly increased the resolution and facilitated the understanding of the transcriptome. Consequently, there has been the discovery of a significantly large number of novel gene loci in the genome. A large number of databases, including the ENCODE consortium, has made available gene annotations for the human genome by integrating data from the systematic explorations [1].

---

* Correspondence: vinods@igib.res.in
[1]GN Ramachandran Knowledge Center for Genome Informatics, CSIR Institute of Genomics and Integrative Biology (CSIR-IGIB), Mathura Road, Delhi 110 025, India
[2]Academy of Scientific and Innovative Research (AcSIR), CSIR-IGIB South Campus, Mathura Road, Delhi 110025, India

The efforts of the GENCODE consortium has been one of the most comprehensive and standardized approach for gene annotation and widely used by the community [1]. The initial efforts of GENCODE in the year 2008 (version 1) annotated 36,247 genes and 83,725 transcripts [2, 3] and subsequent versions of data show the annotations improve over time. The annotations were based on computational analysis, manual annotation, and experimental validation of genes and transcripts. The current release GENCODE Version 24 (V24) released in 2015 for humans has in total 60,554 genes annotated as protein-coding genes (19,815), long noncoding RNA genes (15,941), and small noncoding RNA genes (9882). It is also one of the most comprehensive annotations for long noncoding RNA genes.

Widely used by the community and constantly updated, with an average of three updates every year, we were motivated in understanding how the database evolved in the annotations, as this would provide a snapshot of the dynamic evolution of human gene annotations and specifically the long noncoding RNA annotations. We were interested in exploring both the different classes of annotations and the relative number of genes/transcripts in each annotation version towards understanding how the different gene classes and annotations evolved over time in the last decade.

We systematically analyzed the different annotations of genes/transcripts over different versions of GENCODE, starting with the first release till the latest release (V24) for the Human genome. While GENCODE serves as a major source of long noncoding RNA (lncRNA) annotations and has over time significantly and systematically catalogued the growth of lncRNA annotations, our analysis suggests a significant dynamism in gene annotations, reflective of the evolution and consensus in nomenclature of genes. We also find a number of cases where such dynamism in annotation has contributed to misannotation and in some cases results

which might be highly inconsistent. An overview of the dynamism in annotation and the different facets thereof are presented.

Results

Data compendium of transcripts in the human genome

Through data integration of transcript information from a total of 24 versions of GENCODE from years 2008 to 2015, we assembled a large compendium of a total of 2,51,614 transcripts. The growth of GENCODE has been consistent over the different versions. The initial version started with an annotation of 87,852 transcript annotations of which 43,415 were protein-coding, while 44,437 belonged to other biotypes. The most recent version of GENCODE (V24) annotates 1,99,005 transcripts, out of which 79,865 are protein-coding while 1,19,140 belong to other RNA biotypes. The most recent annotation as per GENCODE V24 estimates approximately 4.18 % of the human genome to be transcribed, significantly up from the estimate of 2.6 % in the first version. The summary of the gene and transcript numbers, the percentage of genome transcribed as annotated in each of the versions, and their growth over the different versions is summarized in Fig. 1.

The compendium of protein-coding and long noncoding RNA annotations

Of the entire compendium of 2,51,614 transcripts, a total of 1,14,114 transcripts were annotated as protein-coding, while a total of 1,20,864 transcripts were annotated as lncRNA biotype, in at least one of the 28 versions of GENCODE. The overlaps between these annotations revealed, a total of 11,069 transcripts had potential moonlighting identities, as shown by clashing annotations in one or the other release of the data resource. The transcripts and their overlapping annotations are summarized in Additional file 1: Figure S1.

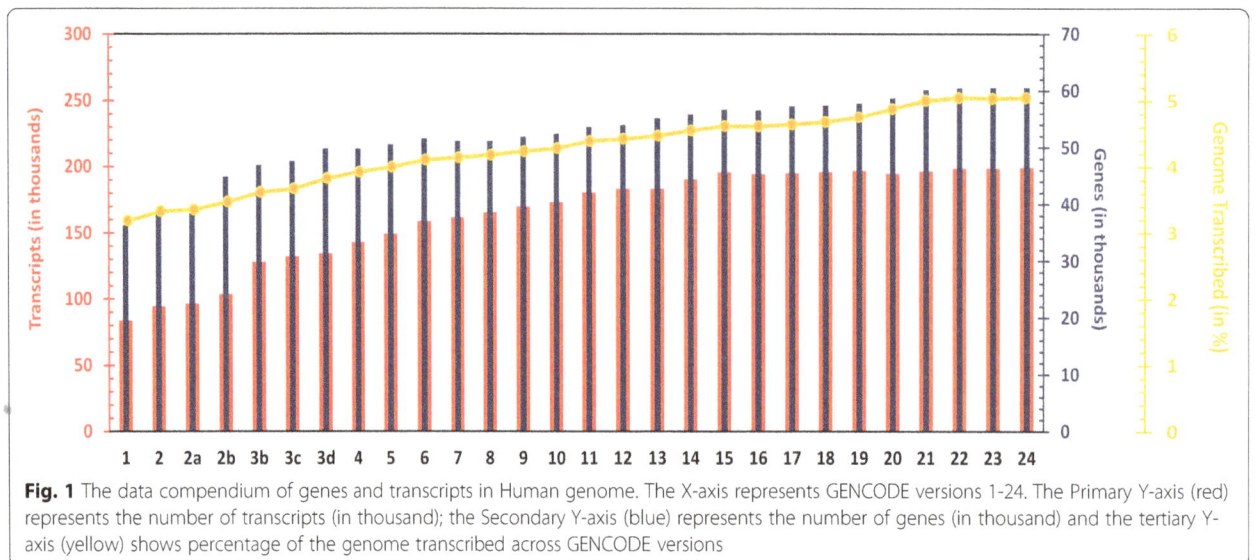

Fig. 1 The data compendium of genes and transcripts in Human genome. The X-axis represents GENCODE versions 1-24. The Primary Y-axis (red) represents the number of transcripts (in thousand); the Secondary Y-axis (blue) represents the number of genes (in thousand) and the tertiary Y-axis (yellow) shows percentage of the genome transcribed across GENCODE versions

Growth of the compendium over time

Over years and versions, the compendium has seen significant addition of transcript annotations, with an average of 6277 additions in every new version. The largest addition to the catalog was with the V3b version in the year 2009, which saw an addition of a whopping 26,715 transcripts to the compendium. This accounted for a significant 20.91 % addition of transcript annotations to the compendium. Of these, a total of 20,499 were protein-coding transcripts, while 3096 were lncRNAs. The update also saw a deletion of 7087 transcript annotations.

While the most significant addition to the protein-coding transcript annotations occurred in V3b, the most significant addition to the lncRNA annotations happened in V4, which saw an addition of 8897 new lncRNA transcript annotations.

The consistent updates to the GENCODE compendium also saw deletion of entries in every update. On an average, 2160 transcript annotations were deleted from the database with every version. The largest deletion of transcript annotations occurred with the V20 update of the compendium in the year 2014. This update accounted for the deletion of 11,410 transcript annotations from the compendium, of which 6727 were protein-coding and 3623 were lncRNAs.

The most significant deletion of protein-coding transcript annotations occurred with V20 which saw the deletion of 6727 transcript annotations, while the most significant deletion of lncRNA annotations occurred in the V4 update which saw the deletion of 4149 transcripts. V20 was close behind with a deletion of 3623 lncRNA transcript annotations. The detail for each version is specified in Table 1.

Consistency in annotations for protein-coding and long noncoding RNAs

Of the total number of transcripts, a total of 54,840 consistently maintained their annotations across all the GENCODE versions. Of these, 32,458 were protein-coding transcripts, while 22,382 belonged to other RNA biotypes. Out of the consistent transcript annotations throughout the versions, 19,520 belonged to lncRNAs. The dynamicity of the GENCODE compendium is summarized in Fig. 2.

Dynamicity of the lncRNA compendium and transformation of annotations

Out of this compendium, a total of 1,37,909 were annotated as noncoding RNA in one of the versions of GENCODE, of which a significant number amounting to 29,512 transcripts were systematically and consistently annotated as lncRNAs in all of the 24 versions. This accounted for 24.41 % of the total lncRNA annotations.

Of the total of 10,718 transcripts which had fleeting identities, a significant number of annotations were from

a protein-coding biotype to lncRNA, which accounted to 6560 transcripts, while the reverse accounted for 5463 transcripts in total. A total of 650 lncRNA transcript annotations reversed back after moonlighting as a protein-coding transcript, while 688 protein-coding transcripts reverted back after moonlighting as an lncRNA.

This dynamic nature of transcript biotypes was consistently observed across all the updates to the GENCODE compendium. The most significant change in the protein-coding transcript annotations happened in V3b leading to 20,499 transformations. In V4, had the most significant change in the lncRNA annotations wherein 10,044 transcripts changed their annotations to lncRNA while simultaneously 4498 lncRNA transcripts mutated their annotations to other biotypes. The largest change from the protein-coding transcripts to other biotypes occurred with V20 update of the compendium in 2014 which accounted for 7212 transcripts. The detail for each version is specified in Table 2.

Differences in the biotypes and annotations between versions of GENCODE

We evaluated the dynamicity in the biotypes under which the transcripts were annotated in different versions of GENCODE. Our analysis revealed a total of 70 biotypes were considered in total for annotation of transcripts. Only a small proportion (17) of their entire compendium of biotypes was systematically used in all the versions of GENCODE. A subset of 9 (Ambiguous ORF, scRNA pseudogene, Mt tRNA pseudogene, snRNA pseudogene, snoRNA pseudogene, rRNA pseudogene, miRNA pseudogene, misc RNA pseudogene) biotypes were dropped after v12, while 12 (ncRNA host, Disrupted domain, TR pseudogene, Artifact, scRNA, TR gene, IG gene, V segment, transcribed pseudogene, J segment, C segment) biotypes were used only in the earlier versions of GENCODE. The presence and absence of all biotypes across various versions of GENCODE are summarized in Fig. 3.

Impact of dynamicity of the lncRNA compendium

We also evaluated the impact of the dynamicity of annotations. Our analysis revealed a total of 1,96,988 transcripts had a dynamic annotation in at least one of the versions of GENCODE. This accounted for a total of 78.29 % of all the transcript annotations in GENCODE.

We closely examined a few candidates which had a significant dynamicity in its annotation (as shown in Additional file 2: Figure S2). We selected candidates which over versions of GENCODE have been dynamically annotated as a protein-coding or long noncoding RNA. One such candidate is C3orf10 (ENST00000256463). C3orf10 gene encodes for a 9-kD protein which plays a role in regulation of actin and microtubule organization. This gene encodes for ENST00000256463 which was annotated as

Table 1 Census of transcripts and their biotypes across all GENCODE versions

S.No	GENCODE versions	Freeze year	No. of Havana transcripts	No. of Ensembl transcripts	Total transcripts	No. of Havana converted to Ensembl ID	Total number of unique transcript IDs which were considered	No. of biotypes	No. of lncRNA biotypes
1	1	2008	67,432	16,293	83,725	66,579	87,852	37	14
2	2	2009	79,899	14,505	94,404	76,890	98,855	36	14
3	2a	2009	83,049	13,352	96,401	81,833	1,01,088	35	14
4	2b	2009	83,049	20,570	10,3619	81,833	1,08,145	39	14
5	v3b	2009	7896	1,19,809	1,27,705	7669	1,27,773	38	14
6	v3c	2009	0	13,2067	1,32,067	0	1,31,891	37	14
7	v3d	2009	0	1,34,266	1,34,266	0	1,34,267	38	15
8	4	2010	0	1,42,637	1,42,637	0	1,42,467	41	15
9	5	2010	0	1,48,880	1,48,880	0	1,48,710	43	15
10	6	2010	0	1,58,489	1,58,489	0	1,58,321	44	16
11	7	2010	0	1,61,375	1,61,375	0	1,61,214	44	16
12	8	2011	0	1,65,067	1,65,067	0	1,64,906	46	18
13	9	2011	0	1,69,419	1,69,419	0	1,69,257	50	20
14	10	2011	0	1,72,975	1,72,975	0	1,72,810	51	20
15	11	2011	0	1,80,272	1,80,272	0	1,80,107	51	19
16	12	2011	0	1,83,086	1,83,086	0	1,82,921	50	19
17	13	2012	0	1,82,967	1,82,967	0	1,82,798	41	18
18	14	2012	0	1,90,051	1,90,051	0	1,89,882	41	18
19	15	2012	0	1,95,433	1,95,433	0	1,95,264	40	17
20	16	2012	0	1,94,034	1,94,034	0	1,93,865	40	17
21	17	2013	0	1,94,871	1,94,871	0	1,94,702	38	15
22	18	2013	0	1,95,584	1,95,584	0	1,95,418	38	14
23	19	2013	0	1,96,520	1,96,520	0	1,96,354	38	14
24	20	2014	0	1,94,334	1,94,334	0	1,94,173	38	14
25	21	2014	0	1,96,327	1,96,327	0	1,96,165	43	17
26	22	2014	0	1,98,442	1,98,442	0	1,98,278	47	17
27	23	2015	0	1,98,619	1,98,619	0	1,98,455	45	16
28	24	2015	0	1,99,169	1,99,169	0	1,99,005	47	18

protein coding in V1 then as an lncRNA in V2-V2a and V3c-V6 and later again annotated as protein coding and further dropped from the database since version 20. In addition to inconsistency to the annotation type, it also had different gene names across versions the name of this transcript also changed: C3orf10 (V1-V8) -> AC034193.5 (V2-V3b) -> BRK1 (V9-V19). There were also few transcripts which had consistently same name such as ENST00 000436930: FER1L5 (V1-V24), ENST00000366438: ATAD 2B (V1-V24) across the entire version with varying annotations. While few transcripts such as ENST00000334998: RP1-163 M9.4 (V1-V2b) -> MST1P9 (V3b-V14) -> MST1L (V15) -> current status does not exist, ENST000 00339140: RP11-167P23.5 (V1-2b) -> FOXR2 (2b-V24), ENST00000408914: RIMKLP (V1-V3d) -> RIMKLB2 (V4-

V5) -> RIMKLBP1 (V6-V24) and had both inconsistent name as well as biotype.

Another example from our analysis is AC074389.6 gene which encodes for a single transcript (ENST00000382528) according to GENCODE annotations. It was annotated as protein coding in V1- 20 and this transcript is annotated as lincRNA from V21. This gene was identified as a novel bioactive peptide in year 2006 derived from precursor proteins which can be used as targets for drug interventions. To identify this new gene, the human genome National Center for Biotechnology Information (NCBI) 33 assembly, July 1, 2003, was used as reference and novelty of peptide sequence was confirmed using Universal Protein Resource (UNIPROT) [4]. Expression profile studies were also conducted to show their presence in various tissues [5].

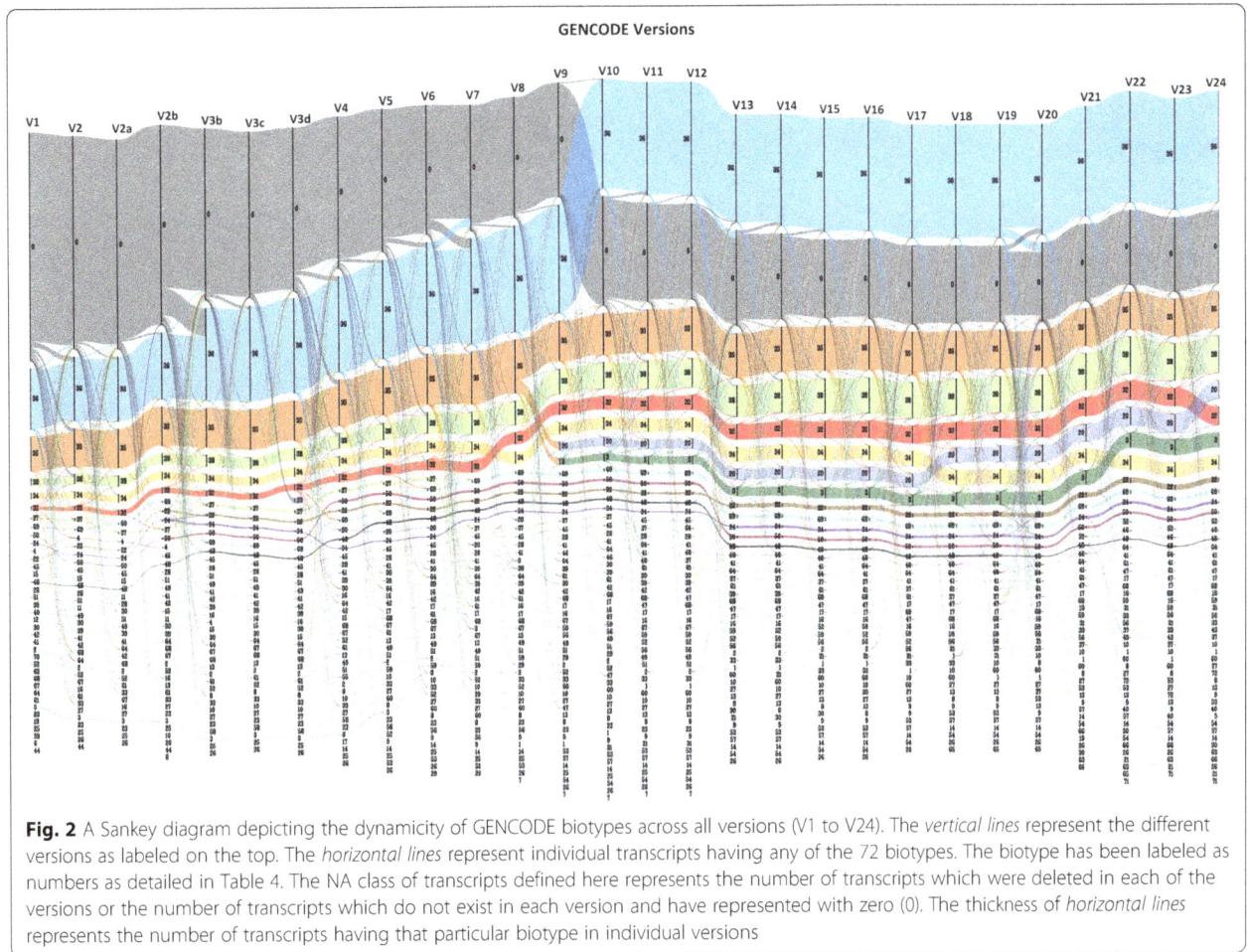

Fig. 2 A Sankey diagram depicting the dynamicity of GENCODE biotypes across all versions (V1 to V24). The *vertical lines* represent the different versions as labeled on the top. The *horizontal lines* represent individual transcripts having any of the 72 biotypes. The biotype has been labeled as numbers as detailed in Table 4. The NA class of transcripts defined here represents the number of transcripts which were deleted in each of the versions or the number of transcripts which do not exist in each version and have represented with zero (0). The thickness of *horizontal lines* represents the number of transcripts having that particular biotype in individual versions

Recently, Wang et al. reported this transcript to be expressed as an Lnc-RI lncRNA, and the same was shown through experimental validation to be ubiquitously expressed [6]. These contrasting reports highlight the genuine concern which arises due to frequent and ever changing landscape of GENCODE annotations.

The transcript ENST00000413529, encoded by the gene SDHAP3, was the most inconsistent transcript across the entire GENCODE compendium, which witnessed a total of nine transitions and was assigned six different biotypes during its short lived journey (V3b-19) Additional file 3: Figure S3.

Using HGNC (The HUGO Gene Nomenclature Committee) [7], one of the largest consortium of the human genes, we wanted to check the existence of the deleted genes in the present GENCODE(V24). The total human gene list extracted from HGNC consisted of 39,777 loci, and there were total of 56,095 GENCODE genes which were present in the earlier GENCODE versions but got eliminated in the current version (V24). When we overlapped the current HGNC genes with the genes deleted in V24, we found 285 genes to be common, out of

which, 35 were lncRNAs. The same is depicted in Additional file 4: Figure S4.

Discussion
The GENCODE compendium of transcript annotations has undoubtedly significantly enhanced the accessibility to a standardized set of genome annotations and accelerated the experimental annotation and understanding of gene functions, especially long noncoding RNA functions. Though there have been a number of databases [8] systematically annotating various aspects of lncRNAs including their functions, interactions etc., all the databases have been lacking continuous updates. GENCODE fills in this gap by covering and integrating the latest in terms of gene and transcript annotation, methodologies, and standards. Notwithstanding the limitations of the resource, which primarily arise from the changing landscape of technologies, definitions and methods for transcriptome analysis, GENCODE still provides one of the most comprehensive and well-accepted compendium of transcript annotations widely used and followed in literature.

Table 2 Details of all the biotypes used in GENCODE and their respective codes as used in our study

Biotype name	Code given
3 prime overlapping ncrna	1
Ambiguous orf	2
Antisense	3
Artifact	4
Bidirectional promoter lncrna	5
C segment	6
Disrupted domain	7
IG C gene	8
IG C pseudogene	9
IG D gene	10
ig gene	11
IG gene	12
IG J gene	13
IG J pseudogene	14
IG pseudogene|ig pseudogene	15
IG V gene	16
IG V pseudogene	17
J segment	18
Known ncrna	19
lincRNA	20
macro lncRNA	21
miRNA	22
miRNA pseudogene	23
misc RNA	24
misc RNA pseudogene	25
Mt rRNA	26
Mt tRNA	27
Mt tRNA pseudogene	28
ncrna host	29
Non-coding	30
Non-stop decay	31
Nonsense-mediated decay	32
Polymorphic pseudogene	33
Processed pseudogene	34
Processed transcript	35
Protein coding	36
Pseudogene	37
Retained intron	38
Retrotransposed	39
Ribozyme	40
rRNA	41
rRNA pseudogene	42
scaRNA	43

Table 2 Details of all the biotypes used in GENCODE and their respective codes as used in our study *(Continued)*

scRNA	44
scRNA pseudogene	45
Sense intronic	46
Sense overlapping	47
snoRNA	48
snoRNA pseudogene	49
snRNA	50
snRNA pseudogene	51
TEC|tec	52
TR C gene	53
TR D gene	54
TR gene	55
TR J gene	56
TR J pseudogene	57
TR pseudogene	58
TR V gene	59
TR V pseudogene	60
Transcribed processed pseudogene	61
Transcribed pseudogene	62
Transcribed unitary pseudogene	63
Transcribed unprocessed pseudogene	64
Translated processed pseudogene	65
Translated unprocessed pseudogene	66
tRNA pseudogene	67
Unitary pseudogene	68
Unprocessed pseudogene	69
V segment	70
Vaultrna	71
sRNA	72

A major limitation of the field has been the inconsistency in the nomenclature of transcript/gene biotypes which significantly adds confusion in the classification and long-term annotation of transcripts, especially lncRNAs. Our analysis of GENCODE suggests that a significant number of 52 biotype annotations were dropped at one point or the other between different versions of GENCODE, which affects a total of 1,96,799 transcript annotations while 17 biotypes remained constant across all GENCODE version for 54,815 transcripts.

In a very dynamic technological and knowledge landscape, it would be imperative for resources to closely integrate the long tail of annotations. It is humanly impossible for organizations to systematically track the growing corpus of literature in the field (Additional file 5: Figure S5), which presently adds over 1000 new publications per year. Therefore, it is imperative to dynamically interlink publications

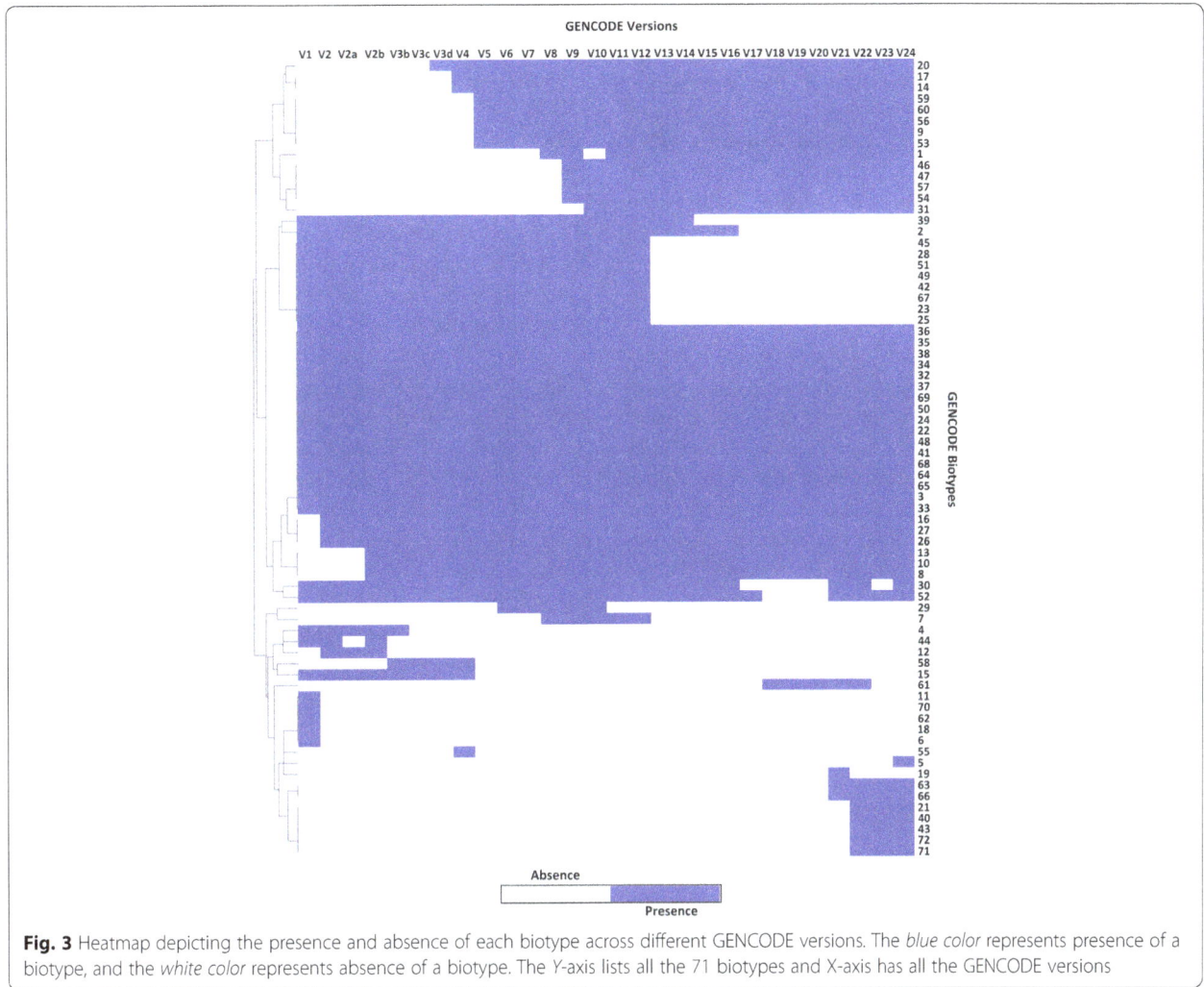

Fig. 3 Heatmap depicting the presence and absence of each biotype across different GENCODE versions. The *blue color* represents presence of a biotype, and the *white color* represents absence of a biotype. The Y-axis lists all the 71 biotypes and X-axis has all the GENCODE versions

and resources related to the field as has been extensively built for protein-coding genes [7].

Another major gap in the field has been the lack of interoperable databases annotating different biological aspects of lncRNAs. Apart from the standard Ensembl IDs followed by GENCODE and used by many other databases, only a small proportion of the lncRNAs 1.46 % of the entire compendium of lncRNAs have also been annotated and provided an HGNC gene symbol. Apart from the standard HGNC gene nomenclature, many publications and resources cite a variety of other nomenclatures, which adds to the confusion and inability to cross-link resources, publications, and analysis results. This major limitation stems from that fact that there has been a lack of standard and consensus standards for nomenclature of lncRNAs. Such standards for nomenclature and annotation of many other noncoding classes including miRNAs have ensured accordance in nomenclature which in turn maintains the compatibility between resources, databases, and citations in publications [7, 9, 10].

A number of resources and databases on lncRNAs have emerged in the recent years and has been comprehensively reviewed by Jalali and co-workers [11]. The resources encompass a variety of biological relationships, interactions, and functionalities. Nevertheless, the integration of the resources into a common platform has been a tedious task due to the variability in annotation standards, version of the annotations used, and lack of interoperability between the resources. The immediate goal would be to enable these complementary resources to be interoperable. The availability of common standards for nomenclature and annotation would enable the resources to be systematically integrated which would in turn enable timely updates. This would facilitate experimental as well as computational biologists wade through the unchartered waters quickly, and effectively.

The update in this ever-growing field has been fast outpacing the efforts by individual groups or laboratories to be able to systematically curate the information in a comprehensive way. Different attempts to fill in the gap of the long tail of bio-curation has emerged in the recent years, including Wiki-based systems for systematic and real-time annotation and curation of biological information. Such resources have been extensively developed not just for model systems but also for noncoding RNA databases. This could be complemented by efforts to automatically tag and annotate data from publications and resources using machine learning approaches developed recently [12].

Conclusion

In summary, our analysis of one of the most comprehensive resource of lncRNAs suggest the dynamic progression of the field in terms of both the number of annotations as well as the changing view of the classification of lncRNAs. While a dynamic change in annotations and a timely release of the updates make the resource unique, popular, and therefore widely used by the community, the dynamicity poses unique challenges to the community. Taking cues from other domains of bio-curation, we propose modalities to mend the gap.

Methods

GENCODE annotation

We downloaded the annotation data in form of Gene Transfer File (GTF) files from the GENCODE database and extracted all the transcript IDs along with their corresponding biotypes across all the versions from V1 to V24. GENCODE consortium has not made available Version 3a publically, hence not included in our study. The census for transcripts and biotypes across versions is detailed in Table 1. There are 28 GENCODE releases in our analyses consisting of genomic elements such as genes, transcripts, Coding sequence (CDS), untranslated regions (UTRS), and Exons annotated by Ensembl and Havana (Human and Vertebrate Analysis and Annotation). These were classified into 71 different biotypes as listed in Table 2 across all versions.

Analysis of consistency of transcripts across GENCODE versions

We extracted all the transcript identifiers comprising of both ENST (Ensembl) and\or OTTHUMT(Havana) IDs along with their transcript type. V1 consisted of only annotations for exons with no separate records for the other genomic elements such as genes, transcripts, or CDS. Hence, we directly used the transcript IDs as assigned to these exons for further analysis.

GENCODE assigned ENSTR/ENSTRR identifiers for pseudo autosomal regions of Y chromosome which are same for the X and Y chromosomes. For our analysis, we replaced all such transcripts with their respective ENST0 IDs in order avoid duplicate entries. We replaced 218 ENST0 IDs with their respective ENSTR /ENSTRR IDs if they had the same ENST identifier and biotype in a particular version.

Moreover, the earlier versions (V1 to 2c) of GENCODE consisted of either OTTHUMT or ENST identifiers for all transcripts. From V3b, GENCODE started to assign both the identifiers to most of the transcripts with an exception of a few which were assigned only IDs prefixed with OTTHUMT. After V3c the OTTHUMT prefixed IDs were systematically phased out as the main identifier, with each transcript having an ENST prefixed ID along with its corresponding OTTHUMT prefixed identifier. 77,193 OTTHUMT prefixed IDs had single ENST prefixed ID throughout their lifetime and hence were replaced with their respective ENST prefixed IDs. While 1982 OTTHUMT prefixed IDs had more than one ENST IDs in the same version therefore such OTTHUMT prefixed IDs were duplicated by assigning them both the Ensembl prefixed IDs while keeping their biotypes intact.

Another set of 3188 OTTHUMT prefixed IDs having more than one ENST prefixed IDs assigned to them across versions were replaced with respective IDs in that version by keeping the biotype of OTTHUMT prefixed ID intact. In addition, for 3272 OTTHUMT prefixed IDs there existed no ENST prefixed ID hence we kept them as it is.

All these transcripts IDs along with their assigned biotypes were organized into compiled record of total annotations. Those transcripts which did not have any biotype assigned to them in GENCODE versions were given a hypothetical code NA (not assigned). All the computation was performed by using custom shell and Perl scripts.

Analysis of consistency of lncRNA transcripts across GENCODE versions

To analyze the distribution and dynamism of lncRNA annotations across the GENCODE versions, we compared the lncRNA biotypes assigned by GENCODE. We made a comprehensive list of all the lncRNA biotypes or transcript biotypes used and dropped across the different versions (as listed in Table 3). While considering lncRNA as a class, we clubbed 23 sub-biotypes, namely 3 prime overlapping ncrna, TEC, Ambiguous orf, Antisense, Bidirectional promoter lncrna, Disrupted domain, Known ncrna, lincRNA, macro lncRNA, misc RNA, ncrna host, Non coding, Processed pseudogene, Processed transcript, Pseudogene, Retained intron, Retrotransposed, Sense intronic, Sense overlapping, Transcribed processed pseudogene, Transcribed unprocessed pseudogene, Unitary pseudogene, and Unprocessed pseudogene. From the compiled record of complete annotations, we extracted the transcripts belonging to these lncRNA subclasses and named it as lncRNA annotations.

Table 3 Number of transcripts added or deleted in each version of GENCODE

GENCODE version	Transcripts added	lncRNAs added	PC transcripts added	Transcripts deleted	lncRNAs deleted	PC transcripts deleted
1	–	–	–	–	–	–
2	13,568	7455	4156	2565	357	1690
2a	5580	3195	1769	3347	1756	1326
2b	7069	1243	1606	12	7	0
v3b	26,715	2924	19,998	7087	1666	1674
v3c	4978	1606	2643	860	169	143
v3d	3581	3481	96	1206	192	967
4	15,138	8897	3786	6937	4149	1481
5	7065	3820	2443	822	323	221
6	10,409	5527	3838	798	156	616
7	11,285	3234	7524	8392	2519	5834
8	5036	2750	1784	1344	61	1268
9	4568	2551	1582	217	67	146
10	3684	2171	1169	131	28	102
11	7817	4801	2290	520	463	56
12	3243	1808	1096	429	237	155
13	6734	3393	1391	6857	120	5272
14	7291	4013	2543	207	118	77
15	5749	3237	2079	367	214	107
16	628	451	132	2027	1052	812
17	1469	1194	206	632	340	185
18	1055	778	158	339	204	109
19	1378	1147	192	442	234	176
20	9229	3676	4238	11,410	3623	6727
21	2218	1709	432	226	119	97
22	2873	1630	751	760	268	320
23	350	212	104	173	117	52
24	758	473	206	208	71	105

Visualization

The distribution of all the transcripts in conjunction with their biotypes across the GENCODE versions from the compiled record for total annotations was visualized using an open web app, RAW [13]. A custom vector-based visualization based D3.js library through an interactable interface was used. The dynamicity of GENCODE annotations across all versions was depicted in form of a Sankey diagram (Fig. 2). In addition, we plotted a Sankey using lncRNA annotations file, as depicted in Fig. 4. Here, we considered four categories, namely lncRNA, protein coding, NA, and others (which included all other biotypes).

We also explored the disparity of biotypes across the GENCODE annotations. Hence, we considered the all the biotypes across different versions and plotted them in form of a heatmap. We observed many biotypes which were eliminated completely while few were retained throughout (Fig. 3).

Comparison across GENCODE versions

We calculated the number of transitions which each transcript went through during their lifetime which has been outlined in the Table 4. We also computed the various biotypes which each transcript was assigned and compiled this information in Table 5.

A compilation of the number of transcripts which were added and deleted in each version of GENCODE was derived from the compiled record of complete annotations. We also did this for both lncRNA and protein-coding transcripts which has been added/deleted, and the same has been outlined in the Table 1.

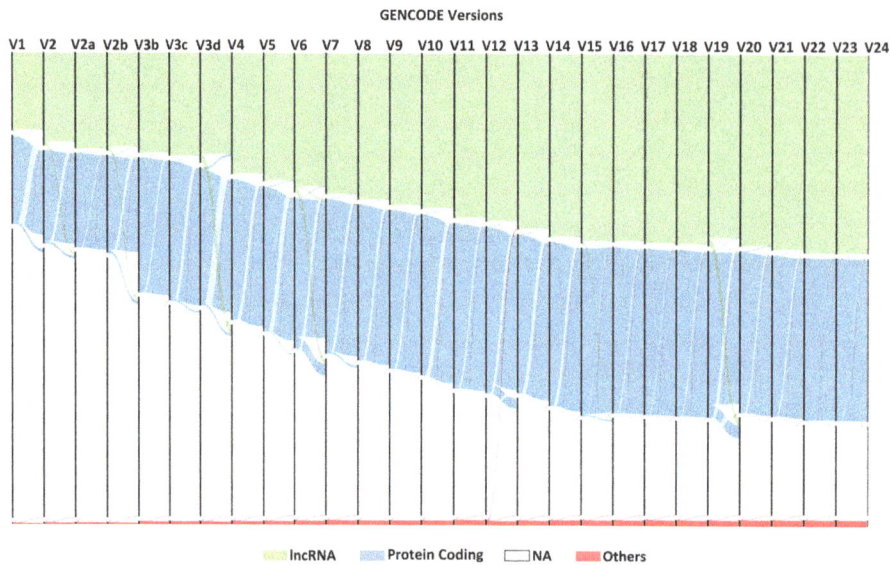

Fig. 4 A Sankey diagram depicting the dynamicity of GENCODE lncRNAs and protein-coding biotypes across all versions (V1 to V24). The lncRNA class considered here covers 23 sub-biotypes which includes 3 prime overlapping ncrna, TEC, Ambiguous orf, Antisense, Bidirectional promoter lncrna, Disrupted domain, Known ncrna, lincRNA, macro lncRNA, misc RNA, ncrna host, Non coding, Processed pseudogene, Processed transcript, Pseudogene, Retained intron, Retrotransposed, Sense intronic, Sense overlapping, Transcribed processed pseudogene, Transcribed unprocessed pseudogene, Unitary pseudogene, Unprocessed pseudogene. The protein-coding class represents the number of transcripts having protein-coding biotype. The NA class of transcripts defined here represents the number of transcripts which were deleted in each of the versions or the number of transcripts which do not exist in each version. While the others category comprises rest of biotypes

While the above table depicted the number of added/deleted transcripts, we also wanted to highlight the different transitions which these protein-coding and lncRNA transcripts went through across the GENCODE versions. Thus, on similar lines, we also produced a table outlining the switching of these transcripts which has been demonstrated in the Table 6.

We also analyzed the abundance of publications for long non coding RNAs over last decade, for which we derived the year wise publication list from Pubmed by searching keyword "lncRNA." The graph shown in Additional file 3: Figure S3 gives a brief layout of the number of publication per year.

Comparison with HGNC

HGNC is the largest and one of the most reliable sources for which assigns unique and standardized nomenclature for human genes created as part of the Human Genome Organization (HUGO) [7]. We wanted to verify whether the genes which do not exist in the present GENCODE version are still present in HGNC. Thus, we extracted all the HGNC genes having approved HGNC IDs (up till last updated: 05/07/16 04:51:01) and checked their presence in last V24.

Table 5 Summary of the number of transitions each transcript went through

No. of transitions	No. of transcripts
0	54,840
1	1,33,630
2	55,951
3	6420
4	1125
5	283
6	95
7	35
8	12
9	3

Table 4 Summary of the number of biotypes assigned to each of the transcripts

No. of biotypes assigned to the transcript	No. of transcripts
1	54,840
2	1,74,779
3	20,528
4	1945
5	256
6	41
7	5

Table 6 Switching of transcripts across versions

GENCODE version	Transcripts added	Transformed to lncRNAs	Transformed to PC transcripts	Transcripts deleted	Transformed from lncRNAs transcripts	Transformed from PC transcripts
1	–	–	–	–	–	–
2	13,568	7781	4336	2565	580	2000
2a	5580	3296	5354	3347	1834	96
2b	7069	1261	1687	12	7	0
v3b	26,715	3096	20,499	7087	2255	2049
v3c	4978	1611	2665	860	194	162
v3d	3581	3722	96	1206	189	1210
4	15,138	10,044	4073	6937	4498	2717
5	7065	4078	2662	822	593	521
6	10,409	6141	4261	798	714	1266
7	11,285	3874	8325	8392	3292	6519
8	5036	2933	1868	1344	155	1508
9	4568	2762	1677	217	178	282
10	3684	2284	1257	131	119	235
11	7817	5028	2530	520	855	322
12	3243	2069	1273	429	469	428
13	6734	4244	1679	6857	557	5660
14	7291	4314	2927	207	631	415
15	5749	3364	2201	367	384	278
16	628	649	311	2027	1271	1019
17	1469	1480	415	632	677	474
18	1055	940	385	339	481	275
19	1378	1289	474	442	582	334
20	9229	4125	4861	11,410	4324	7212
21	2218	2263	535	226	231	618
22	2873	1820	838	760	390	503
23	350	277	195	173	231	112
24	758	527	300	208	197	165

Additional files

Additional file 1: Figure S1. Venn diagram representing the moonlighting of lncRNA and protein-coding transcript annotations.

Additional file 2: Figure S2. Heatmap depicting transitions of the six candidate transcripts from Protein-coding biotype to lncRNA biotype or vice versa over the different versions of GENCODE.

Additional file 3: Figure S3. The transition of ENST00000413529 (SDHAP3) transcript over the various GENCODE versions.

Additional file 4: Figure S4. Common and unique annotated genes of absent in GENCODE V24 and HGNC. Venn diagram shows intersection between genes annotated by GENCODE and HGNC.

Additional file 5: Figure S5. Growth of literature in the field of lncRNAs. The number of publications for each year was retrieved using keyword "lncRNA" from PubMed. The data for 2016 is incomplete at the time of writing the manuscript and therefore marked with dotted lines.

Acknowledgements
The authors also acknowledge constructive criticism and editorial help from Remya Koshy and Ambily Sivadas which significantly improved the readability and perspective of the article.

Funding
The authors acknowledge funding from CSIR India through Grant BSC0123 (GENCODE-C).

Authors' contributions
VS conceptualized the analysis. Data analysis was performed by SJ and SG. SJ and SG prepared the data summaries and visualization. SJ, SG, and VS wrote the manuscript. All authors reviewed the manuscript. All authors read and approved the final manuscript.

Competing interests
The authors declare that they have no competing interests.

Consent for publication
Not applicable.

References

1. ENCODE Project Consortium TEP. The ENCODE (ENCyclopedia Of DNA Elements) Project. Science. 2004;306:636–40.
2. Harrow J, Denoeud F, Frankish A, Reymond A, Chen C-K, Chrast J, et al. GENCODE: producing a reference annotation for ENCODE. Genome Biol. 2006;7:S4.
3. GENCODE Project. GENCODE Data. ftp://ftp.sanger.ac.uk/pub/gencode/ Gencode_human (2015). Accessed 19 Feb 2016.
4. UniProt Consortium TU. UniProt: a hub for protein information. Nucleic Acids Res. 2015;43:D204–12.
5. Jung E, Dittrich W, Scheidler S. Coding genes with a single exon for new bioactive peptides [Internet]. Google Patents; 2008. Available from: http://www.google.com.gt/patents/WO2008074424A3?cl=en.
6. Wang Z-D, Shen L-P, Chang C, Zhang X-Q, Chen Z-M, Li L, et al. Long noncoding RNA lnc-RI is a new regulator of mitosis via targeting miRNA-210-3p to release PLK1 mRNA activity. Sci Rep. 2016;6:25385.
7. Gray KA, Yates B, Seal RL, Wright MW, Bruford EA. Genenames.org: the HGNC resources in 2015. Nucleic Acids Res. 2015;43:D1079–85.
8. Fritah S, Niclou SP, Azuaje F. Databases for lncRNAs: a comparative evaluation of emerging tools. RNA. 2014;20:1655–65.
9. Wright MW, Povey S, Lovering R, Bruford E, Wright M, Lush M, et al. A short guide to long non-coding RNA gene nomenclature. Hum Genomics BioMed Central. 2014;8:7.
10. Genome Information Integration Project And H-Invitational 2 GIIPAH-I, Yamasaki C, Murakami K, Fujii Y, Sato Y, Harada E, et al. The H-Invitational Database (H-InvDB), a comprehensive annotation resource for human genes and transcripts. Nucleic Acids Res. 2008;36:D793–9.
11. Jalali S, Kapoor S, Sivadas A, Bhartiya D, Scaria V. Computational approaches towards understanding human long non-coding RNA biology. Bioinformatics. 2015;31:2241–51.
12. Tarca AL, Carey VJ, Chen X, Romero R, Drăghici S. Machine learning and its applications to biology. PLoS Comput Biol. 2007;3:e116.
13. Caviglia G, Mauri M, Azzi M, Uboldi G: DensityDesign Research Lab, RAW App. http://raw.densitydesign.org/ (2014). Accessed 17 May 2016.

14

Whole-exome sequencing identifies novel candidate predisposition genes for familial polycythemia vera

Elina A. M. Hirvonen[1], Esa Pitkänen[1], Kari Hemminki[2], Lauri A. Aaltonen[1,3] and Outi Kilpivaara[1*]

Abstract

Background: Polycythemia vera (PV), characterized by massive production of erythrocytes, is one of the myeloproliferative neoplasms. Most patients carry a somatic gain-of-function mutation in *JAK2*, c.1849G > T (p.Val617Phe), leading to constitutive activation of JAK-STAT signaling pathway. Familial clustering is also observed occasionally, but high-penetrance predisposition genes to PV have remained unidentified.

Results: We studied the predisposition to PV by exome sequencing (three cases) in a Finnish PV family with four patients. The 12 shared variants (maximum allowed minor allele frequency <0.001 in Finnish population in ExAC database) predicted damaging in silico and absent in an additional control set of over 500 Finns were further validated by Sanger sequencing in a fourth affected family member. Three novel predisposition candidate variants were identified: c.1254C > G (p.Phe418Leu) in *ZXDC*, c.1931C > G (p.Pro644Arg) in *ATN1*, and c.701G > A (p.Arg234Gln) in *LRRC3*. We also observed a rare, predicted benign germline variant c.2912C > G (p.Ala971Gly) in *BCORL1* in all four patients. Somatic mutations in *BCORL1* have been reported in myeloid malignances. We further screened the variants in eight PV patients in six other Finnish families, but no other carriers were found.

Conclusions: Exome sequencing provides a powerful tool for the identification of novel variants, and understanding the familial predisposition of diseases. This is the first report on Finnish familial PV cases, and we identified three novel candidate variants that may predispose to the disease.

Keywords: Myeloproliferative neoplasm, Polycythemia vera, Genetics, Exome sequencing, Familial predisposition

Background

Myeloproliferative neoplasms (MPNs) are a group of hematological malignances with enhanced proliferation of myeloid cells due to an acquired mutation of a single hematopoietic stem cell (HSC) resulting in clonal progeny [1, 2]. Expansion of red blood cells in the peripheral blood and bone marrow is the hallmark of polycythemia vera (PV). PV is a chronic, Philadelphia chromosome-negative myeloproliferative disorder with symptoms including fatigue, pruritus, and splenomegaly. The patients have an increased risk of thrombosis, and the disease may further progress to secondary acute myeloid leukemia (sAML) or myelofibrosis [3].

The global annual incidence rate of PV is approximately 1 per 100,000 persons [4]. Most PV patients (approximately 95%) carry a somatic mutation in exon 14 of a non-receptor tyrosine kinase-coding gene Janus kinase 2 (*JAK2*), c.1849G > T (p.Val617Phe, subsequently referred to as *JAK2V617F*) [5–8]. The major diagnostic criteria for PV according to World Health Organization (2008) are exceptionally high hemoglobin and presence of *JAK2V617F* mutation. Other criteria include consistent bone marrow morphology and low erythropoietin (Epo) levels [9, 10]. When diagnosed with PV, most patients are older than 60 years old, and survival depends on the complications and severity of the disease. In younger patients, life expectancy is reduced when compared to the general population [11].

Janus kinase 2 (JAK2) plays a crucial role in the function and maintenance of HSCs [12], as well as in myelopoiesis through binding to various cytokine receptors including

* Correspondence: outi.kilpivaara@helsinki.fi
[1]Genome-Scale Biology Research Program, Research Programs Unit and Department of Medical and Clinical Genetics, Medicum, University of Helsinki, P.O. Box 6300014 Helsinki, Finland
Full list of author information is available at the end of the article

erythropoietin receptor (Epo-R), thus, contributing to formation of red blood cells [13]. V617F mutation, which is located within the pseudokinase domain of JAK2 protein, leads to constitutive activity that promotes cytokine hypersensitivity and abnormal signaling through Janus kinase-signal transducer and activator of transcription factor (JAK-STAT) pathway [5–8]. In addition, approximately 3% of PV patients, being *JAK2V617F*-negative, harbor mutations in exon 12 of *JAK2*. The mutations in both exons 14 and 12 induce cytokine-independent proliferation of cells expressing Epo-receptors [14]. Also, certain germline mutations in *JAK2* are predicted to represent a mechanism possibly preceding the acquisition of *JAK2V617F* mutation in PV [15].

Although most PV cases appear to be sporadic, familial clustering is also observed in a subset of cases [16–18]. *JAK2V617F*-positive MPNs are strongly associated with *JAK2* 46/1, or 'GGCC' haplotype (rs10974944), which is a common, moderate penetrance predisposition allele [19–22]. Also, a single nucleotide polymorphism (SNP) in the telomerase reverse transcriptase gene (*TERT*), rs2736100, has emerged as another predisposing factor to MPNs [23]. In addition, it has been shown that germline duplication of *ATG2B* and *GSKIP* genes predisposes to familial myeloid malignancies [24], and germline *RBBP6* mutations have been associated with familial MPNs [25]. It is likely that additional inherited genetic factors contribute to PV development as well, since familial clustering of PV suggests the presence of additional susceptibility alleles. High-penetrance predisposition genes to PV have, however, remained unidentified.

Here, we report a Finnish family with four PV patients in two generations (Additional file 1: Figure S1). A family with this many cases is quite exceptional, since PV is a rare disease and usually sporadic. Also, in a familial PV case like here, the age at diagnosis is usually lower, which is an additional indication of the presence of predisposing factors. DNA from all the PV patients in this family was available. The aim of this study was to identify new predisposing gene variants by exome sequencing the DNA of three of the affected family members; the index case and two affected family members (germline) and peripheral blood DNA of the index case. Exome sequencing was not feasible in the fourth PV patient due to very low amount of archived DNA available. The observed, predicted as damaging, germline variants were analyzed in the fourth PV patient of the family, and the variants shared by all four individuals were further screened in six other families with two PV cases in each.

Methods

Patient samples
We investigated a Finnish family with four PV patients (Additional file 1: Figure S1). The index case (1.1) was diagnosed with PV at the age of 36, and with myelofibrosis later at the age of 47. The father (1.2) of the index patient was diagnosed with PV at the age of 48, and the aunt (1.9) with PV and acute leukemia at the age of 91. The uncle (1.10) of the index patient was diagnosed with PV at the age of 83. Two individuals in the family had lymphoma; 1.19 was diagnosed with nodular lymphocyte-predominant Hodgkin lymphoma (NLPHL) at the age of 55, and with diffuse large cell non-Hodgkin lymphoma (NHL) later at the age of 68; 1.6 was diagnosed with differentiated diffuse lymphocytic lymphoma at the age of 89. The healthy daughter of the index case is currently 31 years old.

Both peripheral blood and buccal swab samples were available from the index case. Only formalin-fixed paraffin embedded (FFPE) blocks were available from three other family members diagnosed with PV: 1.2, 1.9, and 1.10. Germline DNA was also available from one of two lymphoma patients (1.19) of the family. The second sample set consisted of FFPE blocks from six other Finnish PV families with two first-degree relative cases in five, and two more distant relatives in the sixth family. The study was approved by the appropriate Ethics Review Committee. Samples were derived either after a signed informed consent or after authorization from the National Supervisory Authority for Welfare and Health. The study was conducted in accordance with the Declaration of Helsinki.

Exome capture and sequencing
DNA was extracted from FFPE blocks with a standard phenol-chloroform method, and the buccal swab sample was extracted with QIAmp DNA Mini kit (Qiagen, Hilden, Germany). DNA from the blood sample was extracted with a standard non-enzymatic TKM buffer-proteinase K method. Sample libraries of gDNA were prepared using NEBNext DNA Library Prep Reagent Set for Illumina (New England Biolabs Ltd. Catalog # E6000), and exomic regions were enriched using Agilent Sure SelectXT Human All Exon V4 + UTRs 50Mb kit (Agilent Technologies, Santa Clara, CA, USA). Paired-end short read sequencing was performed with HiSeq 2000 (Illumina Inc., San Diego, CA, USA) at Karolinska Institutet, Sweden. The DNA library for whole-genome sequencing was prepared with KAPA Hyper Prep Kit (KAPA Biosystems, Wilmington, MA, USA). Paired-end short read sequencing was performed with HiSeq 4000 (Illumina Inc.) at Karolinska Institutet.

Variant analysis
Exome and genome sequencing data was mapped against the human reference genome GRCh37 (1000 Genomes Project reference hs37d5) with Burrow-Wheeler Aligner (BWA MEM, v.0.7.12, https://arxiv.org/abs/1303.3997), and mapped reads refined following the Genome Analysis

Toolkit (GATK) Best Practices workflow including PCR duplicate removal, indel realignment, and base quality score recalibration [26]. Single-nucleotide and short-indel variants were then called using GATK HaplotypeCaller.

The SNV and indel variants in the exome sequencing data were analyzed with an in-house developed analysis and visualization tool (BasePlayer, Katainen et al., manuscript in preparation). A minimum coverage of four reads and the mutated allele present in at least 20% of the reads was required to call a variant. The variants which were present in an in-house control set of 542 Finns [93 whole genome sequenced individuals from the 1000 Genomes Project, 402 whole-genome sequenced individuals from Kuusamo, Finland (Sequencing Initiative Suomi), and 47 uterine leiomyoma patients] were excluded. To exclude common variants, we further filtered the variant set against the Exome Aggregation Consortium (ExAC v.0.3: 3,307 Finns) data [27], setting the maximum allowed minor allele frequency (MAF) in Finnish population to 0.001. Germline variants predicted benign by two independent computational methods PolyPhen-2 [28] and SIFT [29] were excluded.

Direct sequencing

The shared variants found in the exome data and *BCORL1* cDNA were validated by direct Sanger sequencing. Also, the predisposing germline variant rs10974944 in *JAK2* was checked. DNA was extracted from FFPE blocks with a standard phenol-chloroform method or with NucleoSpin® DNA FFPE XS kit (Macherey-Nagel, Düren, Germany). Primers were designed using Primer3Plus software (http://www.bioinformatics.nl/primer3plus). The DreamTaq™ DNA Polymerase (Thermo Scientific, Waltham, MA, USA) or Invitrogen™, Platinum™, SuperFi™ DNA Polymerase (Thermo Scientific) were used in PCR reactions, and PCR products were

purified with A'SAP PCR clean-up method (ArcticZymes, Tromsø, Norway). BigDye Terminator v3.1 sequencing reaction was used in the DNA sequencing, and capillary electrophoresis was performed on an ABI3730xl DNA Analyzer (Applied Biosystems, Foster City, CA, USA) at the Institute for Molecular Medicine Finland (FIMM). The results were analyzed manually using FinchTV v.1.4.0 (Geospiza Inc., Seattle, WA, USA).

For cDNA sequencing, RNA was extracted from index case's whole blood sample with NucleoSpin® RNA kit (Macherey-Nagel), and reverse-transcribed to cDNA with Promega M-MLV Reverse Transcriptase (Thermo Scientific) according to manufacturers' protocols.

Results and discussion
Novel candidate germline variants for PV predisposition

Here, we have studied a Finnish family with four PV cases with the aim of identifying novel PV-predisposing germline variants. Germline DNA exomes from three affected family members and the peripheral blood DNA exome of the index case were sequenced. The average coverage at each base was 58 reads and 88% of the captured regions had a minimum coverage of four reads. We filtered the called variants based on their predicted, damaging effect on the gene product, and their presence in an in-house control set of 542 Finns. We removed the variants that occurred in any of the 542 Finnish controls (MAF < 0.2%, 95% CI [0, 0.05%]), leaving us with 12 shared variants; 1 splice site and 11 possibly pathogenic missense variants (Table 1). We then validated the variants by Sanger sequencing in one additional family member diagnosed with PV (1.10).

From these variants predicted as damaging, 1.10 carried three rare single-nucleotide variants (SNVs): c.1254C > G (p.Phe418Leu) in *ZXDC* (ENST00000389709); c.1931C > G (p.Pro644Arg) in *ATN1* (ENST00000356654);

Table 1 A list of damaging (in silico) germline variants in PV patients detected by exome sequencing (three cases)

Gene	Ensembl gene	Ensembl transcript	Genomic location	Variation, cDNA	Variation, protein
ZXDC*	ENSG00000070476	ENST00000389709	Chr3: 126189754	c.1254C > G	p.Phe418Leu
ATN1*	ENSG00000111676	ENST00000356654	Chr12: 7046361	c.1931C > G	p.Pro644Arg
LRRC3*	ENSG00000160233	ENST00000291592	Chr21: 45877228	c.701G > A, rs148872771	p.Arg234Gln
GNL3	ENSG00000163938	ENST00000418458	Chr3: 52727477	c.1241A > G	p.Tyr414Cys
MDC1	ENSG00000137337	ENST00000376406	Chr6: 30679188	c.2221-1G > T	splice site variant
ITPR3	ENSG00000096433	ENST00000374316	Chr6: 33635026	c.1672C > T, rs780906252	p.Arg558Cys
FAM135A	ENSG00000082269	ENST00000418814	Chr6: 71190668	c.607G > A, rs143901584	p.Val203Met
SLC2A12	ENSG00000146411	ENST00000275230	Chr6: 134312391	c.1756C > T, rs200847615	p.Pro586Ser
WDR86	ENSG00000187260	ENST00000334493	Chr7: 151097265	c.226G > A, rs199824863	p.Asp76Asn
CSMD1	ENSG00000183117	ENST00000537824	Chr8: 3165238	c.3929C > T	p.Ala1310Val
SLC24A2	ENSG00000155886	ENST00000341998	Chr9: 19786283	c.582A > G, rs368590535	p.Ile194Met
ITPKC	ENSG00000086544	ENST00000263370	Chr19: 41224132	c.1092C > G, rs143757004	p.Asp364Glu

The three shared variants in all four cases are marked with an *asterisk*. Genome assembly: GRCh37

and rs148872771, c.701G > A (p.Arg234Gln) in *LRRC3* (ENST00000291592). We screened the three variants in lymphoma patient 1.19, who carried the variant in *LRRC3*. Also, we checked for the three variants in the germlines of eight PV patients from six other Finnish families, but the variants were not observed. It may still be possible, however, that some of the three variants or other variants in these genes may play a role in PV development. We identified also one rare benign (PolyPhen-2, SIFT) missense SNP, rs144332650, c.2912C > G (p.Ala971Gly), and in *BCORL1* (ENST00000540052) in all four PV patients in the family. Mutations in *BCORL1* have been associated with the leukemogenic process in AML [30–32]. PV patients in six other families did not carry the variant.

Zinc-finger X-linked duplicated family member C (ZXDC) belongs to the ZXD family of transcription factors, which has been observed to regulate transcription of major histocompatibility complex (MHC) class I and II genes in antigen presenting cells [33]. In addition to zinc fingers, ZXDC contains a transcriptional activation domain, and a specific domain used for interaction with a transcriptional co-factor class II *trans*-activator (CIITA), which leads to CIITA binding to promoter elements involved in constitutive MHC class II expression [33, 34]. By reducing the expression of ZXDC, CIITA activation of MHC class II gene transcription is significantly reduced [33]. Thus, downregulation of *ZXDC* may contribute to carcinogenesis and malignant progression of tumors by participating in the suppression of MHC class II genes. The mutated site of *ZXDC* identified in our study is the first amino acid of the ninth zinc finger repeat. Zinc fingers are necessary for full activity of cooperation with CIITA [33]. Beyond this role of acting as a co-factor in CIITA function, the *ZXDC* gene function is unknown. ZXDC is enriched in myeloid lineages and has been observed to regulate transcription of key genes during myeloid cell differentiation [35]. During hematopoiesis, it is expressed especially in stem and progenitor cells in the bone marrow, myelocytes, and leukocytes (BloodSpot http://nar.oxfordjournals.org/content/early/2015/10/26/nar.gkv1101.abstract, www.proteinatlas.org). Solely based on gene function, *ZXDC* variant would be the most attractive predisposition candidate of the three, but further studies are warranted.

Atrophin-1 (ATN1) is a nuclear transcriptional corepressor, and aberrant form of ATN1 is associated with neurodegenerative diseases such as dentatorubral-pallidoluysian atrophy (DRPLA), and cancer in humans [36]. It is ubiquitously expressed in neurons [37] and widely in various other tissues, e.g., in hematopoietic cells in the bone marrow (www.proteinatlas.org). ATN1 contains glutamine-repeats, and two arginine-glutamic acid dipeptide-like repeats (RE-repeats) [36]. In neuronal nuclei, ATN1 has been shown to interact with a

transcriptional repressor Eight twenty-one (ETO) protein [38]. In normal hematopoiesis, it is widely expressed in differentiated blood cells, but the expression is lower in stem and progenitor cells (BloodSpot). Thus, it is likely that ATN1 interaction with ETO has no contribution to erythropoiesis or development of myeloproliferative diseases. On the other hand, in patients with AML, ATN1 is expressed more substantially (BloodSpot) compared to normal counterparts. Nevertheless, its normal function is not completely understood.

Little is known about the function of Leucine-rich repeat containing 3 (LRRC3) gene product, but it is widely expressed in different tissues, including the bone marrow. Most malignancies display moderate cytoplasmic-positive staining, and the strongest expression is observed in colorectal cancers (www.proteinatlas.org). One lymphoma patient of the family with DNA available carried the rare variant rs148872771 in *LRRC3*, too, which may indicate the particular variant not being responsible for PV predisposition exclusively.

The SNP in *JAK2*, rs10974944, was identified in all the PV patients in the family. Individuals 1.1, 1.2, and 1.10 were homozygous for the risk variant in their germline (GG genotype), whereas 1.9 was heterozygous (CG genotype). Also, all eight PV patients from the six other families carried the risk variant: two of them were heterozygous, and six were homozygous. The SNP in *TERT*, rs2736100, was also checked from the patients. All the PV patients in the studied family were homozygous for the risk variant (CC).

Germline duplication of *ATG2B* and *GSKIP* has been shown to predispose to familial myeloid malignancies [24]. A possible duplication was checked for in the whole-genome sequence data of the index case by visualizing the depth of coverage in the region. The duplication had not occurred in the index patient's genome.

Detection of somatic variants

We identified the most frequent somatic variation in PV patients, *JAK2V617F*, in index case's blood sample. Loss of heterozygosity (LOH) in the 12 damaging gene variants detected by exome sequencing was looked for in the index case's blood sample, but only the variants c.582A > G (p.Ile194Met) in *SLC24A2* and c.3929C > T (p.Ala1310Val) in *CSMD1*, in addition to *JAK2V617F*, showed clear LOH. In known MPN-associated genes, the index case carried two missense variants, c.680C > T (p.Thr227Met) in *FLT3* and c.5162 T > G (p.Leu1721Trp) in *TET2* (Illumina TruSight® Myeloid Sequencing Panel), predicted as possibly damaging by PolyPhen-2 and SIFT. In addition, we identified possibly damaging missense variants c.3263C > T (p.Ser1088Phe) and c.1235C > T (p.Ala412Val) in *FANCA*, which is one of the genes associated with other myeloid malignancies [39, 40].

BCORL1 gene is located on the X chromosome. The index case being a woman, we checked which of the alleles was expressed; the BCORL1 variant or the wild-type. By Sanger sequencing the cDNA, we identified the expression of BCORL1 variant.

Conclusions

Identification of predisposing genes and mutations is important for families with PV susceptibility and acknowledging family history is essential in order to unveil the genetic background. New hereditary gene defects may lead to screening and genetic counseling of family members and may improve the diagnosis and treatment thus affecting the quality of life. Also, the identification of specific gene mutations gives the possibility to screen individuals at higher risk. This is the first report on Finnish familial PV cases, and we identified three candidate predisposition variants by exome-sequencing. We would like to present these genes as candidates for PV susceptibility and for further validation by the research community.

Acknowledgements
The authors would like to thank Annukka Ruokolainen, Inga-Lill Svedberg, Iina Vuoristo, Sini Nieminen, Marjo Rajalaakso, Heikki Metsola, and Jiri Hamberg for excellent technical assistance. Institute for Molecular Medicine Finland (FIMM) is acknowledged for capillary sequencing. Minna Taipale and Jussi Taipale (Karolinska Institutet, Stockholm, Sweden) are acknowledged for providing excellent exome and genome sequencing services.

Funding
Funding for the work was supported by grants from the Academy of Finland (Finnish Centre of Excellence Program 2012-2017 (#250345), and personal grants for O.K. (#137680, #274474), the Finnish Cancer Society, and Sigrid Juselius Foundation. The funding bodies did not participate in the study design, sample collection, analysis and interpretation of data, or in the writing of the manuscript.

Author's contributions
OK, KH, and LAA designed the study. KH provided the samples. EAMH performed the laboratory experiments and data analysis. EP was responsible for the pipeline. EAMH and OK wrote the manuscript. All authors read and approved the final manuscript.

Competing interests
The authors declare that they have no competing interests.

Consent for publication
Not applicable.

[1]Genome-Scale Biology Research Program, Research Programs Unit and Department of Medical and Clinical Genetics, Medicum, University of Helsinki, P.O. Box 6300014 Helsinki, Finland. [2]Division of Molecular Genetic Epidemiology, German Cancer Research Center (DKFZ), Heidelberg, Germany. [3]Department of Biosciences and Nutrition, Karolinska Institutet, SE-17177 Stockholm, Sweden.

References
1. Adamson JW, Fialkow PJ, Murphy S, Prchal JF, Steinmann L. Polycythemia vera: stem-cell and probable clonal origin of the disease. N Engl J Med. 1976;295:913–6.
2. Jamieson CH, Gotlib J, Durocher JA, Chao MP, Mariappan MR, Lay M, et al. The JAK2 V617F mutation occurs in hematopoietic stem cells in polycythemia vera and predisposes toward erythroid differentiation. Proc Natl Acad Sci U S A. 2006;103:6224–9.
3. Stein BL, Oh ST, Berenzon D, Hobbs GS, Kremyanskaya M, Rampal RK, et al. Polycythemia vera: an appraisal of the biology and management 10 years after the discovery of JAK2 V617F. J Clin Oncol. 2015;33:3953–60.
4. Titmarsh GJ, Duncombe AS, Mcmullin MF, O'rorke M, Mesa R, De Vocht F, et al. How common are myeloproliferative neoplasms? A systematic review and meta-analysis. Am J Hematol. 2014;89:581–7.
5. Baxter EJ, Scott LM, Campbell PJ, East C, Fourouclas N, Swanton S, et al. Acquired mutation of the tyrosine kinase JAK2 in human myeloproliferative disorders. Lancet. 2005;365:1054–61.
6. James C, Ugo V, Le Couedic JP, Staerk J, Delhommeau F, Lacout C, et al. A unique clonal JAK2 mutation leading to constitutive signalling causes polycythaemia vera. Nature. 2005;434:1144–8.
7. Kralovics R, Passamonti F, Buser AS, Teo SS, Tiedt R, Passweg JR, et al. A gain-of-function mutation of JAK2 in myeloproliferative disorders. N Engl J Med. 2005;352:1779–90.
8. Levine RL, Wadleigh M, Cools J, Ebert BL, Wernig G, Huntly BJ, et al. Activating mutation in the tyrosine kinase JAK2 in polycythemia vera, essential thrombocythemia, and myeloid metaplasia with myelofibrosis. Cancer Cell. 2005;7:387–97.
9. Tefferi A, Thiele J, Orazi A, Kvasnicka HM, Barbui T, Hanson CA, et al. Proposals and rationale for revision of the World Health Organization diagnostic criteria for polycythemia vera, essential thrombocythemia, and primary myelofibrosis: recommendations from an ad hoc international expert panel. Blood. 2007;110:1092–7.
10. Tefferi A, Thiele J, Vardiman JW. The 2008 World Health Organization classification system for myeloproliferative neoplasms: order out of chaos. Cancer. 2009;115:3842–7.
11. Stein BL, Saraf S, Sobol U, Halpern A, Shammo J, Rondelli D, et al. Age-related differences in disease characteristics and clinical outcomes in polycythemia vera. Leuk Lymphoma. 2013;54:1989–95.
12. Akada H, Akada S, Hutchison RE, Sakamoto K, Wagner KU, Mohi G. Critical role of Jak2 in the maintenance and function of adult hematopoietic stem cells. Stem Cells. 2014;32:1878–89.
13. Witthuhn BA, Quelle FW, Silvennoinen O, Yi T, Tang B, Miura O, et al. JAK2 associates with the erythropoietin receptor and is tyrosine phosphorylated and activated following stimulation with erythropoietin. Cell. 1993;74:227–36.
14. Scott LM, Tong W, Levine RL, Scott MA, Beer PA, Stratton MR, et al. JAK2 exon 12 mutations in polycythemia vera and idiopathic erythrocytosis. N Engl J Med. 2007;356:459–68.
15. Lanikova L, Babosova O, Swierczek S, Wang L, Wheeler DA, Divoky V, et al. Coexistence of gain-of-function JAK2 germline mutations with JAK2V617F in polycythemia vera. Blood. 2016. doi:10.1182/blood-2016-04-711283.
16. Hemminki K, Jiang Y. Familial polycythemia vera: results from the Swedish Family-Cancer Database. Leukemia. 2001;15:1313–5.
17. Kralovics R, Stockton DW, Prchal JT. Clonal hematopoiesis in familial polycythemia vera suggests the involvement of multiple mutational events in the early pathogenesis of the disease. Blood. 2003;102:3793–6.
18. Cario H, Goerttler PS, Steimle C, Levine RL, Pahl HL. The JAK2V617F mutation is acquired secondary to the predisposing alteration in familial polycythaemia vera. Br J Haematol. 2005;130:800–1.
19. Jones AV, Chase A, Silver RT, Oscier D, Zoi K, Wang YL, et al. JAK2 haplotype is a major risk factor for the development of myeloproliferative neoplasms. Nat Genet. 2009;41:446–9.
20. Kilpivaara O, Mukherjee S, Schram AM, Wadleigh M, Mullally A, Ebert BL, et al. A germline JAK2 SNP is associated with predisposition to the development of JAK2(V617F)-positive myeloproliferative neoplasms. Nat Genet. 2009;41:455–9.
21. Olcaydu D, Harutyunyan A, Jager R, Berg T, Gisslinger B, Pabinger I, et al. A common JAK2 haplotype confers susceptibility to myeloproliferative neoplasms. Nat Genet. 2009;41:450–4.

22. Trifa AP, Cucuianu A, Petrov L, Urian L, Militaru MS, Dima D, et al. The G allele of the JAK2 rs10974944 SNP, part of JAK2 46/1 haplotype, is strongly associated with JAK2 V617F-positive myeloproliferative neoplasms. Ann Hematol. 2010;89:979–83.

23. Oddsson A, Kristinsson SY, Helgason H, Gudbjartsson DF, Masson G, Sigurdsson A, et al. The germline sequence variant rs2736100_C in TERT associates with myeloproliferative neoplasms. Leukemia. 2014;28:1371–4.

24. Saliba J, Saint-Martin C, Di Stefano A, Lenglet G, Marty C, Keren B, et al. Germline duplication of ATG2B and GSKIP predisposes to familial myeloid malignancies. Nat Genet. 2015;47:1131–40.

25. Harutyunyan AS, Giambruno R, Krendl C, Stukalov A, Klampfl T, Berg T, et al. Germline RBBP6 mutations in familial myeloproliferative neoplasms. Blood. 2016;127:362–5.

26. Mckenna A, Hanna M, Banks E, Sivachenko A, Cibulskis K, Kernytsky A, et al. The Genome Analysis Toolkit: a MapReduce framework for analyzing next-generation DNA sequencing data. Genome Res. 2010;20:1297–303.

27. Lek M, Karczewski KJ, Minikel EV, Samocha KE, Banks E, Fennell T, et al. Analysis of protein-coding genetic variation in 60,706 humans. Nature. 2016; 536:285–91.

28. Adzhubei IA, Schmidt S, Peshkin L, Ramensky VE, Gerasimova A, Bork P, et al. A method and server for predicting damaging missense mutations. Nat Methods. 2010;7:248–9.

29. Kumar P, Henikoff S, Ng PC. Predicting the effects of coding non-synonymous variants on protein function using the SIFT algorithm. Nat Protoc. 2009;4:1073–81.

30. Li M, Collins R, Jiao Y, Ouillette P, Bixby D, Erba H, et al. Somatic mutations in the transcriptional corepressor gene BCORL1 in adult acute myelogenous leukemia. Blood. 2011;118:5914–7.

31. Tiacci E, Grossmann V, Martelli MP, Kohlmann A, Haferlach T, Falini B. The corepressors BCOR and BCORL1: two novel players in acute myeloid leukemia. Haematologica. 2012;97:3–5.

32. Rotunno G, Guglielmelli P, Biamonte F, Rumi E, Cazzola M, Vannucchi AM. Mutational analysis of BCORL1 in the leukemic transformation of chronic myeloproliferative neoplasms. Ann Hematol. 2014;93:523–4.

33. Al-Kandari W, Jambunathan S, Navalgund V, Koneni R, Freer M, Parimi N, et al. ZXDC, a novel zinc finger protein that binds CIITA and activates MHC gene transcription. Mol Immunol. 2007;44:311–21.

34. Al-Kandari W, Koneni R, Navalgund V, Aleksandrova A, Jambunathan S, Fontes JD. The zinc finger proteins ZXDA and ZXDC form a complex that binds CIITA and regulates MHC II gene transcription. J Mol Biol. 2007;369:1175–87.

35. Ramsey JE, Fontes JD. The zinc finger transcription factor ZXDC activates CCL2 gene expression by opposing BCL6-mediated repression. Mol Immunol. 2013;56:768–80.

36. Wang L, Tsai CC. Atrophin proteins: an overview of a new class of nuclear receptor corepressors. Nucl Recept Signal. 2008;6:e009.

37. Yazawa I, Nukina N, Hashida H, Goto J, Yamada M, Kanazawa I. Abnormal gene product identified in hereditary dentatorubral-pallidoluysian atrophy (DRPLA) brain. Nat Genet. 1995;10:99–103.

38. Wood JD, Nucifora Jr FC, Duan K, Zhang C, Wang J, Kim Y, et al. Atrophin-1, the dentato-rubral and pallido-luysian atrophy gene product, interacts with ETO/MTG8 in the nuclear matrix and represses transcription. J Cell Biol. 2000;150:939–48.

39. Butturini A, Gale RP, Verlander PC, Adler-Brecher B, Gillio AP, Auerbach AD. Hematologic abnormalities in Fanconi anemia: an International Fanconi Anemia Registry study. Blood. 1994;84:1650–5.

40. Wijker M, Morgan NV, Herterich S, Van Berkel CG, Tipping AJ, Gross HJ, et al. Heterogeneous spectrum of mutations in the Fanconi anaemia group A gene. Eur J Hum Genet. 1999;7:52–9.

15

Genome-wide DNA methylation analysis reveals hypomethylation in the low-CpG promoter regions in lymphoblastoid cell lines

Itsuki Taniguchi[1], Chihiro Iwaya[1], Keizo Ohnaka[2], Hiroki Shibata[1] and Ken Yamamoto[3]* (iD)

Abstract

Background: Epidemiological studies of DNA methylation profiles may uncover the molecular mechanisms through which genetic and environmental factors contribute to the risk of multifactorial diseases. There are two types of commonly used DNA bioresources, peripheral blood cells (PBCs) and EBV-transformed lymphoblastoid cell lines (LCLs), which are available for genetic epidemiological studies. Therefore, to extend our knowledge of the difference in DNA methylation status between LCLs and PBCs is important in human population studies that use these DNA sources to elucidate the epigenetic risks for multifactorial diseases. We analyzed the methylation status of the autosomes for 192 and 92 DNA samples that were obtained from PBCs and LCLs, respectively, using a human methylation 450 K array. After excluding SNP-associated methylation sites and low-call sites, 400,240 sites were subjected to analysis using a generalized linear model with cell type, sex, and age as the independent variables.

Results: We found that the large proportion of sites showed lower methylation levels in LCLs compared with PBCs, which is consistent with previous reports. We also found that significantly different methylation sites tend to be located on the outside of the CpG island and in a region relatively far from the transcription start site. Additionally, we observed that the methylation change of the sites in the low-CpG promoter region was remarkable. Finally, it was shown that the correlation between the chronological age and ageing-associated methylation sites in *ELOVL2* and *FHL2* in the LCLs was weaker than that in the PBCs.

Conclusions: The methylation levels of highly methylated sites of the low-CpG-density promoters in PBCs decreased in the LCLs, suggesting that the methylation sites located in low-CpG-density promoters could be sensitive to demethylation in LCLs. Despite being generated from a single cell type, LCLs may not always be a proxy for DNA from PBCs in studies of epigenome-wide analysis attempting to elucidate the role of epigenetic change in disease risks.

Keywords: DNA methylation, Lymphoblastoid cell lines, Epigenome-wide analysis, Epigenetic epidemiology, Human methylation array

Background

The DNA obtained from EBV-transformed immortalized lymphoblastoid cell lines (LCLs) and peripheral blood cells (PBCs) are commonly used in medical genetic studies. LCLs can be generated from both healthy individuals and patients and supply an unlimited source of genomic DNA. Additionally, LCLs and PBCs have been successfully used for gene expression analyses [1].

DNA methylation is one of the important epigenetic mechanisms regulating gene expression. In addition to sequence variants, it is increasingly accepted that this DNA modification may be implicated in the susceptibility of various multifactorial diseases [2–4]. Recent developments in technology for human genome analysis have enabled us to identify disease-related DNA methylation changes at the genome-wide level. Because it is essential

* Correspondence: yamamoto_ken@med.kurume-u.ac.jp
[3]Department of Medical Biochemistry, Kurume University School of Medicine, 67 Asahi-machi, Kurume, Fukuoka 830-0011, Japan
Full list of author information is available at the end of the article

to use relatively large samples in searching for genes that are susceptible to multifactorial diseases, the DNA sources are limited to LCLs, PBCs, and saliva. However, it is known that DNA methylation status varies between cell types [5]. Therefore, to extend our knowledge of the difference in DNA methylation status between LCLs and PBCs is important in human population studies that use these DNA sources to elucidate the epigenetic risks for multifactorial diseases.

To this end, we designed experiments to compare the DNA methylation status between LCLs and PBCs at an epigenome-wide level using approximately 400,000 methylation data sites from 92 LCL and 192 PBC samples obtained using the Human Methylation 450 K array. We analyzed global differences in methylation profiles and the degree of difference in methylation level of each site in terms of location (inside or outside the CpG island, the distance from transcription start site and promoter type) between LCLs and PBCs. Additionally, the association strength of methylation levels at the ageing-related methylation sites in *FHL2* and *ELOVL2* with chronological age was compared between LCLs and PBCs.

Methods

Subjects

EBV-transformed LCLs derived from 92 healthy Japanese subjects were provided by the Riken Bioresource Center Cell Bank [6]. PBCs were obtained from 192 participants of a baseline survey of the general population from a Fukuoka-based cohort study [7, 8]. This study was performed in accordance with the principles of the Declaration of Helsinki and was approved by the Institutional Review Board at Kyushu University.

DNA methylation chip assay

Genomic DNA was bisulfite-treated using the EZ-96 DNA Methylation Kit (Zymo Research Corporation, Orange, CA), which combines bisulfite conversion and DNA cleanup in a 96-well plate. Genome-wide DNA methylation profiles were obtained using the Illumina HumanMethylation450 BeadChip (Illumina, San Diego, CA) according to the manufacturer's instructions. The GenomeStudio V2011.1 (Methylation Module version 1.9.0) was employed to determine the beta values that reflected the estimated methylation level for each CpG site. The beta value was calculated as: Max(-signal for methylation, 0)/[Max(signal for methylation, 0) + Max(signal for unmethylation, 0) + 100]. Using this metric, the DNA methylation level was represented by a number between 0 (no methylation) and 1 (complete methylation). The signal intensities were

normalized to the internal controls and background prior to beta value calculation.

Selection and classification of DNA methylation sites

Among 473,864 methylation sites on the autosomes, 1305 sites showing low calls (<0.95) were removed for further analyses. To eliminate SNP-associated methylation sites, we screened the nearest SNP for each methylation site using the dbSNP135 database (SNPs categorized in weight = 1 group, http://www.ncbi.nlm.nih.gov/SNP/). We found 72,318 sites in which SNPs were located on the C or G site. Additionally, one methylation site demonstrated an outlier value. After removing these sites; 400,240 methylation sites on the array were available for further analyses. Based on the CpG Islands (CGI) track of the UCSC table browser of the UCSC Genome Bioinformatics database (http://genome.ucsc.edu/index.html), the 400,240 sites on autosomes were classified into two groups, CGI-sites (135,674 sites, inside of CGI) or non-CGI-sites (264,566 sites, outside the CGI). Among the non-CGI sites, 95,625 sites were located near CGI (±2,000 bases) that were classified in a shore group. The distance between the methylation site and the nearest transcription start site (TSS) was calculated using the NCBI RefSeq database. The physical positions on the human genome were based on the Genome Reference Consortium Human Build 37 (GRCh37, http://www.ncbi.nlm.nih.gov/assembly/). Of 400,240 probes, 159,688 demonstrated a TSS between −500 bases and +2,000 bases; among these, 85,700 sites could be classified into high-CpG-density promoters (HCP), intermediate-CpG-density promoters (ICP) and low-CpG-density promoters (LCP), as reported by Mikkelsen et al. [9] (69,836, 10,719, and 5145 in HCP, ICP, and LCP, respectively).

Statistical analysis

To evaluate the difference in methylation level of each site, the data were analyzed using modeling individual Illumina beta values using a generalized linear model (glm) with cell type (LCLs or PBCs), age and sex as the independent variables. P values and the difference in methylation level for each cell type were obtained. The statistical power to detect methylation differences of 0.25 and 0.5 between 192 PBCs and 92 LCLs was estimated to be 50.2 and 97.5%, respectively at a significance level of $P = 0.05$ using G*Power 3.1 software [10]. A principal component analysis (PCA) was performed using the beta values for the 400,240 sites, and the first and second principal component scores for each sample were plotted. The regression analysis was performed using the chronological age of the subjects and the beta values of cg06639320 and cg16867657 for *FHL2* and *ELOVL2*, respectively, with adjustments for sex. These analyses were performed using R (release 2.15.2).

Results

Comparison of global DNA methylation profiles between LCLs and PBCs

To assess the global difference of DNA methylation levels between LCLs and PBCs, we performed a PCA using the methylation data of 400,240 sites on autosomes obtained using the 450 K methylation array. As shown in Fig. 1a, the LCL and PBC groups were clearly distinguished by their first principal component score. Additionally, the PBC samples were distributed within a narrow range, whereas the LCL samples showed a relatively wide range in the second principal component score. These results suggest that there is a global difference in DNA methylation levels between these cell types and that the levels are more diverse in LCLs than in PBCs.

We then examined the difference in methylation level for each site using a glm adjusted for age and sex. As shown in the volcano plot in Fig. 1b, the sites showing lower levels in LCL than in PBC were predominant (low-met-LCL group). The 138,871 sites (34.7% of the total) showed $-\log_{10}(P$ value$) > 10$; among these sites, 85.1% were in the low-met-LCL group. This inclination was observed in each autosome (Additional file 1: Figure S1). Therefore, it was suggested that the main difference in DNA methylation between LCLs and PBCs was hypomethylation in the LCLs and that the change in methylation levels occurred globally in the autosomes.

Hypomethylation observed in the LCLs occurs at sites outside the GpG island

We next assessed the distribution of the difference in methylation levels between LCLs and PBLs in terms of the location of the site (inside or outside the CpG island) (named CGI-site or non-CGI-site). As shown in Fig. 2a,

the distribution of difference was dissimilar between them; the proportion of the sites showing a low P value was larger in the non-CGI-site group (black solid line) than in the CGI-site group (black dashed line). This trend was apparent in the low-met-LCL group (compare the red solid and dashed lines), whereas a dissimilarity of distribution was not observed in the high-met-LCL group (compare the blue solid and dashed lines). These results prompted us to further classify the non-CGI-sites into shore or non-shore groups because the CGI shores were suggested to contribute to tissue-specific DNA methylation [11, 12]. However, we did not find significant differences in the distribution between the shore and non-shore group of the low-met-LCL (Fig. 2b). Taken together, these results suggested that the majority of hypomethylation observed in the LCLs occurred at sites outside the CGIs regardless of shores.

Comparison of the difference in DNA methylation levels observed among LCLs and PBCs in terms of distance from the transcription start site

We further examined the relationship between the distance from the TSS and the difference in DNA methylation levels observed among LCLs and PBCs. We plotted $-\log_{10}(P$ value$)$ for each site against the distance from the nearest TSS (shown in gray dots in Fig. 3a) and indicated a proportion of the site showing $-\log_{10}(P$ value$) > 10$, 25, and 50 in blue, green, and pink dots, respectively (Fig. 3a). The proportion was calculated by dividing the number of the sites meeting the P value criteria by the total number of sites within ±50 bases of window size. We found that the proportion of significantly different sites was lower near the TSS. For instance, approximately 25% of the sites near the TSS

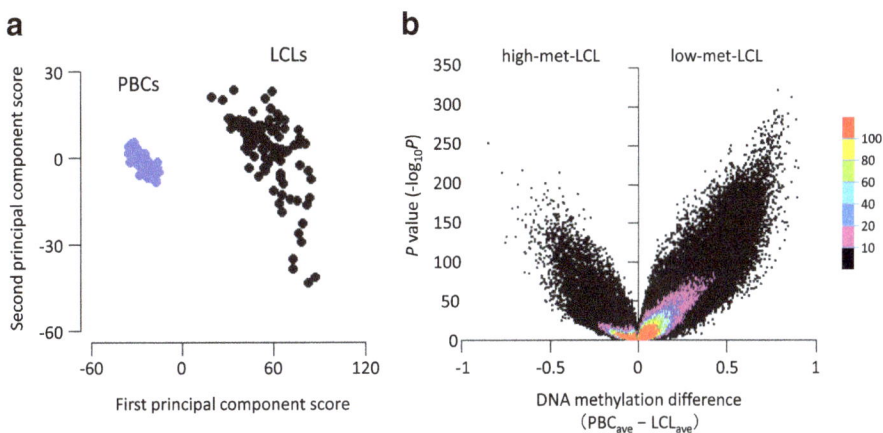

Fig. 1 Global difference in the DNA methylation level between the LCLs and PBCs. **a** Principal component analysis (PCA) plot. PCA was performed using the methylation level of the 400,240 sites on autosomes. The LCL and PBC samples are shown in *black* and *blue dots*, respectively. **b** Volcano plot with the difference of the average of DNA methylation level on the *x*-axis and the *P* value ($-\log_{10}P$) obtained via glm analysis on the *y*-axis. Each *color* shows the dot density ($100 < n$, $80 < n \leq 100$, $60 < n \leq 80$, $40 < n \leq 60$, $20 < n \leq 40$, $10 < n \leq 20$ and $n \leq 10$ per unit area (0.002×1 for *x* and *y*-axis, respectively) in *red, yellow, green, sky blue, blue, pink,* and *black*, respectively)

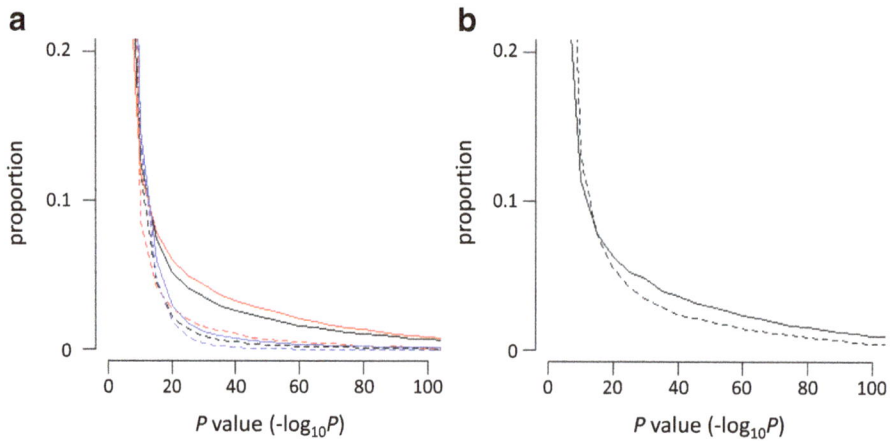

Fig. 2 Distribution of the differences in methylation levels between LCLs and PBLs in terms of CGI. **a** The proportion of *P* values obtained from non-CGI and CGI sites in all samples (*black solid* and *dashed lines*, respectively), non-CGI and CGI sites in the low-met-LCL group (*red solid* and *dashed lines*, respectively), and non-CGI and CGI sites in the high-met-LCL group (*blue solid* and *dashed lines*, respectively) are indicated. **b** The proportion of *P* values obtained from the non-shore and shore sites (*solid* and *dashed lines*, respectively) in the non-CGI sites of the low-met-LCL group are indicated

showed $-\log_{10}(P$ value$) > 10$, whereas this proportion increased to approximately 45% for the sites located approximately ±1000 bases from the TSS in the low-met-LCL group (blue dots, left panel of Fig. 3a). This trend was also observed even in the lower *P* value threshold group (green and pink dots) and in the high-met-LCL group (right panel of Fig. 3a). We then analyzed the sites showing $-\log_{10}(P$-value$) > 10$ separately for CGI- and non-CGI-site groups. As shown in Fig. 3b, the proportion of non-CGI-sites near the TSS was high in both the low- and high-met-LCL groups (red and blue dots, respectively, Fig. 3b). However, the lowest proportion was observed near the TSS in the case of CGI-sites (pink and sky blue dots for low- and high-met-LCL groups, respectively, Fig. 3b). These results

suggested that the low CpG promoter would show a more significant difference in DNA methylation levels than the high CpG promoter.

The methylation sites located in low CpG promoters could be sensitive to demethylation in LCLs

To assess whether the promoter type affects the difference in DNA methylation levels between LCLs and PBCs, the methylation sites located in HCP, LCP and ICP were extracted based on the data set of Mikkelsen et al. [9] (69,836, 10,719, and 5,145, in HCP, ICP, and LCP, respectively), and analyzed the distribution of $-\log_{10}(P$ value$)$ in all, low- and high-met-LCL groups (Fig. 4). It was shown that the proportion of differentially methylated sites was higher in the LCPs than the HCPs.

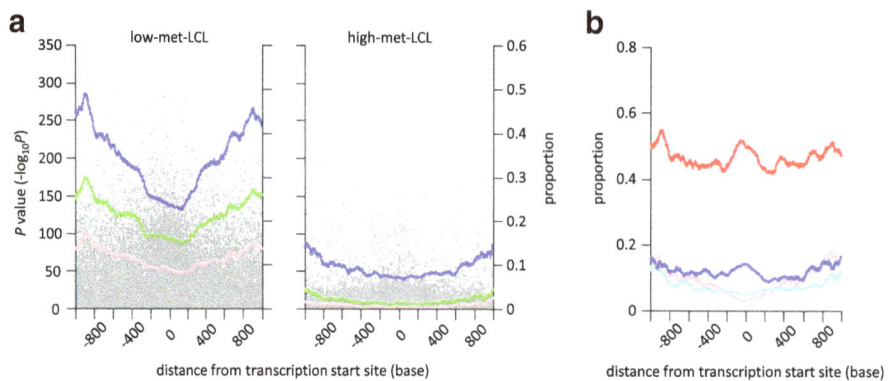

Fig. 3 Distribution of the differences in methylation levels between LCLs and PBLs in terms of TSS. **a** *P* values were plotted against the distance from the nearest TSS (*gray dots*). The proportion of the sites with *P* values ($-\log_{10}P$) greater than 10 (*blue dots*), 25 (*green dots*), and 50 (*pink dots*) in a window size of ±50 bases were plotted. **b** The proportion of the sites with *P* values ($-\log_{10}P$) greater than 10 obtained from non-CGI and CGI sites in the low-met-LCL group (*red* and *pink dots*, respectively), and from non-CGI and CGI sites in the high-met-LCL group (*blue* and *sky blue dots*, respectively) in a window size of ±50 bases were plotted against TSS

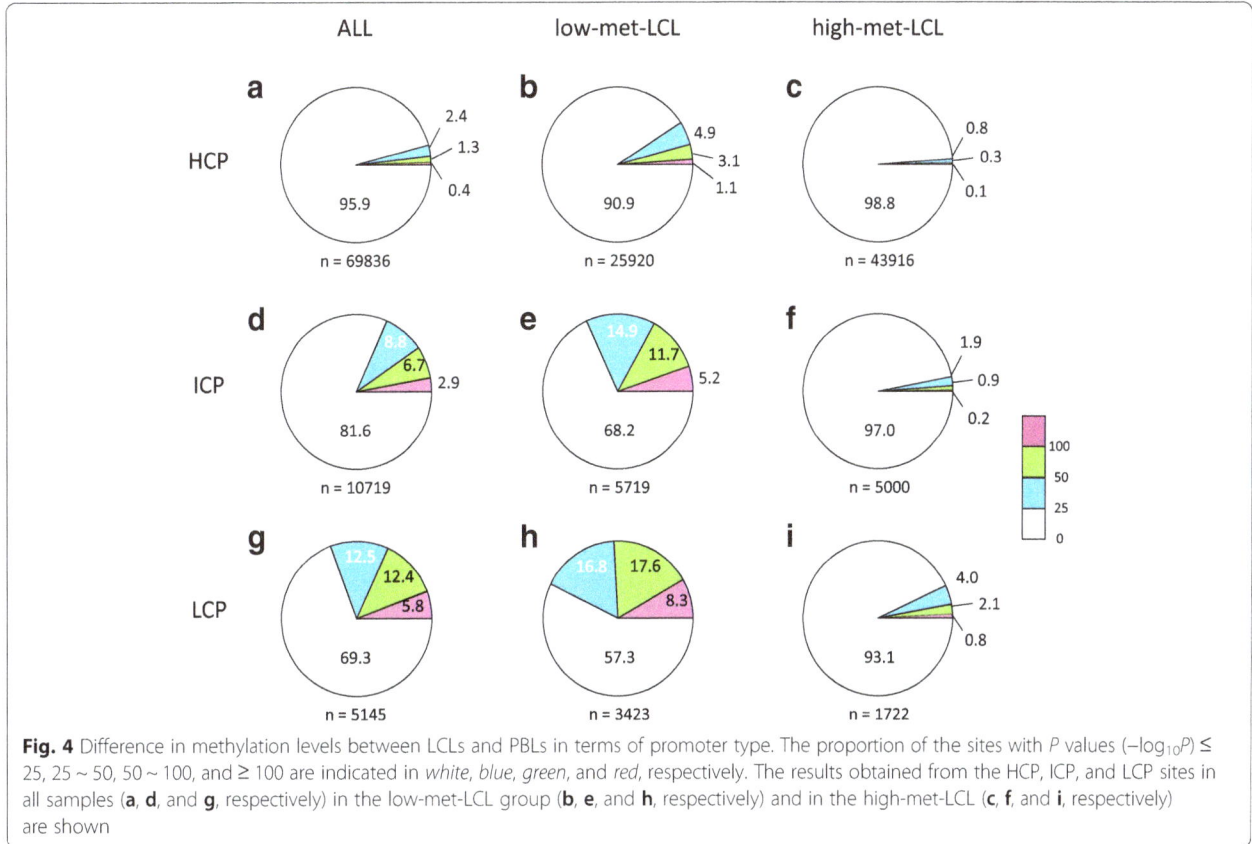

Fig. 4 Difference in methylation levels between LCLs and PBLs in terms of promoter type. The proportion of the sites with P values ($-\log_{10}P$) ≤ 25, 25 ~ 50, 50 ~ 100, and ≥ 100 are indicated in *white*, *blue*, *green*, and *red*, respectively. The results obtained from the HCP, ICP, and LCP sites in all samples (**a**, **d**, and **g**, respectively) in the low-met-LCL group (**b**, **e**, and **h**, respectively) and in the high-met-LCL (**c**, **f**, and **i**, respectively) are shown

In the LCPs, the proportion of the sites showing − $\log_{10}(P$ value$) > 25$ was 30.7%, whereas that in HCPs was 4.1% in all sites (compare Fig. 4a, g). This was more pronounced in the low-met-LCL group (compare Fig. 4b, c, h, i). The sites located in ICPs showed intermediate values between HCPs and LCPs (Fig. 4d–f). These results suggested that the methylation sites located in low CpG promoters could be sensitive to demethylation in LCLs.

To further assess promoter type differences, we compared the HCPs and LCPs methylation level profiles. As shown in Fig. 5, nearly half of the sites in LCPs showed more than 0.6 methylation levels, whereas almost all sites in HCPs were hypomethylated in PBCs. Additionally, it was observed that the methylation levels of highly methylated sites of the LCPs decreased in the LCLs. Therefore, we concluded that highly methylated sites of LCPs caused the difference in DNA methylation levels observed between HCPs and LCPs, especially in the low-met-LCL group.

Comparison between LCLs and PBCs regarding the association between ageing-related CpG sites and chronological age

Using DNA obtained from PBCs, it has been reported that the methylation levels of several CpG sites are associated with chronological age. However, it remains unclear whether LCLs should be utilized for studies on epigenetic ageing biomarkers. To address this issue, we performed a regression analysis for chronological age and known ageing-related CpG sites located in *FHL2* and *ELOVL2* [13, 14]. *FHL2* encodes a member of the four-and-a-half-LIM-only protein family which is suggested to have a role in the assembly of extracellular membranes and in transformation of normal myoblasts to rhabdomyosarcoma cells (OMIN 602633). *ELOVL2* encodes an enzyme which catalyzes the first and rate-limiting reaction of the long-chain fatty acids elongation cycle (OMIM 611814). As shown in Fig. 6a, the methylation level of the PBCs was highly correlated with chronological age (blue dots, $P = 1.7\text{E-}18$ and $r^2 = 0.33$ for *FHL2*, $P = 3.1\text{E-}25$ and $r^2 = 0.44$ for *ELOVL2*). In contrast, the methylation level of the LCLs was varied and the association was weak (black dots, $P = 0.04$ and $r^2 = 0.05$ for *FHL2*, $P = 1.9\text{E-}5$ and $r^2 = 0.18$ for *ELOVL2*). Therefore, these results suggest that DNA obtained from LCLs may not always be an alternative to DNA from PBCs.

Discussion

In this study, we used a 450 K methylation array to investigate the methylation differences between LCLs and PBCs, which are commonly used in genetic epidemiological

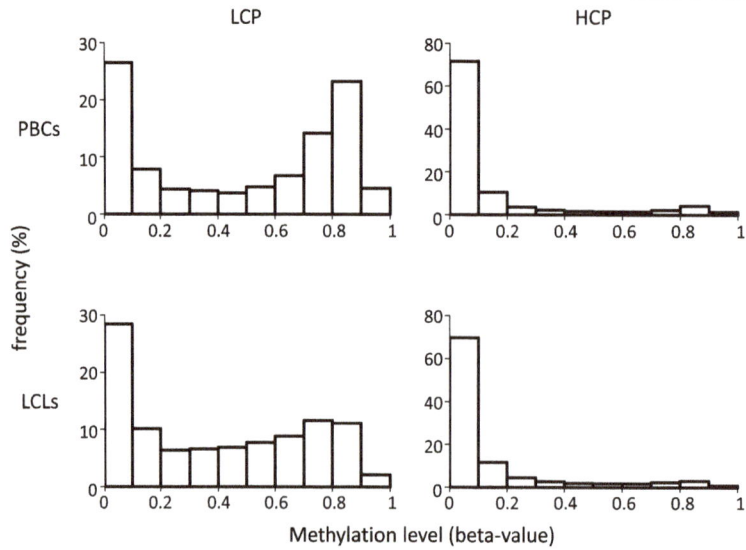

Fig. 5 Distribution of the methylation levels of the sites in the LCPs and HCPs. The frequency in the LCLs and PBCs are shown separately

studies. In all genomes, the majority of the sites in the LCLs showed lower methylation levels than those of the PBCs, and these sites were primarily located in non-CGI regions. Additionally, we found that differentially methylated sites were predominantly located in the LCP region.

Although a relatively small sample number and number of methylation sites were analyzed, previous studies showed that methylation status in LCLs is different from that of PBCs and that the methylation level in LCLs is lower than that of PBCs in the majority of sites [15–20].

Because a large number of samples and more sites were examined, we could investigate the differences in methylation levels between LCLs and PBCs in terms of CGI location, distance from TSS and promoter type as characterized by CG density. We found that a fraction showing a significant difference in methylation level between the LCLs and PBCs was observed near the TSS in the non-CGI sites but not in the CGI sites. This result suggests that the difference in the methylation level of these cell types would be high in the genes in which the promoter shows a low GC content.

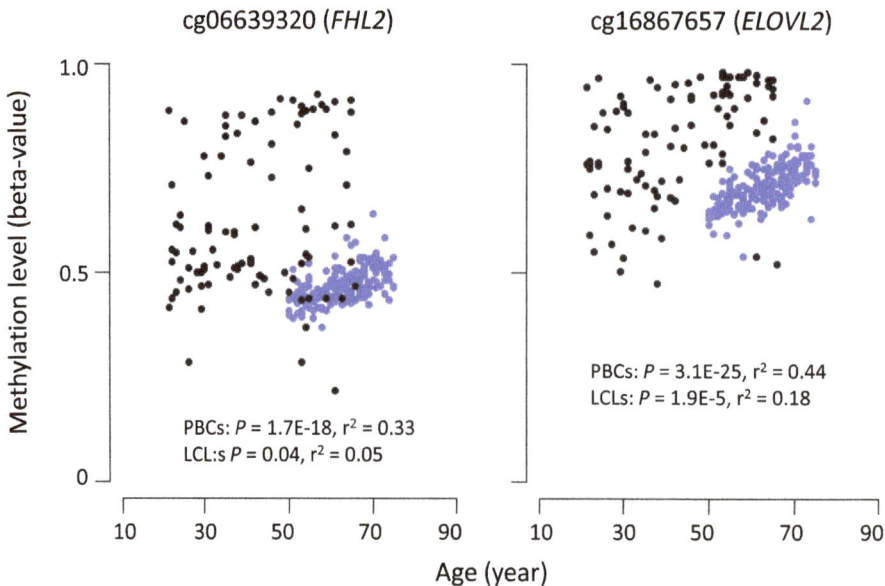

Fig. 6 Regression analyses of the methylation levels and chronological age at the *FHL2* and *ELOVL2* loci. The methylation levels in the LCLs (*black dots*) and PBCs (*blue dots*) were plotted against the age of the donors at the time of providing the specimens. The *P* values and r^2 were obtained by correcting for sex

Genome-wide DNA methylation analysis reveals hypomethylation in the low-CpG promoter regions...

139

We found that significantly different methylation sites were predominant in LCPs but not in HCPs. It has been demonstrated that LCPs are generally associated with tissue-specific genes, whereas HCPs are associated with two classes of genes, including ubiquitous "housekeeping" genes and highly regulated "key developmental" genes [9, 21, 22]. Therefore, our results suggest that the methylation sites located in promoters classified as LCP could have a functional role in distinguishing between LCLs and PBCs by regulating the corresponding gene expression.

The epigenome-wide association studies using human population samples to identify the disease risk loci and epigenomes that are affected by intrinsic or extrinsic factors, such as ageing and smoking, have been progressing [13, 14, 23, 24]. We evaluated the differences in association strength between well-known ageing methylation sites and the chronological age of the samples between LCLs and PBCs and found that the correlation was more significant in PBCs than LCLs. This was due to a larger variance of methylation levels in LCLs than in PBCs. In addition to the differences in cell type, artificial experimental processes, including in vitro culture, culture period and culture freezing, and thawing could cause the large variances in data observed in the LCLs. Therefore, we concluded that DNA obtained from LCLs may not always be a proxy for DNA from PBCs in studies of epigenome-wide analysis attempting to elucidate the role of epigenetic change in disease risks.

Conclusion

There is a global difference in DNA methylation levels between LCLs and PBCs, and the main difference was hypomethylation in the LCLs. The methylation levels of highly methylated sites of the low-CpG-density promoters in PBCs decreased in the LCLs, suggesting that the methylation sites located in low-CpG-density promoters could be sensitive to demethylation in LCLs. The correlation between well-known ageing methylation sites and the chronological age of the samples was more significant in PBCs than LCLs, indicating that despite being generated from a single cell type, LCLs may not always be a proxy for DNA from PBCs in studies of epigenome-wide analysis attempting to elucidate the role of epigenetic change in disease risks.

Abbreviations

CGI: CpG Island; ELOVL2: Elongation of very long chain fatty acids protein 2; FHL2: Four and a half LIM domains 2; Glm: Generalized linear model; HCP: High-CpG-density promoter; ICP: Intermediate-CpG-density promoter; LCL: Lymphoblastoid cell line; LCP: Low-CpG-density promoter; PBC: Peripheral blood cell; PCA: Principal component analysis; TSS: Transcription start site

Acknowledgements
We thank all of the people who have continuously supported the population-based cohort study, the Kyushu University Fukuoka Cohort Study. We also thank Ms. Miki Sonoda for her technical assistance.

Funding
This work was supported by KAKENHI Grant Number 15 K08290 from the Japan Society for the Promotion of Science.

Authors' contributions
KO performed the sample collection KO. KY performed the DNA methylation chip experiments. IT, CI, and KY performed the statistical and bioinformatics analyses. HS supervised the research. All authors wrote and approved the manuscript.

Competing interests
The authors declare that they have no competing interests.

Consent for publication
Not applicable.

Author details
[1]Division of Genomics, Medical Institute of Bioregulation, Kyushu University, 3-1-1 Maidashi, Higashi-ku, Fukuoka 812-8582, Japan. [2]Department of Geriatric Medicine, Graduate School of Medical Sciences, Kyushu University, 3-1-1 Maidashi, Higashi-ku, Fukuoka 812-8582, Japan. [3]Department of Medical Biochemistry, Kurume University School of Medicine, 67 Asahi-machi, Kurume, Fukuoka 830-0011, Japan.

References
1. Powell JE, Henders AK, McRae AF, Wright MJ, Martin NG, Dermitzakis ET, Montgomery GW, Visscher PM. Genetic control of gene expression in whole blood and lymphoblastoid cell lines is largely independent. Genome Res. 2012;22:456–66.
2. Ordovás JM, Smith CE. Epigenetics and cardiovascular disease. Nat Rev Cardiol. 2010;7:510–19.
3. Costenbader KH, Gay S, Alarcón-Riquelme ME, Iaccarino L, Doria A. Genes, epigenetic regulation and environmental factors: which is the most relevant in developing autoimmune diseases? Autoimmun Rev. 2012;11:604–9.
4. Keating ST, El-Osta A. Epigenetic changes in diabetes. Clin Genet. 2013;84:1–10.
5. Ziller MJ, Gu H, Müller F, Donaghey J, Tsai LT, Kohlbacher O, De Jager PL, Rosen ED, Bennett DA, Bernstein BE, et al. Charting a dynamic DNA methylation landscape of the human genome. Nature. 2013;500:477–81.
6. Iwakawa M, Goto M, Noda S, Sagara M, Yamada S, Yamamoto N, Kawakami Y, Matsui Y, Miyazawa Y, Yamazaki H, et al. DNA repair capacity measured by high throughput alkaline comet assays in EBV-transformed cell lines and peripheral blood cells from cancer patients and healthy volunteers. Mutat Res. 2005;588:1–6.
7. Nanri A, Yoshida D, Yamaji T, Mizoue T, Takayanagi R, Kono S. Dietary patterns and C-reactive protein in Japanese men and women. Am J Clin Nutr. 2008;87:1488–96.
8. Yoshida D, Toyomura K, Fukumoto J, Ueda N, Ohnaka K, Adachi M, Takayanagi R, Kono S. Waist circumference and cardiovascular risk factors in Japanese men and women. J Atheroscler Thromb. 2009;16:431–41.
9. Mikkelsen TS, Ku M, Jaffe DB, Issac B, Lieberman E, Giannoukos G, Alvarez P, Brockman W, Kim TK, Koche RP, et al. Genome-wide maps of chromatin state in pluripotent and lineage-committed cells. Nature. 2007;448:553–60.
10. Faul F, Erdfelder E, Lang AG, Buchner A. G*Power 3: a flexible statistical power analysis program for the social, behavioral, and biomedical sciences. Behav Res Methods. 2007;39:175–91.
11. Irizarry RA, Ladd-Acosta C, Wen B, Wu Z, Montano C, Onyango P, Cui H, Gabo K, Rongione M, Webster M, et al. The human colon cancer methylome shows similar hypo- and hypermethylation at conserved tissue-specific CpG island shores. Nat Genet. 2009;41:178–86.
12. Doi A, Park IH, Wen B, Murakami P, Aryee MJ, Irizarry R, Herb B, Ladd-Acosta C, Rho J, Loewer S, et al. Differential methylation of tissue- and cancer-

specific CpG island shores distinguishes human induced pluripotent stem cells, embryonic stem cells and fibroblasts. Nat Genet. 2009;41:1350–53.

13. Garagnani P, Bacalini MG, Pirazzini C, Gori D, Giuliani C, Mari D, Di Blasio AM, Gentilini D, Vitale G, Collino S, et al. Methylation of ELOVL2 gene as a new epigenetic marker of age. Aging Cell. 2012;11:1132–4.

14. Hannum G, Guinney J, Zhao L, Zhang L, Hughes G, Sadda S, Klotzle B, Bibikova M, Fan JB, Gao Y, et al. Genome-wide methylation profiles reveal quantitative views of human aging rates. Mol Cell. 2013;49:359–67.

15. Brennan EP, Ehrich M, Brazil DP, Crean JK, Murphy M, Sadlier DM, Martin F, Godson C, McKnight AJ, van den Boom D, et al. Comparative analysis of DNA methylation profiles in peripheral blood leukocytes versus lymphoblastoid cell lines. Epigenetics. 2009;4:159–64.

16. Sun YV, Turner ST, Smith JA, Hammond PI, Lazarus A, Van De Rostyne JL, Cunningham JM, Kardia SL. Comparison of the DNA methylation profiles of human peripheral blood cells and transformed B-lymphocytes. Hum Genet. 2010;127:651–8.

17. Grafodatskaya D, Choufani S, Ferreira JC, Butcher DT, Lou Y, Zhao C, Scherer SW, Weksberg R. EBV transformation and cell culturing destabilizes DNA methylation in human lymphoblastoid cell lines. Genomics. 2010;95:73–83.

18. Sugawara H, Iwamoto K, Bundo M, Ueda J, Ishigooka J, Kato T. Comprehensive DNA methylation analysis of human peripheral blood leukocytes and lymphoblastoid cell lines. Epigenetics. 2011;6:508–15.

19. Åberg K, Khachane AN, Rudolf G, Nerella S, Fugman DA, Tischfield JA, van den Oord EJ. Methylome-wide comparison of human genomic DNA extracted from whole blood and from EBV-transformed lymphocyte cell lines. Eur J Hum Genet. 2012;20:953–5.

20. Thompson TM, Sharfi D, Lee M, Yrigollen CM, Naumova OY, Grigorenko EL. Comparison of whole-genome DNA methylation patterns in whole blood, saliva, and lymphoblastoid cell lines. Behav Genet. 2013;43:168–76.

21. Weber M, Hellmann I, Stadler MB, Ramos L, Pääbo S, Rebhan M, Schübeler D. Distribution, silencing potential and evolutionary impact of promoter DNA methylation in the human genome. Nat Genet. 2007;39:457–66.

22. Koga Y, Pelizzola M, Cheng E, Krauthammer M, Sznol M, Ariyan S, Narayan D, Molinaro AM, Halaban R, Weissman SM. Genome-wide screen of promoter methylation identifies novel markers in melanoma. Genome Res. 2009;19:1462–70.

23. Breitling LP, Yang R, Korn B, Burwinkel B, Brenner H. Tobacco-smoking-related differential DNA methylation: 27 K discovery and replication. Am J Hum Genet. 2011;88:450–7.

24. Wan ES, Qiu W, Baccarelli A, Carey VJ, Bacherman H, Rennard SI, Agusti A, Anderson W, Lomas DA, Demeo DL. Cigarette smoking behaviors and time since quitting are associated with differential DNA methylation across the human genome. Hum Mol Genet. 2012;21:3073–82.

16

Early-life adversity and long-term neurobehavioral outcomes: epigenome as a bridge?

Alexander M. Vaiserman[*] and Alexander K. Koliada

Abstract

Accumulating evidence suggests that adversities at critical periods in early life, both pre- and postnatal, can lead to neuroendocrine perturbations, including hypothalamic-pituitary-adrenal axis dysregulation and inflammation persisting up to adulthood. This process, commonly referred to as biological embedding, may cause abnormal cognitive and behavioral functioning, including impaired learning, memory, and depressive- and anxiety-like behaviors, as well as neuropsychiatric outcomes in later life. Currently, the regulation of gene activity by epigenetic mechanisms is suggested to be a key player in mediating the link between adverse early-life events and adult neurobehavioral outcomes. Role of particular genes, including those encoding glucocorticoid receptor, brain-derived neurotrophic factor, as well as arginine vasopressin and corticotropin-releasing factor, has been demonstrated in triggering early adversity-associated pathological conditions. This review is focused on the results from human studies highlighting the causal role of epigenetic mechanisms in mediating the link between the adversity during early development, from prenatal stages through infancy, and adult neuropsychiatric outcomes. The modulation of epigenetic pathways involved in biological embedding may provide promising direction toward novel therapeutic strategies against neurological and cognitive dysfunctions in adult life.

Keywords: Biological embedding, DNA methylation, Early-life adversity, Epigenetics, Neurobehavioral outcomes

Background

A growing body of research in recent years highlights the importance of early-life environmental influences in determining the adult health status. On the base of these findings, the Developmental Origin of Adult Health and Disease (DOHaD) hypothesis was proposed postulating that unfavorable environmental early-life conditions can result in "developmental programming" of later-life chronic disease [1, 2]. This hypothesis has been initially focused on the lifelong outcomes of prenatal and neonatal malnutrition. In recent years, however, it became increasingly apparent that non-nutritional impacts such as psychological stress exposure during development can also greatly affect the health status throughout adult life, and epigenetic regulation is considered as a key mechanism mediating these effects.

Most research evidence for the developmental programming by stressful conditions early in life is obtained in rodent models (for reviews, see references [3–6]). In these studies, convincing evidence has been obtained to indicate that early stressful exposures such as perinatal stresses, maternal separation, and inadequate maternal care can cause marked neuroendocrine perturbations persisting up to adulthood and causing impaired cognitive, behavioral, and social functioning during the adult life. A body of studies has demonstrated that prenatal stress and exposure to excess levels of exogenous glucocorticoids can both be related to unfavorable health outcomes including low birth weight, neuroendocrine pathology, and enhanced risk for cardio-metabolic, infectious, and psychiatric disorders throughout the adult life [7–9]. The neuroendocrine effects triggered by prenatal stress have been reported to be associated with depressive- or anxiety-like behavioral phenotypes, including altered levels of physical activity, enhanced immobility throughout a forced swim test, and lowered exploration of novel environments [10]. These effects were shown to be mediated by changes in both maternal and fetal hypothalamic-pituitary-adrenal (HPA) axes causing intrauterine exposure to glucocorticoid excess [11, 12]. A role for in-utero glucocorticoid exposure

* Correspondence: vaiserman@geront.kiev.ua
Laboratory of Epigenetics, Institute of Gerontology, Vyshgorodskaya st. 67, Kiev 04114, Ukraine

induced by maternal stress in rats is evident from research in adult offspring born to either mothers with an intact corticosterone secretion or to intrauterine-stressed adrenalectomized dams [13]. The maternal stress-induced glucocorticoids can pass the placental barrier and thereby disrupt the development of the fetal brain. The stress-related maternal-placental-fetal endocrine and immune/inflammatory candidate mechanisms were proposed as possible candidate mechanisms for long-term effects of the fetal stress exposures on physiological characteristics of the developing organism [14, 15]. Early postnatal stages are another important sensitive period for developmental programming. In rodent research, it was shown that impaired mother-infant interactions (e.g., maternal deprivation/separation) throughout the postnatal period can substantially impair the neuroendocrine regulation, including upregulation of hippocampal glucocorticoid receptor (GR) and hypothalamic corticotropin-releasing factor (CRF), and also can lead to enhancement of the adrenocorticotropic hormone and corticosterone levels [16–18]. Such early adversity-induced neuroendocrine changes may lead to behavioral issues during adulthood, including the impaired learning, memory, and also depressive- and anxiety-like behaviors [19]. The long-term effects caused by variation in postnatal maternal care in rodents (e.g., low or high levels of licking and grooming, LG) are most studied in this context to date. The offspring of high-LG mothers exhibited lower levels of stress responsivity, better performance on cognitive tasks, and exploratory behavior in a novel environment during adulthood than the offspring that have been reared by low-LG dams [20, 21]. In these studies, the physiological and biochemical changes induced by adversities in early life were accompanied by substantial alterations on the level of epigenetic control of gene expression and related changes in patterns of DNA methylation, histone modifications, and microRNA regulation.

The evidence for long-lasting effects of adversities in early life from human studies are rather scarce and mainly limited to change in DNA methylation level which is thought to be the most stable form of epigenetic modification. This review is mostly focused on the results obtained from human studies highlighting the causative role of epigenetic pathways in mediating the link between the adversity in early development from prenatal stages through infancy and adult neuropsychiatric outcomes.

Search strategy

In this review, we searched the PubMed database (http://www.ncbi.nlm.nih.gov/pubmed/) to find all published studies on the epigenetic links (both at the genome-wide and candidate gene levels) between early-life adversity and long-term neurobehavioral outcomes in humans. In our search, we used combinations of the following search terms: "biological embedding," "early-life adversity," "epigenetic,"

"epigenome," "DNA methylation," "neurobehavioral," and "neuropsychiatric." The time period of the search covered articles published from 1994 to 2017 with no language restrictions, although only English language studies were eventually included. There was no restriction on the type of study design; therefore, all clinical, epidemiological, and quasi-experimental studies satisfying the search criteria were included. Several relevant experimental studies closely related to the topic under discussion were also eligible for inclusion. We used these papers to determine whether there is a coherence of effects across humans and non-human species and to examine the contribution of epigenetic mechanisms in biological embedding of adverse early-life exposures.

Biological embedding of adverse experiences in early life

Currently, the regulation of gene activity by epigenetic mechanisms (mitotically or meiotically heritable changes in gene expression that occur without any change in DNA sequence) is suggested to be a key player in mediating the link between stressful events early in life and adult neurobehavioral outcomes [22, 23]. DNA is known to maintain stability during the whole life cycle (except for mutations that occur randomly), while the epigenetic marks are dramatically changed throughout the early developmental stages to initiate distinguished patterns of expression among different developing tissues. The main mechanisms of epigenetic regulation in mammals are covalent modification of DNA by methylation, post-translational modifications (including acetylation, phosphorylation, methylation, and ubiquitination) of the histone proteins, as well as regulation by non-coding RNAs (ncRNAs) [24].

There are numerous lines of evidence indicating that the mammalian epigenome (i.e., the totality of epigenetic marks across the whole genome) is the most labile and, thereby, most sensitive to various environmental and hormonal cues, at specific stages of early development [25]. In mammals, a global demethylation of DNA followed by remethylation was shown to occur throughout the development of germ cells. A second genome-wide demethylation wave takes place in early embryogenesis, and patterns of methylation are re-established after implantation of the blastocyst [26]. The phases of post-fertilization demethylation and remethylation are likely playing a role in the removal of epigenetic information acquired by the parental generation [27, 28]. Once established throughout early development, epigenetic marks are stable maintained through cell division.

The epigenome thereby seems particularly susceptible to unfavorable environmental conditions during the stages of gametogenesis and early embryogenesis [29]. In mammals, including humans, the period of maximal epigenetic plasticity continues from before birth until weaning [30]. Various environmental cues in early life, particularly severe stresses or trauma, can cause lifelong epigenetic modifications which may, in turn, set the organism off on phenotypic trajectories to health or disease [31, 32]. There are numerous evidences from animal studies that environmental adversities and/or psychosocial stresses early in life can trigger epigenetic modifications with significant functional consequences for brain plasticity and behavior and subsequently lead to a variety of cognitive dysfunctions and psychiatric disorders in adult life [33–36]. The crucial role of epigenetic machinery in the biological embedding of stressful exposures in early life has been demonstrated in a number of rodent models, where considerable variations in both DNA methylation and histone modification have been reported in offspring exposed to different prenatal stresses, inappropriate maternal care, maternal deprivation/separation, as well as to juvenile social enrichment/isolation [3–6].

Remarkably, the earlier the organism is affected by stressful experiences during the intrauterine period, the more pronounced long-term consequences are usually observed, suggesting a causative role of epigenetic processes in pathways to adult-life pathological conditions. One good example for that is a mouse study by Mueller and Bale [37], where exposure to various stressful events during fetal development resulted in elevated stress sensitivity in adulthood, which has been manifested in modified expression of GR and corticotropin-releasing hormone, and also in the enhanced responsivity of HPA axis in prenatally affected animals. These effects have been accompanied by changes in methylation levels and in expression of GR and CRF genes. The period of early gestation was identified in this study as a particularly sensitive stage, suggesting a strong evidence for epigenetic involvement in developmental programming of neuroendocrine functions.

Postnatal exposures, such as neonatal handling (an experimental procedure in which animals are briefly separated from the dam and handled for the first 10 postnatal days), were also shown to have a profound impact on epigenetic profiles. This is a widespread experimental procedure used to understand how adversity early in life can affect neurobehavioral development of animals and place them on a pathway to disease. It has been found that neonatal handling can induce a persistent increase in the transcription level of the nuclear receptor subfamily 3, group C, member 1 (NR3C1) gene encoding the GR [38]. The epigenetic changes, such as those triggered by handling, have been observed in offspring raised by high-LG mothers. Among them, there was a reduced level of DNA methylation in the promoter region of the hippocampal NR3C1 gene [39]. These effects occurred throughout the first week of postnatal life, and they have been shown to be reversed by cross-fostering, persisted into adulthood, and associated

with changed histone acetylation and transcription factor (NGFI-A) binding to the GR promoter. Another gene that was found to be expressed in the adult prefrontal cortex in rats in response to an adversity in early life is the brain-derived neurotrophic factor (BDNF) gene playing a crucial role in the neural and behavioral plasticity and in development of various psychiatric disorders related to adversity early in life, such as depression, bipolar disorder, autism, and schizophrenia [40]. These expression changes have been accompanied by corresponding changes in the methylation levels.

Findings from human studies suggesting a role of epigenetic mechanisms in long-lasting effects of adversities in early life are more limited compared to those obtained from animal models due to the restricted access to suitable biological materials, but they clearly demonstrate that these mechanisms can also operate in man. In the succeeding subsections, findings from human studies are summarized and discussed.

Maternal adversity in pregnancy

It has been well documented in many human studies that maternal exposure to adverse conditions during pregnancy, including an unfavorable social environment, anxiety, depression, starvation, and pain (all known to increase the intrauterine level of glucocorticoids), can be linked to a variety of cognitive and behavioral problems during the adulthood [41–43]. There is consistent evidence that antenatal stress or anxiety has a programming effect on the fetus which can persist up to adulthood and result in an elevated risk of psychiatric and behavioral pathological conditions, including autism, schizophrenia and anxiety/depression-related behaviors, and also impaired cognitive performance in later life [44].

Recent studies highlighted the role of epigenetic mechanisms in mediating long-lasting outcomes of maternal adversities during pregnancy. Pre- and early postnatal dysregulation of epigenetic pathways resulted in genome-wide modulating gene expression in different tissues including the brain, by that influencing the functioning and connectivity of neural circuitry and affecting the risk for neurobehavioral impairments in later life [41]. In a methylome-wide association study (MWAS), maternal depression-associated changes in the DNA methylation levels were revealed in neonatal T lymphocytes; these alterations were found to persist to adult age in the hippocampal tissues [45]. A strong association was observed between maternal depressive symptoms during pregnancy and increased level of NR3C1 exon 1F methylation in male infants, and also lowered methylation level of another gene responsible for these associations, BDNF IV, in both male and female infants [46]. Gestational exposure to maternal depressed or anxious mood in the third trimester of prenatal development caused an enhanced methylation in the CpG-rich region of the promoter and exon1F of the GR gene (NR3C1) in the newborn cord blood, and these effects have been demonstrated to be persistent throughout the infancy [47]. Surprisingly, these epigenetic effects were revealed in offspring but not in maternal blood samples. The methylation levels of NR3C1 gene in cord blood have been associated with the levels of stress response in 3-month-old infants (as measured by salivary cortisol levels), assuming functional consequences of these epigenetic variations for the HPA stress responsiveness. Radtke et al. [48] revealed that maternal exposure to intimate partner violence throughout pregnancy affected the methylation level of NR3C1 gene in the whole blood DNA of 10–19-year-old adolescent offspring. As in the case of the maternal depression during pregnancy, these epigenetic effects were observed in affected offspring, but not in maternal blood. The same effects have been seen in cord blood as a result of maternal pregnancy-associated anxiety [49]. Similar findings were also reported on the SLC6A4 gene encoding the serotonin transporter. The methylation levels of SLC6A4 have been shown to be associated with a number of prenatal and postnatal adverse exposures, including maternal depression during pregnancy, as well as childhood trauma and abuse [50]. Prenatal exposure to a maternal depressed mood throughout the 2nd trimester of gestation has been revealed to be associated with a decreased methylation level in promoter region of SLC6A4 gene, in leukocytes from maternal peripheral blood and in umbilical cord leukocytes obtained from neonates at birth, while no such effects regarding the BDNF gene were observed [51].

In several studies, importance of epigenetic regulation of placental genes, playing a crucial role in maternal-fetal interactions, in long-lasting outcomes of maternal adversity has been reported. Prenatal exposure to maternal anxiety and/or depression has been shown to adversely influence the neurobehavioral development of newborns. These unfavorable neurodevelopmental outcomes have been demonstrated to be linked to the increased methylation levels of the NR3C1 and 11β-hydroxysteroid dehydrogenase type 2 (11β-HSD2) placental genes and to significant perturbations of the HPA axis [52]. The expression levels of the placental human SLC6A4 gene were found to be substantially elevated in placentas from mothers who had untreated mood disorders during pregnancy in comparison with control women [53]. An association between the maternal mood throughout the pregnancy and downregulation of placental 11β-HSD2 gene encoding the cortisol-metabolizing enzyme was revealed in the study by O'Donnell et al. [54]. The infants whose mothers were exposed to higher socioeconomic adversity levels in their pregnancy have also been found to have the lowest methylation levels in the placental 11β-HSD2 gene [55]. The authors have suggested that such methylation patterns of this gene indicate that

cues from environment transmitted from mother to fetus during gestation may program the response to potentially unfavorable environment in postnatal life via lesser exposure to cortisol throughout prenatal development. Overall, the findings from these investigations indicate that placental genes can be implicated in intrauterine programming of neurological functioning.

A schematic representation of hypothetical mechanisms linking maternal adversity in pregnancy to neurobehavioral and cognitive dysfunction in offspring is given in Fig. 1.

Adversity in childhood

In addition to in-utero developmental stage, early infancy is another critical stage of epigenetic plasticity. Early postnatal development is characterized by very rapid growth of various organs and organ systems including the brain and the rest of the nervous system, and epigenetic regulation is regarded as a crucial process through which the formation of specific synapses occurs throughout critical developmental periods [56, 57]. Therefore, in addition to adversities throughout the gestational development, early postnatal adversity can also have a potential for long-term epigenetic programming.

There is increasing evidence that unfavorable conditions in early infancy related to maltreatment, poor quality parenting or loss of parents, parental psychiatric disorders, exposures to physical, sexual, psychological, or emotional abuse, etc. can lead a number of adverse neurobehavioral and cognitive outcomes in adulthood [58–60]. Since the central nervous system interacts with the immune system via the HPA axis and autonomic nervous system (ANS), the immune dysregulation is regarded as a core component of these programming effects. Adversity in early life has been shown to be related to alterations in neural development (particularly of the hippocampus, amygdala, and prefrontal cortex), ANS and HPA axis dysregulation [61], and enhanced levels of inflammatory mediators [62]. The impaired neural development is believed to be a central pathway by which adversity early in life can increase the inflammation level and thereby the risk for adverse psychophysical health outcomes. Moreover, the exposure to chronic stressful conditions in infancy can lead to failure or depletion of normal physiologic processes ("allostatic load hypothesis") and thereby impair the physiological response to stress and other health outcomes in adult life through a process called biological embedding [62–65]. Adult subjects having a history of adversity in their childhood showed the decreased volumes of prefrontal cortex and hippocampus, enhanced level of activation of HPA axis in response to stress, and elevated inflammation levels compared to nonmaltreated persons [63].

Long-lasting emotional and cognitive dysfunctions caused by adversities in early life are thoroughly studied in rodent models. Typically, animals exposed to a postnatal

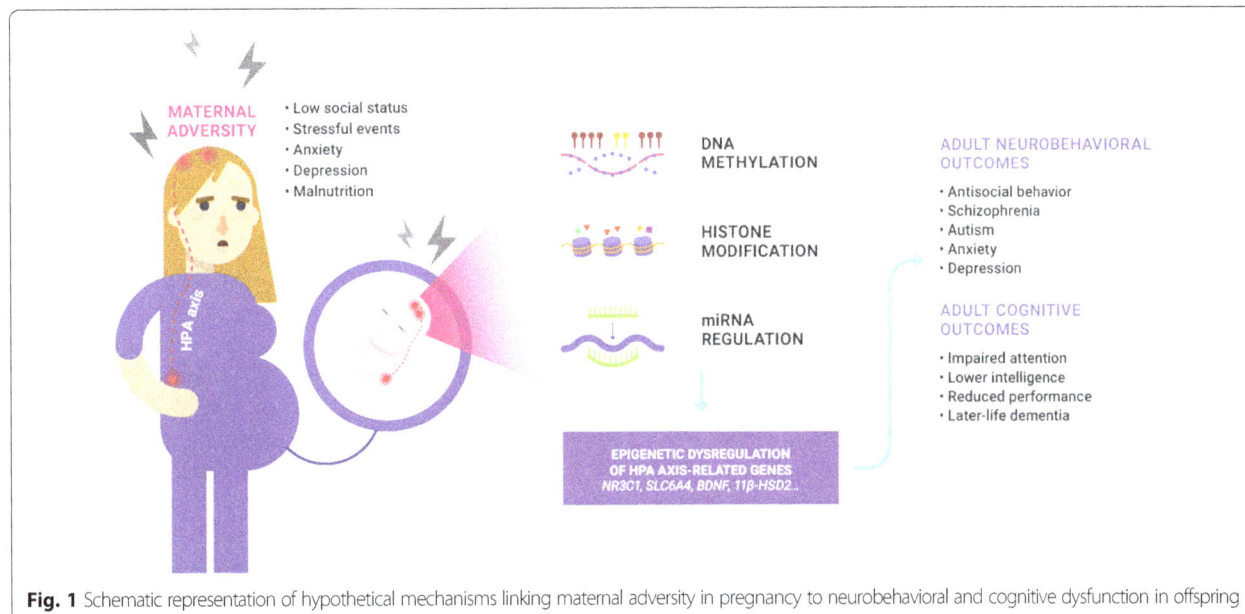

Fig. 1 Schematic representation of hypothetical mechanisms linking maternal adversity in pregnancy to neurobehavioral and cognitive dysfunction in offspring

maternal deprivation demonstrate elevated neuroendocrine response to stress, cognitive impairment, and enhanced levels of anxiety and depressive-like behavior [66–68]. Several phenotypes reported in these models of early-life adversities were likely to share common neurobiological mechanisms. So, there is evidence for impaired glucocorticoid negative-feedback control of the HPA axis, reduced hippocampal neurogenesis, and altered glutamate neurotransmission in both prenatally stressed rats and those animals that experienced inadequate maternal care [68].

These findings from animal models have been extended to humans by highlighting associations between adversities in early life and modified epigenetic patterns in adulthood [69–72]. Substantial epigenetic effects were revealed for the assortment of genes involved in the etiology of conversion disorders, aggressive and suicidal behaviors, and callous-unemotional traits [73, 74]. Among these genes, most are involved in mediating the HPA axis, brain development, immune response, neurotransmission, serotonin synthesis, and other processes. In some studies, psycho-emotional trauma in childhood has been demonstrated to be a potential risk factor for developing depressive symptoms later in life, particularly in response to additional trigger stressful events [75, 76]. For instance, women having a history of childhood abuse and actual diagnosis of major depression showed a sixfold higher level of adrenocorticotropic hormone stress response than the age-matched control individuals, suggesting that there may be permanent changes in set-points for HPA activity in response to stress among those persons who were exposed to early-life stressful conditions [77]. On the basis of these findings, Heim et al. suggested that trauma early in life is related to the sensitization of neuroendocrine stress responses, immune activation,

enhanced central CRF activity, glucocorticoid resistance, and lower hippocampal volume throughout the adult life [75]. These neuroendocrine changes triggered by stresses in early life can likely affect the risk of developing depression in response to stress during adulthood. Data from recent studies highlight the critical role of epigenetic regulation in the linkage between trauma in childhood and depression in adult life [76].

It should be noted, however, that the link between adverse conditions in early life and unfavorable neurobehavioral outcomes in adulthood can be dependent not only on epigenetic processes per se, but also on the genetic background of affected individuals. In particular, different combinations of functional polymorphisms in dopamine and serotonin pathway genes can result in both responder and non-responder phenotypes in the wide range from adverse to advantageous early-life circumstances. For example, a functional polymorphism in the promoter gene of monoamine oxidase A (MAOA), a mitochondrial enzyme that degrades the neurotransmitters including serotonin, norepinephrine, and dopamine, was demonstrated to mediate the association between adversities early in life and enhanced risk for violence and antisocial behavior in adulthood. In the research by Frazzetto et al., the MAOA genotype was found to moderate the link between traumatic events experienced from birth up to the age of 15 years and physical aggression in adult life, as assessed by the Aggression Questionnaire [78]. In this study, scores of physical aggression were shown to be higher in those adult men who have been exposed to traumatic events early in life and who carried the low MAOA activity allele (MAOA-L). These findings were confirmed by later studies. The interaction of

MAOA genotype and childhood adversity on antisocial outcomes was examined in a meta-analysis of 27 studies conducted by Byrd and Manuck [79]. Across 20 male cohorts, adversity in early life has been demonstrated to be a stronger predictor of adult antisocial outcomes for a low-activity, compared to a high-activity, MAOA genotype. Similar, but less consistent, findings were reported in 11 female cohorts studied.

An association between the adversities early in life and long-lasting changes in processes of epigenetic regulation at the whole-genome level was demonstrated repeatedly. Many of such studies used low socioeconomic status (SES) as indicator of early stressful conditions. Low SES, generally accompanied by an enhanced stress load due to a poor quality of nutrimental intake, infections, and higher load of physical work, has been found to strongly predict a number of psycho-emotional pathologies such as schizophrenia and depression in adult life [67]. Disadvantaged early SES was related to profiles of adult blood DNA methylation [80]. Most of the genes differentially methylated in association with low early-life SES are known to be functionally implicated in metabolic and cell signaling pathways. Genome-wide transcriptional profiling demonstrated that in healthy adults with low-SES childhood background, genes bearing response elements for CREB/ATF family of transcription factors transmitting adrenergic signals to leukocytes were substantially upregulated, whereas genes with response elements for the GR, regulating the secretion of cortisol and transducing the anti-inflammatory signals to the immune system, were significantly downregulated [81]. Individuals exposed to low-SES conditions in early life also exhibited raised cortisol levels, elevated expression of pro-inflammatory transcription factor NF-kappaB, as well as increased production of the pro-inflammatory cytokine interleukin 6. On the basis of data obtained, the authors suggested that "low early-life SES programs a defensive phenotype characterized by resistance to glucocorticoid signaling, which in turn facilitates exaggerated adrenocortical and inflammatory responses. Although these response patterns could serve adaptive functions during acute threats to well-being, over the long term, they might exact an allostatic toll on the body that ultimately contributes to the chronic diseases of aging." Chen et al. have revealed that the unfavorable effects of the low-SES conditions in early life on immune system functioning and inflammatory processes in adult life can be at least partly prevented by the high-level maternal warmth [82]. These alterations have been accompanied by changes in genome-wide transcription profiles. Those individuals who had low SES level early in life and whose mothers demonstrated high warmth toward them showed reduced Toll-like receptor-stimulated production of interleukin 6 and lowered activity of immune activating transcription factor (AP-1) and NF-kappaB in comparison with those subjects who had low SES level early

in life but have experienced lower maternal warmth. These findings suggest that disadvantageous effect of low socioeconomic environment in early life might be buffered by a supportive family climate.

Similar lasting effects have been obtained for the unfavorable experiences such as sexual/physical abuse or neglect in early life [83]. Three hundred sixty-two differentially methylated promoters have been identified by a genome-wide analysis in the hippocampal neurons isolated from postmortem brain samples in subjects with a history of heavy abuse throughout infancy in comparison with control persons [84]. Among them, those genes implicated in a cellular or neural plasticity were shown to be the most differentially methylated. Nine hundred ninety-seven gene promoters were identified as being differentially methylated in association with abuse throughout childhood in a whole-blood DNA from adult subjects [85]. Most of these genes are involved in important pathways of cellular signaling associated with development and regulation of transcription. Four hundred forty-eight gene promoters were differentially methylated in T cells from adult male individuals exposed to parental physical aggression in the age of 6 to 15 years relative to a control group [86]. Most these genes are known to play an important role in aggressive behavior.

Long-lasting social and behavioral problems were also observed in persons who experienced parental neglect through the institutionalization in early life [87]. Differential patterns of whole-genome DNA methylation were found in blood samples from institutionalized children and children reared by their biological parents [88]. Most of these differentially methylated genes are related to immune and cellular signaling pathways, including those responsible for development and functioning of the brain as well as in neural communication. One hundred seventy-three genes were differentially methylated among subjects with and without the placement into the foster care system during their childhood [69]. Most of these genes are involved in the ubiquitin-mediated proteolysis pathway, which plays an important role in immune/inflammatory responses, in antigen processing and presentation pathways, and also in some important cellular processes. Moreover, 72 genes known to be related to the control of apoptosis and transcriptional regulation exhibited increased methylation levels in those individuals who had a history of foster care placement, while 101 genes involved in protein catabolic processes and in the control of posttranslational protein modification exhibited lowered levels of methylation in comparison with control subjects. Summary of evidence on the epigenetic link between adverse early-life events and adult-life neurobehavioral outcomes obtained from epigenome-wide association studies (EWAS) is given in Table 1. As we can see from Table 1, most of these EWAS data have been obtained from small samples; therefore, one must use some caution in interpreting these

Table 1 Summary of evidence on the link between adverse early-life events and adult neurobehavioral outcomes from epigenome-wide association studies

Condition /exposure	Stage at exposure	Age at detection	Tissue/cells	Population, sample size (n)	DMRs or up/downregulated genes, n	Function/pathway	Ref.
Maternal depression	Prenatal	Adult	Hippocampal tissue samples	Male postmortem samples with (n = 12) or without (n = 50) a history of maternal depression	294 DMRs associated with 234 genes	Immune system functions	[45]
Low SES	Childhood	45 years	Blood	40 British adults	586 hypermethylated and 666 hypomethylated gene promoters	Cell signaling pathways	[80]
	Childhood	25–40 years	Blood	103 healthy adults	73 upregulated and 37 downregulated genes	Raised cortisol levels; increased IL-6 production	[81]
	Childhood	25–40 years	Blood	53 healthy adults with a history of low early-life SES	330 upregulated and 161 downregulated genes in participants who grew up with high maternal warmth	Immune activation and systemic inflammation; diminishing these outcomes by supportive family climate	[82]
Child neglect /abuse	Childhood	Adult	Hippocampal neurons	25 French-Canadian men with a history of severe childhood abuse and 16 control subjects	248 hypermethylated DMRs; 114 hypomethylated DMRs	Cellular/neuronal plasticity	[84]
	Childhood	45 years	Blood	12 British men with a history of childhood abuse and 28 control subjects	311 hypermethylated and 686 hypomethylated gene promoters	Development, regulation of transcription	[85]
	Childhood	Adult	T lymphocytes	8 subjects with a history of physical aggression from age 6 to 15 years and 57 controls	171 hypermethylated and 277 hypomethylated gene promoters	Aggressive behavior	[86]

DMRs, differentially methylated regions; SES, socioeconomic status

results. Further studies with larger samples are clearly required in order to allow more reliable conclusions.

In addition to EWAS, consistent evidence for the importance of epigenetic regulation in mediating long-term effects of early adversity was also provided from some candidate gene research. While a full-genome analysis allows to generate hypotheses on the underlying molecular mechanisms, the candidate gene approach allows to determine whether a specific gene of interest makes a contribution in each particular case. For example, in the McGowan et al. study, epigenetic differences in the brain loci substantially involved in the pathophysiology of suicide have been observed [89]. Specifically, by studying the postmortem hippocampal brain samples from suicidal individuals, the history of neglect/abuse in early childhood was shown to be associated with a lowered hippocampal volume and with severe cognitive impairments. Moreover, the gene encoding ribosomal RNA (rRNA) was significantly hypermethylated throughout the promoter and 5′ regulatory region in the brains of suicide victims, consistent with the decreased level of expression of rRNA gene in the hippocampus. Subsequently, McGowan et al. [90] examined epigenetic differences in a neuron-specific promoter of NR3C1 gene among postmortem hippocampal tissues from suicide completers with or without the child abuse history. The NR3C1 gene was selected for analysis since the decreased level of GRs

within the hippocampus is believed to lead to elevated HPA stress response and thereby might account for an enhanced risk of psychopathology and poorer emotional regulation in those subjects who were abused in childhood. In that study, the levels of expression of NR3C1 gene were considerably decreased in suicide victims having a history of childhood abuse compared to non-abused suicide victims or control individuals; no differences, however, were revealed among non-abused suicide victims and control subjects. The essential effect on the expression of transcripts from the exon 1F NR3C1 promoter has been also observed. Labonté et al. also reported the enhanced methylation levels in the promoter of the 1F NR3C1 and decreased expression of this gene in the hippocampus of suicide completers having a history of abuse compared to either suicide completers with no abuse history or to control subjects [91]. In a more recent research by Bustamante et al., the childhood maltreatment assessed by a retrospective self-report questionnaire has been significantly associated with methylation levels in the NR3C1 promoter region in whole blood of adult persons [92]. Tyrka et al. also demonstrated that adversity in childhood can be linked to the risk for adult psychiatric disorders via epigenetic regulation of glucocorticoid signaling genes such as NR3C1 and gene coding for FK506 binding protein 51 (FKBP5) [93]. In another recent study by the same authors, the reduced methylation levels of NR3C1 gene were significantly associated with maltreatment in childhood and anxiety, depressive and substance-use disorders in adulthood [94]. An enhanced risk for development of stress-associated psychiatric disorders in adulthood was shown to be associated with childhood trauma-dependent, allele-specific demethylation of the functional glucocorticoid response elements of FKBP5 gene playing an important role in regulation of the HPA axis [95]. The demethylation of FKBP5 gene has been associated with elevated stress-dependent gene transcription followed by lasting dysregulation of the stress hormone system and by global effect on immune functioning and areas of the brain related to stress regulation.

One important issue in these studies is limited access to human neural tissues. Thereby, most candidate gene studies examining role of epigenetic variations in developmental programming of adult behavioral and cognitive dysfunctions are based on samples from peripheral tissues. In the above-mentioned study by Bick et al., significant negative correlation was observed between the mothers' parenting reports and methylation levels of NR3C1 gene and also a macrophage migration inhibitory factor gene functionally implicated in the expression of NR3C1 and immune response in offspring blood 5 to 10 years after assessing the maternal caregiving quality [69].

Childhood-adversity caused epigenetic modifications in NR3C1 gene in the human blood samples have also been reported in the Tyrka et al. research [96]. In this study, enhanced levels of NR3C1 methylation were observed in leukocyte DNA from healthy adult individuals exposed to inadequate nurturing or maltreatment during their childhood. The elevated methylation levels of the exon 1F NR3C1 promoter in the peripheral blood of individuals suffering from borderline personality disorder or major depressive disorder have been revealed to be associated with the severity of childhood maltreatment [97]. No such changes have been found, however, in bulimic women exposed to abuse in their childhood [98]. Similar findings have been obtained for a serotonin system playing a crucial role in the brain development (including a region important in stress-regulation such as the hippocampus) and in the etiology of depression [99]. In this study, the childhood trauma, along with male gender and smaller hippocampal volume, has been independently associated with higher levels of peripheral serotonin transporter methylation in adulthood.

In some candidate gene researches, low SES has been used as a reliable indicator of adverse early-life conditions. In the research by Miller and Chen, the adolescent subjects whose families owned homes throughout their early childhood demonstrated higher levels of NR3C1 expression and lower levels of toll-like receptor 4 (TLR4) gene expression in leukocytes from peripheral blood compared to individuals with low SES in early life [100]. Data from this study indicated that low SES early in life may trigger a pro-

inflammatory phenotype in later life. Similar findings have been obtained in African-American men who often have low SES in their childhood. In the study by Witek-Janusek et al., higher levels of childhood trauma and indirect exposure to neighborhood violence have been shown to be related to a greater acute stress-induced IL-6 response and also to a reduced methylation of the IL-6 promoter and lower cortisol response in adulthood [101]. Summary of evidence on the epigenetic link between adverse early-life events and adult-life neurobehavioral outcomes obtained from candidate gene studies is given in Table 2.

Overall, these research findings highlight the mechanisms implicated in early adversity-induced impairment of neuroendocrine pathways associated with stress reactivity and adult social behavior. One of the most significant health outcomes of the childhood adversity is lasting neuroendocrine disturbance caused by adversity-induced alterations in the methylation levels of NR3C1 gene, thereby leading to changed cortisol production and various pathological conditions in adulthood [39, 102].

A schematic representation of hypothetical mechanisms linking childhood adversity to later-life neurobehavioral and cognitive dysfunction is given in Fig. 2.

Natural experiment-based evidence

Causal relationship between stressful events in early life and health problems later in life is evident from a body of quasi-experimental states ("natural experiments"), referring to any kind of naturally occurring circumstances in which subsets of the population have different levels of exposure to a supposed causal factor [103]. Currently, a man-made famine, where dietary insults and chronic stress tend to co-occur in exposed populations, is typically used in quasi-experimental design [104]. The nutritional status might be affected by stressful events at different levels including self-selection of dietary components, intake of calories, and utilization of metabolic wastes for energy production, whereas nutritional factors can, in turn, affect stress response through influencing both peripheral and central mechanisms of stress reactivity [14, 105]. Thereby, famine has numerous features that can be beneficial for its use as a natural experiment in studying lasting outcomes of stressful events in early life [104, 106], although using such research design could confound attempts to distinguish the (intergenerational) effects of nutrition and stress.

In a number of quasi-experimental statues, impaired cognitive and behavioral functioning, as well as psychiatric illness in adulthood, has been reported in cohorts exposed in their early life to natural disasters or stressful historic events, such as World War II [107], Holocaust [108–110], Israeli-Arab war of 1967 [111, 112], Chinese Famine of 1959–1961 [113–117], and Dutch famine of 1944–1945 [118–120]. In some recent quasi-experimental studies, evidence for the epigenetic embedding of stressful historic

Table 2 Summary of evidence on the epigenetic link between adverse early-life events and adult neurobehavioral outcomes from candidate gene studies

Condition/ exposure	Stage at exposure	Age at detection	Tissue/cells	Population, sample size (n)	Function/pathway	Gene/element	Epigenetic outcome	Ref.
Low SES	0–5 years	25–40 years	Saliva	103 adults	Decreased glucocorticoid and increased pro-inflammatory signaling	CREB/ATF gene family NR3C1	Upregulation Downregulation	[81]
Child maltreatment	Early childhood	Adulthood	Postmortem hippocampus	18 male suicide subjects, 12 controls	Impaired ribosomal functioning	RRNa promoter	Hypermethylation	[89]
	Childhood	Adulthood	Postmortem hippocampus	12 abused and 12 non-abused suicide subjects, 12 controls	Impaired stress reactivity	NR3C1	Downregulation	[90]
	Childhood	Adulthood	Postmortem hippocampus	21 abused and 21 non-abused suicide subjects, 14 controls	Impaired stress reactivity	NR3C1 1F	Hypermethylation	[91]
	Childhood	Adulthood	Whole blood	74 maltreated and 73 control subjects	Impaired stress reactivity	NR3C1 1F promoter region	Hypermethylation	[92]
	Childhood	18–59 years	Leukocytes	58 female and 41 male subjects	Impaired stress reactivity	NR3C1 promoter	Hypermethylation	[93]
	Childhood	18–65 years	Leukocytes	213 female and 127 male subjects	Impaired stress reactivity	NR3C1 promoter	Hypomethylation	[94]
	Childhood	19–59 years	Peripheral blood cells	30 subjects with and 46 without the history of child trauma	Immune functioning, stress reactivity	Glucocorticoid response elements of FKBP5 gene	Allele-specific demethylation	[95]
	Childhood	18–59 years	Leukocytes	58 female and 41 male subjects	Impaired stress reactivity	NR3C1 promoter	Hypermethylation	[96]
	Childhood	Adulthood	Peripheral blood	200 subjects with different rate of child maltreatment	Impaired stress reactivity	NR3C1 1F promoter	Hypermethylation	[97]
	Childhood	18–65 years	Peripheral blood	33 maltreated subjects, 36 controls	Stress-related psychopathology	Serotonin transporter gene promoter	Hypermethylation	[99]
	Childhood	18–25 years	Peripheral blood	34 African American men	Higher proinflammatory response to stress	IL-6 gene promoter	Hypomethylation	[101]

events was obtained. Although no relationship between the exposure to the Dutch famine throughout the intra-uterine period and whole-genome DNA methylation level in adulthood was observed [121], an association between the famine exposure during early gestation and changed patterns of methylation at CpG dinucleotides in genes

responsible for growth, development, and metabolism in the whole blood of adult individuals was found in the Tobi et al. study [122]. The gene-specific differences in the patterns of DNA methylation associated with in-utero exposure to the Dutch famine have been indicated in several studies including those of Heijmans et al. [123] and Tobi et al. [124]. The

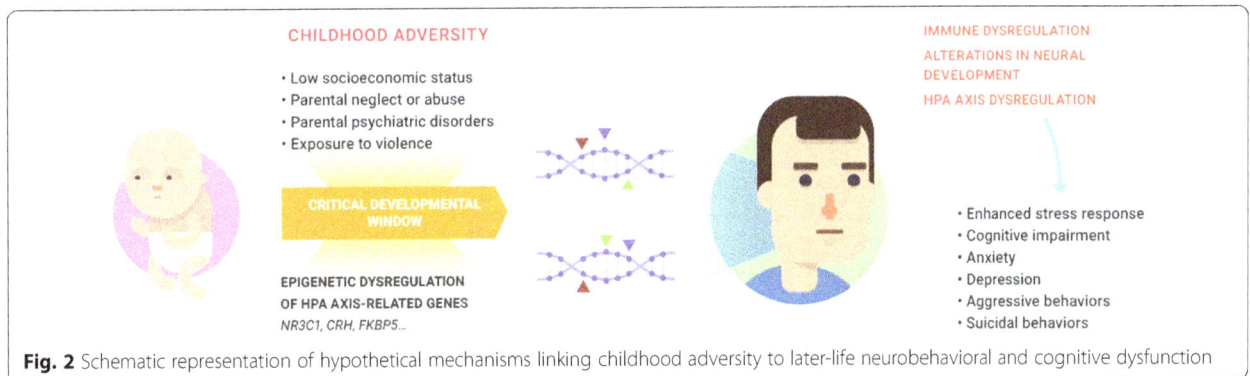

CHILDHOOD ADVERSITY

· Low socioeconomic status
· Parental neglect or abuse
· Parental psychiatric disorders
· Exposure to violence

CRITICAL DEVELOPMENTAL WINDOW

EPIGENETIC DYSREGULATION OF HPA AXIS-RELATED GENES
NR3C1, CRH, FKBP5...

IMMUNE DYSREGULATION
ALTERATIONS IN NEURAL DEVELOPMENT
HPA AXIS DYSREGULATION

· Enhanced stress response
· Cognitive impairment
· Anxiety
· Depression
· Aggressive behaviors
· Suicidal behaviors

Fig. 2 Schematic representation of hypothetical mechanisms linking childhood adversity to later-life neurobehavioral and cognitive dysfunction

differences in DNA methylation levels in HPA axis-associated genes, such as corticotropin-releasing hormone (CRH) and NR3C1 genes, however, were non-significant in the latest study. Importantly, the effect of prenatal famine exposure has been demonstrated to strongly depend on the exposure period, with differences being most pronounced when the famine exposure occurred throughout the periconceptional period rather than throughout the late gestational developmental stage. These findings demonstrate that early intrauterine period is the most sensitive stage in human ontogenesis [125, 126]. Since epigenome is demonstrated to be most plastic during this ontogenetic period, these data are suggestive for the role of epigenetic modifications in driving life-lasting effects of exposures to famine and/or other disasters in early life.

Transgenerational transmission of childhood trauma

Evidence has been also obtained that early adversity-induced neuronal and behavioral effects can be transgenerationally transmitted via epigenetic mechanisms to subsequent generations [45, 127]. In the study by Yehuda et al. [128], the Holocaust survivor offspring were studied which are known to have altered GR sensitivity and vulnerability to psychiatric disorders. In this study, adult offspring with both maternal and paternal Holocaust-induced posttraumatic stress disorder (PTSD) demonstrated decreased levels of the 1F NR3C1 promoter methylation, while offspring with paternal PTSD only exhibited higher methylation levels of 1F NR3C1 promoter in the peripheral blood mononuclear cells compared to participants without parental Holocaust exposure. Similar transgenerational effects were found for another historical event, such as the Tutsi genocide. In the Perroud et al. study [129], mothers exposed to the Tutsi genocide as well as their offspring had higher mineralocorticoid receptor levels and lower cortisol and GR levels in comparison with non-exposed mothers and their children. Furthermore, the exposed mothers and their progeny had higher levels of methylation of the 1F NR3C1 exon than non-exposed subjects.

To examine the mechanisms underlying such transgenerational effects of paternal trauma, Gapp et al. [130] using a mouse model of unpredictable maternal separation and maternal stress have demonstrated that postnatal trauma changes coping behavior in adverse conditions in exposed males when adult and in their adult male offspring. These behavioral changes have been accompanied by elevated levels of NR3C1 expression and reduced methylation of the NR3C1 promoter in the hippocampus. The DNA methylation levels were also lowered in sperm cells of exposed males when adult. Interestingly, the transgenerational transmission of neurobehavioral symptoms has been shown to be prevented by paternal environmental enrichment, and this effect was linked to the reversal of changes in DNA methylation and expression of NR3C1 gene in the hippocampus of the male offspring.

Conclusion

Accumulating evidence assume that adversity early in life can be associated with later neuropsychiatric, cognitive, and behavioral outcomes. Recent findings suggest that childhood adversity can have greater impact on the later health status than stressful exposures in adulthood, as assumed by phenomenon of the biological embedding of early experience. In several studies, credible evidence is obtained that adversity early in life can reach far into the later adulthood partly due to cellular aging, as evident from recent data indicating that severe traumatic and social exposures as well as institutional care history in childhood can be embedded at the molecular level through accelerated telomere shortening [131–133].

It is increasingly clear that epigenetic control of gene expression plays a central role in these effects. Some authors hypothesized that such early-life "epigenetic tuning" may likely prepare particular genes for responses to subsequent triggers [134]. By this mechanism, the functional performance of different tissues and organs may be established well before they are actually challenged. In evolutionary terms, such epigenetic fine-tuning of the expression of responsible genes enables the organism to adapt to varying environmental conditions [135], but it can enhance the risk for disorders, including neuropsychiatric ones, later in life [136]. Thus, epigenetic studies can provide insight into the mechanisms mediating the relationship between early-life adversities, aberrant neuroplastic interactions, and adult health outcomes. The role of particular genes such as hippocampal NR3C1 as well as genes coding arginine vasopressin and CRF in the neurons of the paraventricular nucleus has been demonstrated in mediating the effects of early adversity-associated pathological conditions [70, 137].

In the past decade, significant advances have been achieved in the emerging field of neurobehavioral epigenetics. Some research challenges, however, must be addressed for further progress in the field. For example, it is still not clear how stable are modifications of epigenetic marks which are induced by adversities in early life. Recent findings suggest that such epigenetic modifications can be long-term or even life-long and may persist up to the highest age categories [138–140]. These data, however, are rather scarce; therefore, it requires further investigation. Another issue is that epigenetic patterns can be specific not only for distinct cell types but also for specific neuronal pathways in the same brain regions [141]. Thus, focus of further research will likely be shifted from particular candidate genes to particular candidate gene pathways that may be epigenetically labile in response to adverse conditions in early life. Moreover, as significant epigenetic modifications originate both within and among different types of tissues, one more potentially important issue is applicability of samples from peripheral blood for determining epigenetic modifications in human studies. Indeed, though widespread epigenetic changes might be induced by traumatic experiences in early life, these effects may greatly vary in magnitude and direction in neuronal tissues relative to non-neuronal tissues including peripheral blood. This may limit the opportunity to studying the long-term neurobehavioral impacts of early trauma basing only on peripheral blood samples or buccal swabs. Therefore, animal models, which provide an opportunity to simultaneously measure the epigenetic profiles in both peripheral and brain tissues, can be useful in highlighting epigenetic pathways underlying the biological embedding of early adversity. The use of animal models certainly raises questions concerning the specificity of such pathways across different mammalian species and about similarities and differences among these pathways in various animal species and humans. In addition, the potential impact of subsequent exposure to traumatic events during adolescence and adulthood on the process of epigenetic embedding of early experiences also needs to be considered as the findings from some investigations suggest that epigenome continues to be labile during adulthood [142].

A genuine incorporation of novel knowledge about the mechanisms underlying the process of epigenetic embedding of adverse experiences during sensitive developmental periods into the current paradigm on the causation of adult neurobehavioral and cognitive dysfunction will certainly move the focus of efforts targeted toward prevention of adult psychopathological conditions from the later life stages to early developmental stages from conception to weaning. Indeed, reducing or eliminating risk factors early in life would likely have a potential to prevent neurological dysfunctions and psychopathologies in adult life. In this context, an important point is that epigenetic states, in contrast to the relatively stable genetic information, are reversible and can be modified by environmental factors [143]. Therefore, modulation of epigenetic pathways involved in biological embedding may provide promising new direction toward novel therapeutic strategies against neurological and cognitive dysfunctions in adult life.

Over the last years, the therapeutic potential of pharmaceuticals targeted at chromatin modifying enzymes, such as histone deacetylase (HDAC) inhibitors, in treatment of cognitive and behavioral impairments as well as psychiatric disorders such as anxiety, depression, fear, and schizophrenia has been repeatedly demonstrated [144–146]. In a number of preclinical animal models, the convincing evidence is obtained that treatment with HDAC inhibitors can be effective in the prevention and therapy of experimentally induced cognitive and behavioral abnormalities. Treatment with the HDAC inhibitor sodium butyrate (SB)

reversed the abnormal hyperactive behavior in a rat model of D-amphetamine-induced mania-like behavior [147]. Furthermore, SB and other HDAC inhibitor, valproate, abolished manic-like behaviors and protected the rat brain from metabolic disturbances induced by the metabolic poison, ouabain [148]. These findings were subsequently confirmed in models of depressive- and manic-like behaviors induced by chronic mild stress or maternal deprivation [149]. In the same models, SB treatment improved the recognition memory and reversed the stress-induced decrease of hippocampal neurotrophic factors including BDNF, nerve growth factor, and glial cell line-derived neurotrophic factor [150], and also abolished the maternal deprivation- and chronic mild stress-induced dysfunction in the striatum of rats [151]. In a mouse model of valproic acid (VPA)-induced autism, chronicadministration of SB resulted in attenuating the experimentally induced deficits in a novel object recognition and loss of hippocampal dendritic spine, as well as significantly increased level of acetylation of the histone H3 in hippocampus [152]. In the same mouse model of VPA-induced autism, SB attenuated autism-like deficits in social behavior and modified transcription levels of many behavior-associated genes in the prefrontal cortex, in particular, genes implicated in neuronal excitation or inhibition [153]. In a mouse model of isoflurane-induced cognitive deficits, treatment with SB attenuated the repression of contextual fear memory, apparently by promoting histone acetylation and expression of histone acetylation-mediated genes [154]. In the 6-hydroxydopamine-induced Parkinson's disease rat model, administration of SB resulted in a substantial attenuation of motor deficits and also in an increase of striatal dopamine, BDNF, and global H3 histone acetylation levels [155]. The cognition-protective effect of SB was revealed in a rat model of chronic cerebral hypoperfusion; this effect was, at least in part, mediated through enhancing histone acetylation and facilitating the transcription of Nrf2 downstream genes in the hippocampus [156]. HDAC inhibitor phenylbutyrate (PBA) was shown to be able to attenuate hippocampal neuronal loss and reverse the Alzheimer's disease-like phenotype in a mouse model of Alzheimer's disease [157]. In rats with neonatal ventral hippocampal lesions which are commonly used for modeling neurodevelopmental aspects of schizophrenia, treatment with PBA reversed the unfavorable behavioral consequences of these lesions in the ventral hippocampus [158]. In a maternal separation rat model, treatment with another HDAC inhibitor, suberoylanilide hydroxamic acid (SAHA), reversed early-life stress-induced visceral hypersensitivity and anxiety behavior [159]. Taken together, these results suggest that targeting epigenome by specific pharmacological interventions can be a promising therapeutic option in treatment of neuropsychiatric and cognitive impairments, including those related to biological embedding of early life exposures. There are certainly many important issues which need to be addressed before implementation of such interventions in clinical practice, including their effective dose levels, administration frequency, safety, and potential side effects. These issues remain to be addressed in future clinical trials.

Abbreviations
BDNF: Brain-derived neurotrophic factor; CRF: Hypothalamic corticotropin-releasing factor; CRH: Corticotropin-releasing hormone; EWAS: Epigenome-wide association studies; GR: Glucocorticoid receptor; HDAC: Histone deacetylase; HPA: Hypothalamic-pituitary-adrenal; LG: Licking/grooming; MAOA: Monoamine oxidase A; MWAS: Methylome-wide association study; NR3C1: Nuclear receptor subfamily 3 group C member; PBA: Phenylbutyrate; PTSD: Posttraumatic stress disorder; SAHA: Suberoylanilide hydroxamic acid; SB: Sodium butyrate; SES: Socioeconomic status; VPA: Valproic acid

Acknowledgements
The authors would like to thank Oksana Zabuga for the helpful assistance in preparing the manuscript.

Funding
Not applicable.

Authors' contributions
AMV conceived the idea for the manuscript and produced the first draft. AKK was involved in creating the figures and also in the critical review. Both authors read and approved the final manuscript.

Authors' information
AV is a Professor and Head of the Laboratory of Epigenetics, Institute of Gerontology, Kiev, Ukraine. AK is a Research associate of the Laboratory of Epigenetics, Institute of Gerontology, Kiev, Ukraine.

Consent for publication
Not applicable.

Competing interests
The authors declare that they have no competing interests.

References
1. Hanson MA, Gluckman PD. Developmental origins of health and disease—global public health implications. Best Pract Res Clin Obstet Gynaecol. 2015;29(1):24–31.
2. Eriksson JG. Developmental origins of health and disease—from a small body size at birth to epigenetics. Ann Med. 2016;48(6):456–67.
3. Champagne FA. Interplay between social experiences and the genome: epigenetic consequences for behavior. Adv Genet. 2012;77:33–57.
4. Doherty TS, Roth TL. Insight from animal models of environmentally driven epigenetic changes in the developing and adult brain. Dev Psychopathol. 2016;28:1229–43.
5. Curley JP, Champagne FA. Influence of maternal care on the developing brain: mechanisms, temporal dynamics and sensitive periods. Front Neuroendocrinol. 2016;40:52–66.
6. Kim DR, Bale TL, Epperson CN. Prenatal programming of mental illness: current understanding of relationship and mechanisms. Curr Psychiatry Rep. 2015;17(2):5.
7. Fowden AL, Valenzuela OA, Vaughan OR, Jellyman JK, Forhead AJ. Glucocorticoid programming of intrauterine development. Domest Anim Endocrinol. 2016;56:S121–32.
8. Silberman DM, Acosta GB, Zorrilla Zubilete MA. Long-term effects of early life stress exposure: role of epigenetic mechanisms. Pharmacol Res. 2016;109:64–73.
9. Maccari S, Polese D, Reynaert ML, Amici T, Morley-Fletcher S, Fagioli F. Early-life experiences and the development of adult diseases with a focus on mental illness: the human birth theory. Neuroscience. 2017;342:232–51.
10. Weinstock M. The long-term behavioural consequences of prenatal stress. Neurosci Biobehav Rev. 2008;32:1073–86.
11. Reynolds RM, Jacobsen GH, Drake AJ. What is the evidence in humans that DNA methylation changes link events in utero and later life disease? Clin Endocrinol. 2013a;78:814–22.
12. Reynolds RM, Labad J, Buss C, Ghaemmaghami P, Räikkönen K. Transmitting biological effects of stress in utero: implications for mother and offspring. Psychoneuroendocrinology. 2013b;38:1843–9.
13. Barbazanges A, Piazza PV, Le Moal M, Maccari S. Maternal glucocorticoid secretion mediates long-term effects of prenatal stress. J Neurosci. 1996;16:3943–9.
14. Entringer S, Buss C, Wadhwa PD. Prenatal stress and developmental programming of human health and disease risk: concepts and integration of empirical findings. Curr Opin Endocrinol Diabetes Obes. 2010;17:507–16.
15. Entringer S, Buss C, Wadhwa PD. Prenatal stress, development, health and disease risk: a psychobiological perspective—2015 Curt Richter Award Paper. Psychoneuroendocrinology. 2015;62:366–75.
16. Lippmann M, Bress A, Nemeroff CB, Plotsky PM, Monteggia LM. Long-term behavioural and molecular alterations associated with maternal separation in rats. Eur J Neurosci. 2007;25:3091–8.
17. Nishi M, Horii-Hayashi N, Sasagawa T. Effects of early life adverse experiences on the brain: implications from maternal separation models in rodents. Front Neurosci. 2014;8:166.
18. Tractenberg SG, Levandowski ML, de Azeredo LA, Orso R, Roithmann LG, Hoffmann ES, et al. An overview of maternal separation effects on behavioural outcomes in mice: evidence from a four-stage methodological systematic review. Neurosci Biobehav Rev. 2016;68:489–503.
19. Braun K, Champagne FA. Paternal influences on offspring development: behavioural and epigenetic pathways. J Neuroendocrinol. 2014;26(10):697–706.
20. Korosi A, Baram TZ. Plasticity of the stress response early in life: mechanisms and significance. Dev Psychobiol. 2010;52(7):661–70.
21. Perry R, Sullivan RM. Neurobiology of attachment to an abusive caregiver: short-term benefits and long-term costs. Dev Psychobiol. 2014;56(8):1626–34.
22. Szyf M, Tang YY, Hill KG, Musci R. The dynamic epigenome and its implications for behavioral interventions: a role for epigenetics to inform disorder prevention and health promotion. Transl Behav Med. 2016;6(1):55–62.
23. Cowan CS, Callaghan BL, Kan JM, Richardson R. The lasting impact of early-life adversity on individuals and their descendants: potential mechanisms and hope for intervention. Genes Brain Behav. 2016;15(1):155–68.
24. Canovas S, Ross PJ. Epigenetics in preimplantation mammalian development. Theriogenology. 2016;86(1):69–79.
25. Vaiserman A. Epidemiologic evidence for association between adverse environmental exposures in early life and epigenetic variation: a potential link to disease susceptibility? Clin Epigenetics. 2015;7:96.
26. Vaiserman AM, Koliada AK, Jirtle RL. Non-genomic transmission of longevity between generations: potential mechanisms and evidence across species. Epigenetics Chromatin. 2017;27:38.
27. Lee HJ, Hore TA, Reik W. Reprogramming the methylome: erasing memory and creating diversity. Cell Stem Cell. 2014;14:710–9.
28. Trerotola M, Relli V, Simeone P, Alberti S. Epigenetic inheritance and the missing heritability. Hum Genomics. 2015;9:17.
29. Vickaryous N, Whitelaw E. The role of the early embryonic environment on epigenotype and phenotype. Reprod Fertil Dev. 2005;17:335–40.
30. Hochberg Z, Feil R, Constancia M, Fraga M, Junien C, Carel JC. Child health, developmental plasticity, and epigenetic programming. Endocr Rev. 2011;32:159–224.
31. Feil R, Fraga MF. Epigenetics and the environment: emerging patterns and implications. Nat Rev Genet. 2012;13(2):97–109.
32. Boyce WT, Kobor MS. Development and the epigenome: the 'synapse' of gene–environment interplay. Dev Sci. 2015;18(1):1–23.
33. McGowan PO, Roth TL. Epigenetic pathways through which experiences become linked with biology. Dev Psychopathol. 2015;27(2):637–48.
34. Provencal N, Binder EB. The neurobiological effects of stress as contributors to psychiatric disorders: focus on epigenetics. Curr Opin Neurobiol. 2015;30:31–7.
35. Isles AR. Neural and behavioral epigenetics; what it is, and what is hype. Genes Brain Behav. 2015;14(1):64–72.

36. Halldorsdottir T, Binder EB. Gene × environment interactions: from molecular mechanisms to behavior. Annu Rev Psychol. 2017;68:215–41.

37. Mueller BR, Bale TL. Sex-specific programming of offspring emotionality after stress early in pregnancy. J Neurosci. 2008;28:9055–65.

38. O'Donnell D, Larocque S, Seckl JR, Meaney MJ. Postnatal handling alters glucocorticoid, but not mineralocorticoid messenger RNA expression in the hippocampus of adult rats. Brain Res Mol Brain Res. 1994;26:242–8.

39. Weaver IC, Cervoni N, Champagne FA, D'Alessio AC, Sharma S, Seckl JR, et al. Epigenetic programming by maternal behavior. Nat Neurosci. 2004;7:847–54.

40. Kundakovic M, Gudsnuk K, Herbstman JB, Tang D, Perera FP, Champagne FA. DNA methylation of BDNF as a biomarker of early-life adversity. Proc Natl Acad Sci U S A. 2015;112(22):6807–13.

41. Monk C, Spicer J, Champagne FA. Linking prenatal maternal adversity to developmental outcomes in infants: the role of epigenetic pathways. Dev Psychopathol. 2012;24:1361–76.

42. Lewis AJ, Austin E, Knapp R, Vaiano T, Galbally M. Perinatal maternal mental health, fetal programming and child development. Healthcare (Basel). 2015;3(4):1212–27.

43. Newman L, Judd F, Olsson CA, Castle D, Bousman C, Sheehan P, et al. Early origins of mental disorder—risk factors in the perinatal and infant period. BMC Psychiatry. 2016;16:270.

44. Babenko O, Kovalchuk I, Metz GA. Stress-induced perinatal and transgenerational epigenetic programming of brain development and mental health. Neurosci Biobehav Rev. 2015;48:70–91.

45. Nemoda Z, Massart R, Suderman M, Hallett M, Li T, Coote M, et al. Maternal depression is associated with DNA methylation changes in cord blood T lymphocytes and adult hippocampi. Transl Psychiatry. 2015;5:e545.

46. Braithwaite EC, Kundakovic M, Ramchandani PG, Murphy SE, Champagne FA. Maternal prenatal depressive symptoms predict infant NR3C1 1F and BDNF IV DNA methylation. Epigenetics. 2015;10:408–17.

47. Oberlander TF, Weinberg J, Papsdorf M, Grunau R, Misri S, Devlin AM. Prenatal exposure to maternal depression, neonatal methylation of human glucocorticoid receptor gene (NR3C1) and infant cortisol stress responses. Epigenetics. 2008;3:97–106.

48. Radtke KM, Ruf M, Gunter HM, Dohrmann K, Schauer M, Meyer A, et al. Transgenerational impact of intimate partner violence on methylation in the promoter of the glucocorticoid receptor. Transl Psychiatry. 2011;1:1–6.

49. Hompes T, Izzi B, Gellens E, Morreels M, Fieuws S, Pexsters A. Investigating the influence of maternal cortisol and emotional state during pregnancy on the DNA methylation status of the glucocorticoid receptor gene (NR3C1) promoter region in cord blood. J Psychiatr Res. 2013;47:880–91.

50. Provenzi L, Giorda R, Beri S, Montirosso R. SLC6A4 methylation as an epigenetic marker of life adversity exposures in humans: a systematic review of literature. Neurosci Biobehav Rev. 2016;71:7–20.

51. Devlin AM, Brain U, Austin J, Oberlander TF. Prenatal exposure to maternal depressed mood and the MTHFR C677T variant affect SLC6A4 methylation in infants at birth. PLoS One. 2010;5:e12201.

52. Conradt E, Lester BM, Appleton AA, Armstrong DA, Marsit CJ. The roles of DNA methylation of NR3C1 and 11β-HSD2 and exposure to maternal mood disorder in utero on newborn neurobehavior. Epigenetics. 2013;8:1321–9.

53. Ponder KL, Salisbury A, McGonnigal B, Laliberte A, Lester B, Padbury JF. Maternal depression and anxiety are associated with altered gene expression in the human placenta without modification by antidepressant use: implications for fetal programming. Dev Psychobiol. 2011;53:711–23.

54. O'Donnell KJ, Bugge Jensen A, Freeman L, Khalife N, O'Connor TG, Glover V. Maternal prenatal anxiety and downregulation of placental 11β-HSD2. Psychoneuroendocrinology. 2012;37:818–26.

55. Appleton AA, Armstrong DA, Lesseur C, Lee J, Padbury JF, Lester BM, et al. Patterning in placental 11-B hydroxysteroid dehydrogenase methylation according to prenatal socioeconomic adversity. PLoS One. 2013;8:e74691.

56. Bale TL. Epigenetic and transgenerational reprogramming of brain development. Nat Rev Neurosci. 2015;16(6):332–44.

57. Qiao Y, Yang X, Jing N. Epigenetic regulation of early neural fate commitment. Cell Mol Life Sci. 2016;73(7):1399–411.

58. Varese F, Smeets F, Drukker M, Lieverse R, Lataster T, Viechtbauer W, et al. Childhood adversities increase the risk of psychosis: a meta-analysis of patient-control, prospective- and cross-sectional cohort studies. Schizophr Bull. 2012;38:661–71.

59. Brent DA, Silverstein M. Shedding light on the long shadow of childhood adversity. JAMA. 2013;309:1777–8.

60. Strüber N, Strüber D, Roth G. Impact of early adversity on glucocorticoid regulation and later mental disorders. Neurosci Biobehav Rev. 2014;38:17–37.

61. Chiang JJ, Taylor SE, Bower JE. Early adversity, neural development, and inflammation. Dev Psychobiol. 2015;57(8):887–907.

62. Ehrlich KB, Ross KM, Chen E, Miller GE. Testing the biological embedding hypothesis: is early life adversity associated with a later proinflammatory phenotype? Dev Psychopathol. 2016;28(4pt2):1273–83.

63. Danese A, McEwen BS. Adverse childhood experiences, allostasis, allostatic load, and age-related disease. Physiol Behav. 2012;106:29–39.

64. Remmes J, Bodden C, Richter SH, Lesting J, Sachser N, Pape HC, et al. Impact of life history on fear memory and extinction. Front Behav Neurosci. 2016;10:185.

65. Rubin LP. Maternal and pediatric health and disease: integrating biopsychosocial models and epigenetics. Pediatr Res. 2016;79(1–2):127–35.

66. Meaney MJ, Szyf M, Seckl JR. Epigenetic mechanisms of perinatal programming of hypothalamic–pituitary–adrenal function and health. Trends Mol Med. 2007;13:269–77.

67. Hackman DA, Farah MJ, Meaney MJ. Socioeconomic status and the brain: mechanistic insights from human and animal research. Nat Rev Neurosci. 2010;11:651–9.

68. Maccari S, Krugers HJ, Morley-Fletcher S, Szyf M, Brunton PJ. The consequences of early-life adversity: neurobiological, behavioural and epigenetic adaptations. J Neuroendocrinol. 2014;26(10):707–23.

69. Bick J, Naumova O, Hunter S, Barbot B, Lee M, Luthar SS, et al. Childhood adversity and DNA methylation of genes involved in the hypothalamus–pituitary–adrenal axis and immune system: whole-genome and candidate-gene associations. Dev Psychopathol. 2012;24:1417–25.

70. Vialou V, Feng J, Robison AJ, Nestler EJ. Epigenetic mechanisms of depression and antidepressant action. Annu Rev Pharmacol Toxicol. 2013;53:59–87.

71. Smart C, Strathdee G, Watson S, Murgatroyd C, McAllister-Williams RH. Early life trauma, depression and the glucocorticoid receptor gene––an epigenetic perspective. Psychol Med. 2015;45(16):3393–410.

72. Turecki G, Meaney MJ. Effects of the social environment and stress on glucocorticoid receptor gene methylation: a systematic review. Biol Psychiatry. 2016;79(2):87–96.

73. DeLisi M, Vaughn MG. The vindication of Lamarck? Epigenetics at the intersection of law and mental health. Behav Sci Law. 2015;33(5):607–28.

74. Frodl T. Do (epi)genetics impact the brain in functional neurologic disorders? Handb Clin Neurol. 2017;139:157–65.

75. Heim C, Newport DJ, Mletzko T, Miller AH, Nemeroff CB. The link between childhood trauma and depression: insights from HPA axis studies in humans. Psychoneuroendocrinology. 2008;33:693–710.

76. Heim C, Binder EB. Current research trends in early life stress and depression: review of human studies on sensitive periods, gene-environment interactions, and epigenetics. Exp Neurol. 2012;233:102–11.

77. Heim C, Newport DJ, Heit S, Graham YP, Wilcox M, Bonsall R, et al. Pituitary-adrenal and autonomic responses to stress in women after sexual and physical abuse in childhood. JAMA. 2000;284:592–7.

78. Frazzetto G, Di Lorenzo G, Carola V, Proietti L, Sokolowska E, Siracusano A, et al. Early trauma and increased risk for physical aggression during adulthood: the moderating role of MAOA genotype. PLoS One. 2007;2(5):e486.

79. Byrd AL, Manuck SB. MAOA, childhood maltreatment, and antisocial behavior: meta-analysis of a gene–environment interaction. Biol Psychiatry. 2014;75(1):9–17.

80. Borghol N, Suderman M, McArdle W, Racine A, Hallett M, Pembrey M, et al. Associations with early-life socio-economic position in adult DNA methylation. Int J Epidemiol. 2012;41:62–74.

81. Miller GE, Chen E, Fok AK, Walker H, Lim A, Nicholls EF, et al. Low early-life social class leaves a biological residue manifested by decreased glucocorticoid and increased proinflammatory signaling. Proc Natl Acad Sci U S A. 2009;106:14716–21.

82. Chen EE, Miller GE, Kobor MS, Cole SW. Maternal warmth buffers the effects of low early-life socioeconomic status on pro-inflammatory signaling in adulthood. Mol Psychiatry. 2011;16:729–37.

83. Hornung OP, Heim CM. Gene–environment interactions and intermediate phenotypes: early trauma and depression. Front Endocrinol (Lausanne). 2014;5:14.

84. Labonté B, Suderman M, Maussion G, Navaro L, Yerko V, Mahar I, et al. Genome-wide epigenetic regulation by early-life trauma. Arch Gen Psychiatry. 2012;69.722–31.

85. Suderman M, Borghol N, Pappas JJ, Pinto Pereira SM, Pembrey M, Hertzman C, et al. Childhood abuse is associated with methylation of multiple loci in adult DNA. BMC Med Genet. 2014;7:13.

86. Provençal N, Suderman MJ, Guillemin C, Vitaro F, Côté SM, Hallett M, et al. Association of childhood chronic physical aggression with a DNA methylation signature in adult human T cells. PLoS One. 2014;9:e89839.

87. Julian MM. Age at adoption from institutional care as a window into the lasting effects of early experiences. Clin Child Fam Psychol Rev. 2013;16(2):101–45.

88. Naumova O, Lee M, Koposov R, Szyf M, Dozier M, Grigorenko EL. Differential patterns of whole-genome DNA methylation in institutionalized children and children raised by their biological parents. Dev Psychopathol. 2012;24:143–55.

89. McGowan PO, Sasaki A, Huang TC, Unterberger A, Suderman M, Ernst C, et al. Promoter-wide hypermethylation of the ribosomal RNA gene promoter in the suicide brain. PLoS One. 2008;3:e2085.

90. McGowan PO, Sasaki A, D'Alessio AC, Dymov S, Labonté B, Szyf M, et al. Epigenetic regulation of the glucocorticoid receptor in human brain associates with childhood abuse. Nature Neurosci. 2009;12:342–8.

91. Labonté B, Yerko V, Gross J, Mechawar N, Meaney MJ, Szyf M, et al. Differential glucocorticoid receptor exon 1(B), 1(C), and 1(H) expression and methylation in suicide completers with a history of childhood abuse. Biol Psychiatry. 2012b;72:41–8.

92. Bustamante AC, Aiello AE, Galea S, Ratanatharathorn A, Noronha C, Wildman DE, et al. Glucocorticoid receptor DNA methylation, childhood maltreatment and major depression. J Affect Disord. 2016;206:181–8.

93. Tyrka AR, Ridout KK, Parade SH. Childhood adversity and epigenetic regulation of glucocorticoid signaling genes: associations in children and adults. Dev Psychopathol. 2016;28(4pt2):1319–31.

94. Tyrka AR, Parade SH, Welch ES, Ridout KK, Price LH, Marsit C, et al. Methylation of the leukocyte glucocorticoid receptor gene promoter in adults: associations with early adversity and depressive, anxiety and substance-use disorders. Transl Psychiatry. 2016;6(7):e848.

95. Klengel T, Mehta D, Anacker C, Rex-Haffner M, Pruessner JC, Pariante CM, Pace TW, et al. Allele-specific FKBP5 DNA demethylation mediates gene-childhood trauma interactions. Nat Neurosci. 2013;16(1):33–41.

96. Tyrka AR, Price LH, Marsit C, Walters OC, Carpenter LL. Childhood adversity and epigenetic modulation of the leukocyte glucocorticoid receptor: preliminary findings in healthy adults. PLoS One. 2012;7:e30148.

97. Perroud N, Paoloni-Giacobino A, Prada P, Olié E, Salzmann A, Nicastro R, et al. Increased methylation of glucocorticoid receptor gene (NR3C1) in adults with a history of childhood maltreatment: a link with the severity and type of trauma. Transl Psychiatry. 2011;1:e59.

98. Steiger H, Labonté B, Groleau P, Turecki G, Israel M. Methylation of the glucocorticoid receptor gene promoter in bulimic women: associations with borderline personality disorder, suicidality, and exposure to childhood abuse. Int J Eat Disord. 2013;46:246–55.

99. Booij L, Szyf M, Carballedo A, Frey EM, Morris D, Dymov S, et al. DNA methylation of the serotonin transporter gene in peripheral cells and stress-related changes in hippocampal volume: a study in depressed patients and healthy controls. PLoS One. 2015;10(3):e0119061.

100. Miller G, Chen E. Unfavorable socioeconomic conditions in early life presage expression of proinflammatory phenotype in adolescence. Psychosom Med. 2007;69:402–9.

101. Witek-Janusek L, Tell D, Gaylord-Harden N, Mathews HL. Relationship of childhood adversity and neighborhood violence to a proinflammatory phenotype in emerging adult African American men: an epigenetic link. Brain Behav Immun. 2017;60:126–35.

102. Davidson RJ, McEwen BS. Social influences on neuroplasticity: stress and interventions to promote well-being. Nat Neurosci. 2012;15:689–95.

103. Vaiserman A. Early-life origin of adult disease: evidence from natural experiments. Exp Gerontol. 2011;46(2–3):189–92.

104. Mill J, Heijmans BT. From promises to practical strategies in epigenetic epidemiology. Nat Rev Genet. 2013;14:585–94.

105. Entringer S, Wadhwa PD. Developmental programming of obesity and metabolic dysfunction: role of prenatal stress and stress biology. Nestle Nutr Inst Workshop Ser. 2013;74:107–20.

106. Steiger H, Thaler L. Eating disorders, gene–environment interactions and the epigenome: roles of stress exposures and nutritional status. Physiol Behav. 2016;162:181–5.

107. Kesternich I, Siflinger B, Smith JP, Winter JK. The effects of World War II on economic and health outcomes across Europe. Rev Econ Stat. 2014;96:103–18.

108. Yehuda R, Bierer LM. Transgenerational transmission of cortisol and PTSD risk. Prog Brain Res. 2008;167:121–35.

109. Bercovich E, Keinan-Boker L, Shasha SM. Long-term health effects in adults born during the Holocaust. Isr Med Assoc J. 2014;16:203–7.

110. Keinan-Boker L, Shasha-Lavsky H, Eilat-Zanani S, Edri-Shur A, Shasha SM. Chronic health conditions in Jewish Holocaust survivors born during World War II. Isr Med Assoc J. 2015;17(4):206–12.

111. Malaspina D, Corcoran C, Kleinhaus KR, Perrin MC, Fennig S, Nahon D, et al. Acute maternal stress in pregnancy and schizophrenia in offspring: a cohort prospective study. BMC Psychiatry. 2008;8:71.

112. Kleinhaus K, Harlap S, Perrin M, Manor O, Margalit-Calderon R, Opler M, et al. Prenatal stress and affective disorders in a population birth cohort. Bipolar Disord. 2013;15(1):92–9.

113. St Clair D, Xu M, Wang P, Yu Y, Fang Y, Zhang F, et al. Rates of adult schizophrenia following prenatal exposure to the Chinese famine of 1959–1961. JAMA. 2005;294(5):557–62.

114. MQ X, Sun WS, Liu BX, Feng GY, Yu L, Yang L, et al. Prenatal malnutrition and adult schizophrenia: further evidence from the 1959–1961 Chinese famine. Schizophr Bull. 2009;35(3):568–76.

115. Song S, Wang W, Hu P. Famine, death, and madness: schizophrenia in early adulthood after prenatal exposure to the Chinese Great Leap Forward Famine. Soc Sci Med. 2009;68(7):1315–21.

116. Huang C, Phillips MR, Zhang Y, Zhang J, Shi Q, Song Z, et al. Malnutrition in early life and adult mental health: evidence from a natural experiment. Soc Sci Med. 2013;97:259–66.

117. Wang C, An Y, Yu H, Feng L, Liu Q, Lu Y, et al. Association between exposure to the Chinese famine in different stages of early life and decline in cognitive functioning in adulthood. Front Behav Neurosci. 2016;10:146.

118. de Rooij SR, Painter RC, Phillips DI, Osmond C, Tanck MW, Bossuyt PM, et al. Cortisol responses to psychological stress in adults after prenatal exposure to the Dutch famine. Psychoneuroendocrinology. 2006;31(10):1257–65.

119. de Rooij SR, Veenendaal MV, Räikkönen K, Roseboom TJ. Personality and stress appraisal in adults prenatally exposed to the Dutch famine. Early Hum Dev. 2012;88:321–5.

120. Susser E, St Clair D. Prenatal famine and adult mental illness: interpreting concordant and discordant results from the Dutch and Chinese Famines. Soc Sci Med. 2013;97:325–30.

121. Lumey LH, Terry MB, Delgado-Cruzata L, Liao Y, Wang Q, Susser E, et al. Adult global DNA methylation in relation to pre-natal nutrition. Int J Epidemiol. 2012;41:116–23.

122. Tobi EW, Slieker RC, Stein AD, Suchiman HE, Slagboom PE, van Zwet EW, et al. Early gestation as the critical time-window for changes in the prenatal environment to affect the adult human blood methylome. Int J Epidemiol. 2015;44(4):1211–23.

123. Heijmans BT, Tobi EW, Stein AD, Putter H, Blauw GJ, Susser ES, et al. Persistent epigenetic differences associated with prenatal exposure to famine in humans. Proc Natl Acad Sci U S A. 2008;105:17046–9.

124. Tobi EW, Lumey LH, Talens RP, Kremer D, Putter H, Stein AD, et al. DNA methylation differences after exposure to prenatal famine are common and timing- and sex-specific. Hum Mol Genet. 2009;18:4046–53.

125. Heijmans BT, Tobi EW, Lumey LH, Slagboom PE. The epigenome: archive of the prenatal environment. Epigenetics. 2009;4:526–31.

126. Roseboom TJ, Painter RC, van Abeelen AF, Veenendaal MV, de Rooij SR. Hungry in the womb: what are the consequences? Lessons from the Dutch famine. Maturitas. 2011;70:141–5.

127. Gröger N, Matas E, Gos T, Lesse A, Poeggel G, Braun K, et al. The transgenerational transmission of childhood adversity: behavioral, cellular, and epigenetic correlates. J Neural Transm (Vienna). 2016;123(9):1037–52.

128. Yehuda R, Daskalakis NP, Lehrner A, Desarnaud F, Bader HN, Makotkine I, et al. Influences of maternal and paternal PTSD on epigenetic regulation of the glucocorticoid receptor gene in Holocaust survivor offspring. Am J Psychiatry. 2014;171(8):872–80.

129. Perroud N, Rutembesa E, Paoloni-Giacobino A, Mutabaruka J, Mutesa L, Stenz L, et al. The Tutsi genocide and transgenerational transmission of maternal stress: epigenetics and biology of the HPA axis. World J Biol Psychiatry. 2014;15(4):334–45.

130. Gapp K, Bohacek J, Grossmann J, Brunner AM, Manuella F, Nanni P, et al. Potential of environmental enrichment to prevent transgenerational effects of paternal trauma. Neuropsychopharmacology. 2016;41(11):2749–58.

131. Puterman E, Gemmill A, Karasek D, Weir D, Adler N.E, Prather AA, et al. Lifespan adversity and later adulthood telomere length in the nationally representative US Health and Retirement Study. Proc Natl Acad Sci U S A 2016;113(42):E6335–E6342.

132. Humphreys KL, Esteves K, Zeanah CH, Fox NA, Nelson CA 3rd, et al. Accelerated telomere shortening: tracking the lasting impact of early institutional care at the cellular level. Psychiatry Res. 2016;246:95–100.

133. Mitchell C, Hobcraft J, McLanahan SS, Siegel SR, Berg A, Brooks-Gunn J, et al. Social disadvantage, genetic sensitivity, and children's telomere length. Proc Natl Acad Sci U S A. 2014;111(16):5944–9.

134. Scott BR, Belinsky SA, Leng S, Lin Y, Wilder JA, Damiani LA. Radiation-stimulated epigenetic reprogramming of adaptive-response genes in the lung: an evolutionary gift for mounting adaptive protection against lung cancer. Dose-Response. 2009;7:104–31.

135. Barouki R, Gluckman PD, Grandjean P, Hanson M, Heindel JJ. Developmental origins of non-communicable disease: implications for research and public health. Environ Health. 2012;11:42.

136. Godfrey KM, Costello PM, Lillycrop KA. The developmental environment, epigenetic biomarkers and long-term health. J Dev Orig Health Dis. 2015;6(5):399–406.

137. Palma-Gudiel H, Córdova-Palomera A, Leza JC, Fañanás L. Glucocorticoid receptor gene (NR3C1) methylation processes as mediators of early adversity in stress-related disorders causality: a critical review. Neurosci Biobehav Rev. 2015;55:520–35.

138. Turecki G, Ota VK, Belangero SI, Jackowski A, Kaufman J. Early life adversity, genomic plasticity, and psychopathology. Lancet Psychiatry. 2014;1(6):461–6.

139. Vaiserman AM. Early-life nutritional programming of longevity. J Dev Orig Health Dis. 2014;5:325–38.

140. Kundakovic M, Jaric I. The epigenetic link between prenatal adverse environments and neurodevelopmental disorders. Genes (Basel). 2017;8(3):104.

141. McGowan PO. Epigenetic clues to the biological embedding of early life adversity. Biol Psychiatry. 2012;72:4–5.

142. Talens RP, Christensen K, Putter H, Willemsen G, Christiansen L, Kremer D, et al. Epigenetic variation during the adult lifespan: cross-sectional and longitudinal data on monozygotic twin pairs. Aging Cell. 2012;11:694–703.

143. Sen P, Shah PP, Nativio R, Berger SL. Epigenetic mechanisms of longevity and aging. Cell. 2016;166(4):822–39.

144. Penney J, Tsai LH. Histone deacetylases in memory and cognition. Sci Signal. 2014;7(355):re12.

145. Fuchikami M, Yamamoto S, Morinobu S, Okada S, Yamawaki Y, Yamawaki S. The potential use of histone deacetylase inhibitors in the treatment of depression. Prog Neuro-Psychopharmacol Biol Psychiatry. 2016;64:320–4.

146. Qiu X, Xiao X, Li N, Li Y. Histone deacetylases inhibitors (HDACis) as novel therapeutic application in various clinical diseases. Prog Neuro-Psychopharmacol Biol Psychiatry. 2017;72:60–72.

147. Moretti M, Valvassori SS, Varela RB, Ferreira CL, Rochi N, Benedet J, et al. Behavioral and neurochemical effects of sodium butyrate in an animal model of mania. Behav Pharmacol. 2011;22(8):766–72.

148. Lopes-Borges J, Valvassori SS, Varela RB, Tonin PT, Vieira JS, Gonçalves CL, et al. Histone deacetylase inhibitors reverse manic-like behaviors and protect the rat brain from energetic metabolic alterations induced by ouabain. Pharmacol Biochem Behav. 2015, 128:89–95.

149. Resende WR, Valvassori SS, Réus GZ, Varela RB, Arent CO, Ribeiro KF, et al. Effects of sodium butyrate in animal models of mania and depression: implications as a new mood stabilizer. Behav Pharmacol. 2013;24(7):569–79.

150. Valvassori SS, Varela RB, Arent CO, Dal-Pont GC, Bobsin TS, Budni J, et al. Sodium butyrate functions as an antidepressant and improves cognition with enhanced neurotrophic expression in models of maternal deprivation and chronic mild stress. Curr Neurovasc Res. 2014;11(4):359–66.

151. Valvassori SS, Resende WR, Budni J, Dal-Pont GC, Bavaresco DV, Réus GZ, et al. Sodium butyrate, a histone deacetylase inhibitor, reverses behavioral and mitochondrial alterations in animal models of depression induced by early- or late-life stress. Curr Neurovasc Res. 2015;12(4):312–20.

152. Takuma K, Hara Y, Kataoka S, Kawanai T, Maeda Y, Watanabe R, et al. Chronic treatment with valproic acid or sodium butyrate attenuates novel object recognition deficits and hippocampal dendritic spine loss in a mouse model of autism. Pharmacol Biochem Behav. 2014;126:43–9.

153. Kratsman N, Getselter D, Elliott E. Sodium butyrate attenuates social behavior deficits and modifies the transcription of inhibitory/excitatory genes in the frontal cortex of an autism model. Neuropharmacology. 2016;102:136–45.

154. Zhong T, Qing QJ, Yang Y, Zou WY, Ye Z, Yan JQ, et al. Repression of contexual fear memory induced by isoflurane is accompanied by reduction in histone acetylation and rescued by sodium butyrate. Br J Anaesth. 2014; 113(4):634–43.

155. Sharma S, Taliyan R, Singh S. Beneficial effects of sodium butyrate in 6–OHDA induced neurotoxicity and behavioral abnormalities: modulation of histone deacetylase activity. Behav Brain Res. 2015;291:306–14.

156. Liu H, Zhang JJ, Li X, Yang Y, Xie XF, Hu K. Post-occlusion administration of sodium butyrate attenuates cognitive impairment in a rat model of chronic cerebral hypoperfusion. Pharmacol Biochem Behav. 2015;135:53–9.

157. Cuadrado-Tejedor M, Ricobaraza AL, Torrijo R, Franco R, Garcia-Osta A. Phenylbutyrate is a multifaceted drug that exerts neuroprotective effects and reverses the Alzheimer's disease-like phenotype of a commonly used mouse model. Curr Pharm Des. 2013;19(28):5076–84.

158. Sandner G, Host L, Angst MJ, Guiberteau T, Guignard B, Zwiller J. The HDAC inhibitor phenylbutyrate reverses effects of neonatal ventral hippocampal lesion in rats. Front Psychiatry. 2011;1:153.

159. Moloney RD, Stilling RM, Dinan TG, Cryan JF. Early-life stress-induced visceral hypersensitivity and anxiety behavior is reversed by histone deacetylase inhibition. Neurogastroenterol Motil. 2015;27(12):1831–6.

17

Molecular characterization of exonic rearrangements and frame shifts in the *dystrophin* gene in Duchenne muscular dystrophy patients in a Saudi community

Nasser A. Elhawary[1,2*], Essam H. Jiffri[3], Samira Jambi[4], Ahmad H. Mufti[1], Anas Dannoun[1], Hassan Kordi[1], Asim Khogeer[5], Osama H. Jiffri[3], Abdelrahman N. Elhawary[6] and Mohammed T. Tayeb[1]

Abstract

Background: In individuals with Duchenne muscular dystrophy (DMD), exon skipping treatment to restore a wild-type phenotype or correct the frame shift of the mRNA transcript of the *dystrophin* (*DMD*) gene are mutation-specific. To explore the molecular characterization of *DMD* rearrangements and predict the reading frame, we simultaneously screened all 79 *DMD* gene exons of 45 unrelated male DMD patients using a multiplex ligation-dependent probe amplification (MLPA) assay for deletion/duplication patterns. Multiplex PCR was used to confirm single deletions detected by the MLPA.

Results: There was an obvious diagnostic delay, with an extremely statistically significant difference between the age at initial symptoms and the age of clinical evaluation of DMD cases (*t* value, 10.3; 95% confidence interval 5.95–8.80, $P < 0.0001$); the mean difference between the two groups was 7.4 years. Overall, we identified 147 intragenic rearrangements: 46.3% deletions and 53.7% duplications. Most of the deletions (92.5%) were between exons 44 and 56, with exon 50 being the most frequently involved (19.1%). Eight new rearrangements, including a mixed deletion/duplication and double duplications, were linked to seven cases with DMD. Of all the cases, 17.8% had duplications with no hot spots. In addition, confirmation of the reading frame hypothesis helped account for new *DMD* rearrangements in this study. We found that 81% of our Saudi patients would potentially benefit from exon skipping, of which 42.9% had a mutation amenable to skipping of exon 51.

Conclusions: Our study could generate considerable data on mutational rearrangements that may promote future experimental therapies in Saudi Arabia.

Keywords: Duchenne muscular dystrophy, *Dystrophin* gene, Large rearrangements, Frame shift, MLPA, Saudi community

Background

Dystrophinopathies are the most common form of muscular dystrophy in childhood. They are caused by mutations in the *dystrophin* gene (*DMD*; OMIM #300377) [1, 2]. Duchenne muscular dystrophy (DMD; OMIM #310200) is a severe form of muscular dystrophy, with an incidence of 1 in 3600–5000 male births [3]. Becker muscular dystrophy (BMD) is a milder form of DMD, with an incidence of 1 in 20,000 male births (BMD; OMIM # 300376) [4].

DMD is characterized by rapidly progressive degeneration and necrosis of the proximal muscles and calf pseudo-hypertrophy. Most *DMD* patients show muscle weakness at age 2 or 3, but it may be seen as early as infancy. Patients commonly lose independent ambulation by the age of 12 and die of dilated cardiomyopathy around the second or third decade. In comparison, patients with BMD exhibit relatively minor pathological

* Correspondence: naelhawary@uqu.edu.sa
[1]Department of Medical Genetics, Medicine College, Umm Al-Qura University, P.O. Box 57543, Mecca 21955, Saudi Arabia
[2]Department of Molecular Genetics, Faculty of Medicine, Ain Shams University, Cairo 11566, Egypt
Full list of author information is available at the end of the article

symptoms, slower progression, later onset, and longer survival. Patients with an intermediate form of the disease, intermediate muscular dystrophy (IMD), may continue to walk until they are 16 years of age [4, 5].

The *DMD* gene is one of the largest known genes in humans, with 79 exons (approximately 2.4 Mb of genomic DNA) [1] expressing a 427-kDa muscular protein that plays a fundamental role in stabilizing the sarcolemma. It does so by using a complex of glycoproteins associated with dystrophin to link actin filaments within the cytoskeleton and the extracellular matrix. Lack of dystrophin breaks these connections, altering the plasma membrane and finally producing myofiber degeneration and necrosis [6]. Thus, according to the reading frame hypothesis [7], *DMD* mutations that destroy the reading frame result in a truncated, non-functional dystrophin protein associated with a "DMD" phenotype. These mutations frequently generate a premature stop codon that activates nonsense-mediated mRNA decay [8]. On the other hand, mutations that maintain the reading frame can permit semi-functional dystrophin protein and thus give rise to a "BMD" phenotype [4, 9]. Together, these two phenotype-genotype correlations explain more than 92% of all cases [7].

Different types of mutations have been reported in patients with DMD and BMD. These are mainly large rearrangements (deletions in approximately 60–70% of patients and duplications in approximately 7–10%), with the remaining being point mutations (mainly nonsense mutations) and small deletions or insertions [10, 11]. Most gross deletions can be detected by multiplex PCR (mPCR) [12, 13] and are clustered in the proximal and central hot spot regions [14, 15]. Although a large proportion of the duplications were reported many years ago [16, 17], most laboratories do not systematically screen for these rearrangements. Duplication analysis and the determination of at-risk carrier status of the *DMD* gene require quantitative investigation, which is laborious and technically demanding [18, 19]. Previous studies have applied Southern blotting [16, 20], pulsed-field gel electrophoresis, quantitative mPCR [21–23], multiplex amplifiable probe hybridization [24], and comparative genomic hybridization microarray [25].

Given that deletions and duplications of one or more exons are found in the majority (70%) of patients, it is most cost-efficient and labor-efficient to check for these mutations first. A reliable and rapid technique, multiplex ligation-dependent probe amplification (MLPA), has been applied to cover the whole *DMD* gene to detect deletions and duplications and to identify exactly which exons are involved in deletions or duplications [18, 26–29]. This approach reveals whether a given exon is present and allows the copy number of each exon to be calculated by comparing relative peak heights. MLPA can detect both

deletions and duplications in patients as well as in female carriers. Compared to array comparative genomic hybridization, MLPA is a low-cost and technically uncomplicated method.

Although several studies have investigated exonic deletions in different populations [11, 30–36], it is unknown where these deletions occur in the Saudi population. Although a few studies have described the molecular diagnosis of DMD in Saudi patients, the large deletions associated with disease were examined in only some exons, and the studies were limited by small sample sizes [37, 38]. Molecular characterization of the large *DMD* gene has been proposed to address large intragenic rearrangements in the whole exons of the *DMD* gene using an MLPA strategy covering nearly 75% of whole gene mutations. Thus, accurate molecular diagnosis may provide information on eligibility for mutation-specific treatments. A plausible frame shift hypothesis suggests how one might reduce disease severity via exon skipping, for example, by correcting the fidelity of the translational reading frame with large *DMD* deletions or restoring the wild type with large *DMD* duplications.

To our knowledge, the present study is the first study using the MLPA strategy to identify genotype-phenotype correlations in *DMD* patients in a Saudi community. Our results will add valuable data on de novo mutations in this population and to the databases of different DMD web pages as well.

Methods

Ethics statement and participants

All participants were enrolled under a protocol approved by the Institutional Biomedical Ethics Committee at Umm Al-Qura University (ref. #HAPO-02-K-012). Parents of all participants gave written consent after being informed about the aim of the study.

The study included 45 unrelated male patients with DMD selected from 65 families from the western region of the Kingdom of Saudi Arabia (KSA), including Jeddah, Mecca, Taif, and Hada. Twenty additional eligible male patients did not enroll because their parents refused to share their clinical data, their clinical profiles were incomplete, or their creatine phosphokinase assessments were missing. For each patient included in the study, a clinical data sheet was recorded in the database of the Molecular Genetics Laboratory in the Department of Medical Genetics at Umm Al-Qura University. Clinical information was independent of any molecular DNA data for the *DMD* gene or its protein. *Dystrophin* probands were diagnosed by clinical geneticists or pediatricians based on strict criteria including a clinical presentation expected for DMD, family history of X-linked muscular dystrophy, or muscle biopsy with a dystrophin analysis performed using immunohistochemistry. Clinical diagnosis of dystrophin probands included age at onset, age at clinical evaluation, calf pseudohypertrophy, age at wheelchair confinement, cardiac function, and motor function. A histopathological study was performed before molecular DNA analysis if muscle biopsies were available. To avoid bias, we included only one case for each family. We categorized patients according to age at loss of ambulation: DMD ≤ 12 years, IMD 12–16 years, and BMD > 16 years. Cases with a family history of autosomal recessive inheritance or with normal dystrophin protein were excluded.

DNA isolation

Genomic DNA was isolated from buccal cells using the Oragene DNA-OGR-575 kit (DNA Genotek Inc., Ottawa, ON, Canada) according to the manufacturer's protocol with some modifications. Briefly, the full buccal cells were collected within 30 min, and the Oragene tube was capped immediately. The cells were incubated with the OGR-lysis buffer in a water bath at 53 °C to release the DNA, which was then precipitated by ethanol and dissolved in elution buffer [39].

Multiplex polymerase chain reaction

The genomic DNA of all *DMD* patients was subjected to multiplex PCR (mPCR) to screen for *DMD* deletions using 15 primer sets (Additional file 1: Table S1). The oligonucleotides included flanking sequences of exons 4, 8, 12, 17, 19, 44, 45, 48, and 51 [12] and of exons 6, 13, 47, 50, 52, and 60 [13]. We made some modifications to Chamberlain's mPCR set by not adding dimethylsulfoxide, which could result in a lower PCR yield. However, PCR cycling was programmed as initial denaturing at 95 °C for 6 min (1 round), then 94 °C for 30 s, annealing at 53 °C for 30 s, 65 °C for 4 min (repeated for 23 rounds), and final elongation at 65 °C for 7 min [12]. Hot-start mPCR was performed using Beggs' PCR program [13]: 95 °C for 6 min (1 round) and 25 subsequent cycles including DNA denaturing at 95 °C for 30 s, annealing at 56 °C for 1 min, and elongation at 68 °C for 4 min. Amplification reactions were carried out on thermal cycler Engine Dyad (Bio-Rad Laboratories Inc., Hercules, CA). PCR products (10–15 μl) were separated on 3% NuSieve agarose (BMA Bioproducts, Rockland, ME). The gels were viewed using the Gel Documentation and Analysis System (G-Box, SynGene, Frederick, MD, USA).

Multiplex ligation-dependent probe amplification

We analyzed all DMD cases for large deletions and large duplications using MLPA SALSA P034/P035 DMD kits (http://www.mrc-holland.com) following the manufacturer's instructions. In brief, denaturation, hybridization, ligation, and amplification steps were performed on a DNA Engine Dyad thermal cycler (Bio-Rad Laboratories

Inc., Hercules, CA). Finally, PCR amplification was performed using SALSA MLPA PCR primers labeled with the FAM dye. A mixture of 0.7 μl of PCR product, 0.2 μl of 600 LIZ GS size-standard, and 9.0 μl of Hi-Di formamide was incubated for 3 min at 86 °C and cooled at 4 °C for 2 min. The MLPA product mix was separated on a POP7 polymer (Applied Biosystems Inc., Life Technologies, Foster City, CA) at 60 °C with the setting of 1. 6 kV for injection voltage, 18 s for injection time, 15 kV for run voltage, and 1800 s for run time.

Data analysis

The raw data were analyzed using GeneMapper Software 5 (Applied Biosystems Inc., Life Technologies, Foster City, CA). The DNA of cases with single-exon deletions were re-examined using conventional PCR. Initial analysis was performed with the naked eye to look for missed exon-specific peaks. For the remaining samples, the peak height of each exon was divided by the two nearest control peaks. The median ratio across all samples for each peak was calculated and used as a reference for one copy. For the sake of accuracy, any normalized ratio below 0.3 was considered a possible deletion. A duplication was considered if a normalized ratio was 1.8–2.0. If any single-exon deletion was identified, conventional PCR amplification was carried out to validate this deletion using primer sets and PCR conditions given in the Leiden Muscular Dystrophy pages (http://www.dmd.nl).

Databases and confirming mutations for the DMD gene

We checked all mutations recorded in this study according to available databases established by the Leiden Muscular Dystrophy pages (http://www.dmd.nl) [40], the Leiden Open Variation Database 3.0 (http://www.lovd.nl/3.0/home) [41], and UMD-DMD (http://www.umd.be/DMD/) [31, 32]. Databases for exon skipping to restore the DMD reading frame were found on sites developed by Leiden University Medical Center (http://www.exonskipping.nl/?s=exon+skipping&submit=Go) and CureDuchenne (https://www.cureduchenne.org/cure/edystrophin/).

Statistical analysis

Hardy-Weinberg equilibrium (HWE) deviation was examined for X-linked *DMD* cases in this study using the Online Encyclopedia for Genetic Epidemiology studies software (http://www.oege.org/software/hwe-mr-calc.shtml). We used the G*Power Software (http://www.psycho.uni-duesseldorf.de/abteilungen/aap/gpower3/download-and-register/) to estimate power analysis to determine adequate sample sizes to achieve an 80% power for t testing of point biserial model. "*Priori*" sample size and "*post hoc*" power estimations were tested knowing our DMD sample size, a probability of $\alpha = 0.05$, and the effect size index "r" (the absolute value of the correlation coefficient in the population, 0 < "r" < 1). We used paired *t* test analysis to compare the significant difference between the age at onset and the age of clinical evaluation for each *DMD* case. A two-sided *P* value less than 0.05 was considered to indicate statistical significance and 95% confidence interval (CI) for all analyses.

Results
Clinical profile
Among 45 unrelated patients, 21 were diagnosed with DMD, 10 with IMD, and 5 with BMD. The unassigned patients were defined as not determined (ND), as they were too young to permit a definitive diagnosis (*n* = 9). The median age at onset was 3.5 years (range 1.0–7.0 years), while the median presenting age was 11.5 years (1.5–20 years) (Fig. 1). Most of the patients reported initial symptoms between 1 and 3 years of age (71.1%, 32/45), followed by those reporting symptoms at 4–5 years (24.4%, 11/45). The age at clinical evaluation was most frequently between 10 and 12 years (35.6%, 16/45). We found an extremely statistically significant difference between the age at initial symptoms and the age at clinical evaluation of DMD cases (*t* value, 10.3; 95% CI 5.95–8.8, *P* < 0.0001). The mean difference in age between the two groups was 7.4 years.

Hardy-Weinberg equilibrium
All affected males were in HWE at the *DMD* gene deletions/duplications ($\chi^2 = 1.00$, *P* = 0.317), where the heterozygotes were absent in such X-linked recessive mode of inheritance.

Large-scale rearrangements
Using mPCR, we identified 55 large deletions in the 45 unrelated *DMD* patients. MLPA detected 147 intragenic rearrangements, 68 (46.3%) of which were large

deletions and 79 (53.7%) of which were large duplications. All deletions identified by mPCR were confirmed by the MLPA-based screening. The utility of MLPA assay for all exons is clear, as 13 (19%) of 68 deletions were detected using MLPA but were not detected by conventional mPCR analysis. The percentage of cases with deletions and duplications were 46.7% (21/45) and 17.8% (8/45), respectively. Table 1 includes the large rearrangements that were identified in the present study and had been previously described.

New mutations in the *DMD* gene
We also identified seven previously undescribed large *DMD* rearrangements from eight Saudi cases. These new mutations, based on the *DMD* databases [31, 32, 40, 41], included one mixed rearrangement (del 45–52 + dup 21–23), one large deletion (del 45–56), two large duplications (8–30 and 17–24), and three double duplications (dup 2–4 + dup 18–19, dup 13 + dup 21–24, and dup 56–58 + dup 62–64). These large mutations were from eight (17.8%) of the Saudi patients (Table 2).

Distribution of rearrangements
In this study, deletions did not have a random distribution. We found that 92.5% (63/68) of hot spot deletions were linked to exons 44–56 (central region), whereas 7.5% (5/68) of deletions were related to exons 10–20. Exon 50 was most frequently involved in deletions (19.1%, 13/68), followed by exons 48 and 49 (each 11.8%, 8/68). The rate of deletions increased from a minimum in exon 44 to a maximum in exon 50 and then decreased until the 3′ end of the *DMD* gene, with no deletion in exons 57–79 (distal region) (Fig. 2). Moreover, we found that the number of cases with a deletion of *only* one exon was lower than the number with deletions of more than one exon (9/21, 42.8% versus 12/21, 57.1%). About half of the deletions (44.4%) were detected only once, in

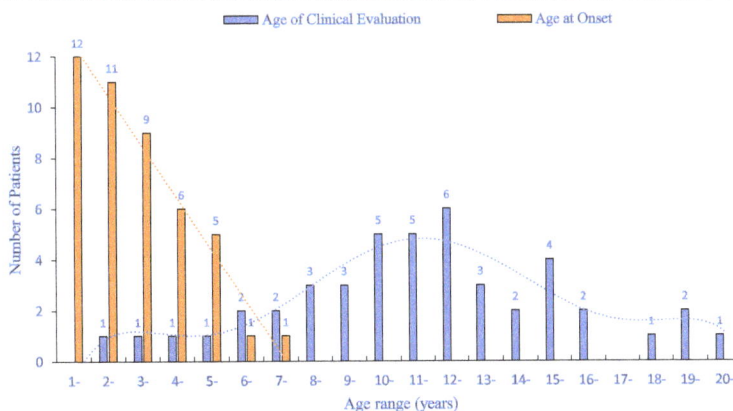

Fig. 1 The age at onset and the age of clinical evaluation of DMD patients in this study. The analysis of DMD cases showed an apparent diagnostic delay

Table 1 Previously described large rearrangements identified in this study and their reading frame shifts

Family no.	Phenotype	Multiplex PCR	MLPA del/dup	Exon(s) del/dup	Codons del/dup	Frame shift	Amino acid change[a]	cDNA[a]
DS-23	DMD	No del	Del 10–11	371	123 [2/3]	Stop at 323	p.His321PhefsX3	c.961_1331del
DS-1	DMD	Del 19	Del 18–20	454	151 [1/3]	−1	p.Arg723Lys874	c.2169_2622del
DS-37	IMD	Del 44	Del 44	148	49 [1/3]	Stop at 2113	p.Arg2098AsnfsX16	c.6291_6438del
DS-34	ND	Del 44–48	Del 44–48	808	269 [1/3]	−1	p.Arg2098Gln2366del	c.6291_7098del
DS-24	DMD	Del 45	Del 45	176	58 [2/3]	Stop at 2163	p.Glu2147AlafsX17	c.6439_6614del
DS-8	IMD	Del 45–50	Del 45–50	871	290 [1/3]	Stop at 2155	p.Glu2147LeufsX9	c.6439_7309del
DS-38	IMD	Del 45–52	Del 45–52	1222	407 [1/3]	Stop at 2168	p.Glu2147LeufsX22	c.6439_7660del
DS-29	DMD	Del 47–50	Del 47–50	547	182 [1/3]	Stop at 2263	p.Val2257LeufsX7	c.6763_7309del
DS-20	ND	Del 47–50	Del 47–50	547	182 [1/3]	Stop at 2263	p.Val2257LeufsX7	c.6763_7309del
DS-48	BMD	Del 48	Del 48	186	62	In-frame	p.Val2305Gln2366del	c.6913_7098del
DS-30	DMD	Del 50	Del 49–50	211	70 [1/3]	Stop at 2375	p.Glu2367LeufsX9	c.7099_7309del
DS-31	IMD	Del 50	Del 49–50	211	70 [1/3]	Stop at 2375	p.Glu2367LeufsX9	c.7099_7309del
DS-36	DMD	Del 50	Del 50	109	36 [1/3]	Stop at 2409	p.Arg2401LeufsX9	c.7201_7309del
DS-32	IMD	Del 50	Del 50	109	36 [1/3]	Stop at 2409	p.Arg2401LeufsX9	c.7201_7309del
DS-33	ND	Del 50	Del 50	109	36 [1/3]	Stop at 2409	p.Arg2401LeufsX9	c.7201_7309del
DS-35	ND	Del 50	Del 50	109	36 [1/3]	Stop at 2409	p.Arg2401LeufsX9	c.7201_7309del
DS-27	ND	Del 50–52	Del 50–52	460	153 [1/3]	Stop at 2422	p.Arg2401LeufsX22	c.7201_7660del
DS-12	DMD	Del 51	Del 51	233	77 [2/3]	Stop at 2469	p.Ser2437CysfsX33	c.7310_7542del
DS-18	DMD	No del	Del 55	190	63 [1/3]	Stop at 2700	p.Val2677ThrfsX24	c.8028_8217del
DS-11	ND	No del	Dup 50–51	343	114	In-frame	p.Arg2401Lys2514dup	c.7201-?_7542+?dup

DMD Duchenne muscular dystrophy, *IMD* intermediate muscular dystrophy, *BMD* Becker muscular dystrophy, *ND* not determined
[a]These data are based on the Leiden Muscular Dystrophy Pages (http://www.dmd.nl/) and the UMD-DMD (http://www.umd.be/DMD/)

agreement with the high allelic heterogeneity of the *DMD* gene.

Duplications were distributed in the proximal (68/79, 86.1%), central (5/79, 6.3%), and distal regions (6/79, 7.6%) (Fig. 2). Unlike deletions, duplicated exons were more frequent in the proximal region (26 duplications) than in the central and distal regions (13 duplications). The most frequent duplications were of exons 21, 22, and 23 (7.6%, 6/79 each), followed by exons 18 and 19 (5.1%, 4/79 each). We did not find any duplications in exons 31–49 (central region) or exons 69–79 (distal region) within our cases (Fig. 2). Similar to deletions, 42.5% of exonic duplications (17/40) were observed only once, revealing a considerable heterogeneity of duplications.

Reading frame shift and phenotype correlation

Gene rearrangements (deletions and duplications) were correlated with clinical phenotypes in 28 unrelated cases: 11 (39.3%) with DMD, 7 (25%) with IMD, 3 (10.7%) with BMD, and 7 (25%) with ND. We also predicted the translational reading frame in 28 DMD cases with rearrangements identified in this study, using the Leiden Muscular Dystrophy pages (http://www.dmd.nl). Applying the reading frame rule revealed consistency with the frame shift rule for 90.9% (10/11) of the individuals with

Table 2 New *DMD* mutational rearrangements identified in this study and their predicted reading frame shifts

Case no.	Phenotype	Multiplex PCR	MLPA del/dup	Exon(s) del/dup	Codons del/dup	Frame shift	Amino acid change[a]
DS-2	DMD	No del	dup 2–4 and dup 18–19	233; 212	77 [2/3] 70 [2/3]	+ 2 + 2	p.Val89MetfsX15 [b] p.Lys724Gly795fsX1
DS-14	DMD	No del	dup 8–30	3679	1226 [1/3]	+ 1	p.Val218K1412fsX
DS-15	ND	No del	dup 8–30	3679	1226 [1/3]	+ 1	p.Val218K1412fsX
DS-52	BMD	No del	dup 13 and dup 21–24	120 654	40 218	0 0	p.Val495Val535dup p.Asp875Lys1093dup
DS-50	BMD	No del	dup 17–24	1248	428	0	p.Ile665Lys1093dup
DS-53	IMD	del 45–51	del 45–52 and dup 21–23	1222; 540	407 [1/3] 180	− 1[c] 0	p.Glu2147LeufsX22 [c] p.Asp875Asn1055dup
DS-25	IMD	del 45–51	del 45–56	1952	650 [2/3]	− 2	p.Glu2147Ser2798
DS-22	DMD	No del	dup 56–58 and dup 62–64	451 198	150 [1/3] 66	+ 1 0	p.Ser2798Lys2891fsX4 p.Ser3056Asp3122

DMD Duchenne muscular dystrophy, *IMD* intermediate muscular dystrophy, *BMD* Becker muscular dystrophy, *ND* not determined

[a]Theoretical amino acid change based on the database of the Leiden Muscular Dystrophy Pages (http://www.dmd.nl/)

[b]Previously described duplication at cDNA (c.6439_7660del) showing amino acid change (p.Glu2147LeufsX22) resulting in a termination transcript at codon 2168

[c]Previously described deletion at cDNA (c.32_265dup) showing amino acid change (p.Val89MetfsX15) resulting in a termination transcript at codon 103

DMD phenotypes and 100% (7/7) of the individuals with IMD phenotypes. Likewise, the *DMD* genes in all cases with BMD phenotypes had in-frame functional effects on the DMD protein (cases #DS-48, #DS-50, and #DS-52) (Tables 1 and 2). All previously described rearrangements we detected gave rise to a stop codon and thus a truncated protein, except for case #DS-48 with a BMD phenotype and case #DS-11 with ND, which gave rise to in-frame predictions (Table 2). Two cases identified in this study (#DS-53 and #DS-22) reflected both in-frame and reading frame shift predictions (Fig. 3). The complex rearrangement of case #DS-53 (del 45–52 + dup 21–23) was associated with an IMD phenotype, with translational reading frame predictions with in-frame and frame shift patterns (Fig. 3). This phenotype may have occurred from the addition of exons 21–23 to the mRNA transcript lessening the damaging effect of del 45–52 on the functional protein. On the contrary, case #DS-22 could not have corrected for the harmful dup 56–58, giving rise to a DMD phenotype (Table 2).

Discussion

The present study used a facile, reliable, and time-consuming MLPA strategy to identify large rearrangements covering all 79 exons of the *DMD* gene. Our results showed the prevalence of 46.7 and 17.8%, respectively, for large deletions and large duplications in 45 Saudi patients with DMD. Unlike the hot spot deletions in exons 44–56 (92.5%), the hot spot deletions near the 5′ end of the gene were not distinctive, and no large hot spot duplications were found anywhere along the *DMD* gene. The presence of an unusual MLPA pattern in our Saudi sample, including non-contiguous duplications as well as contiguous deletions combined with

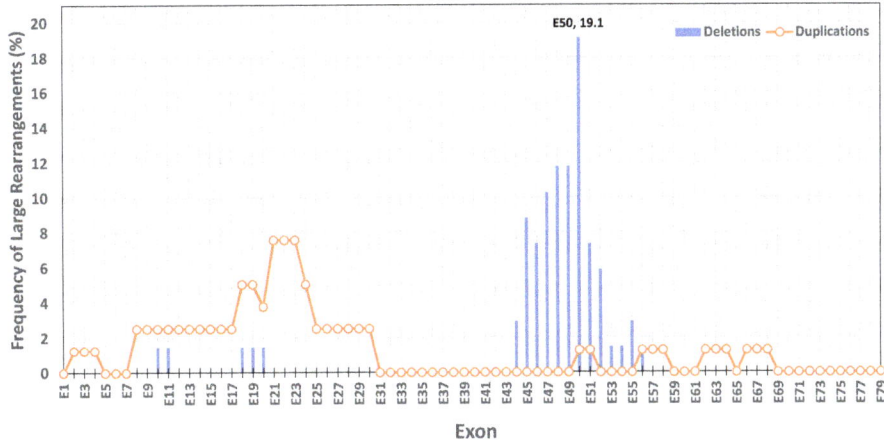

Fig. 2 The frequency of large mutational rearrangements for each exon of the *DMD* gene. A region with a high frequency of deletions was found in exons 44–56. No such region of frequency was detected for large duplications

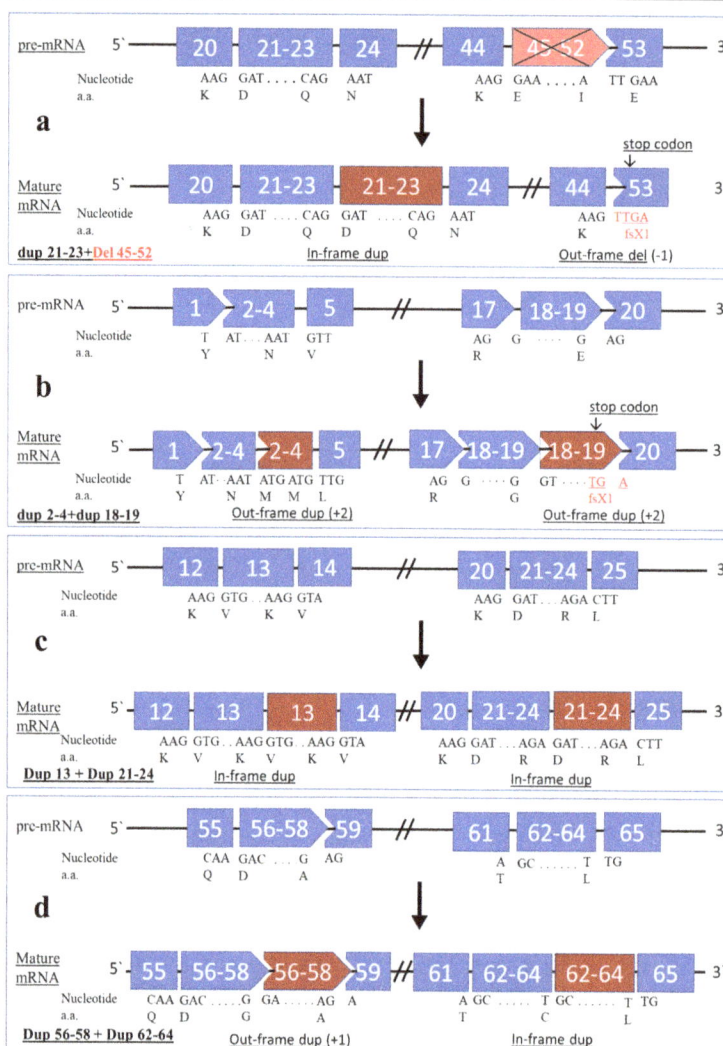

Fig. 3 A schematic overview of new complex large rearrangements in the *DMD* gene. **a** The case #DS-53 with an unusual mixed rearrangement (dup 21–23 + del 45–52) leads to an out-of-frame shift giving rise to a severe DMD phenotype. **b** The case #DS2 with a double duplication (dup 2–4 + dup 18–19) results in out-of-frame shifts with a DMD phenotype. **c** The case #DS-52 with two *in-frame* shift due to double duplications (dup 13 + dup 21–24) giving rise to a BMD phenotype. **d** The case #DS-22 showed two double duplications within the mature mRNA giving an out-frame (dup 56–58), in-frame (dup 62–64) mutations giving rise to a DMD phenotype

non-contiguous duplications, suggests complex rearrangements. Our findings regarding double, separate duplications and complex rearrangements are consistent with some previous reports in Serbian and South African patients [18, 42].

Results from MLPA-related studies among different ethnic populations are conflicting in terms of rates of large rearrangements within the *DMD* gene. Studies have found rates of deletions (and duplications) of 71.8–79.0% (16.4–19.8%) in Chinese [29], 79.5% (6.5%) in Indian [43], 60% (10.0%) in Japanese [44], 45.5–71.8% (16.7%) in Korean [35], and 28.2% (20.5%) in Taiwanese [45] populations. When compared with our sample, a Turkish sample has also been shown to have a relatively higher rate of deletions within the *DMD* gene (63.7%)

[46], likely because of admixture with other European ethnicities. Rates of *DMD* deletions and duplications in some other Middle Eastern populations are more similar to what we found: Egyptian (51.3% deletions) [47], Iranian (51% deletions) [48], Moroccan (51% deletions) [49], and Syrian (49.0% deletions; 9.8% duplications) [50]. The majority of the reported *DMD* gene mutations in our Saudi data showed translational reading frame shifts (94.4%), while 5.6% of the mutations did not follow the reading frame rule. This latter outcome is relatively consistent with the corresponding values in the TREAT-NMD DMD Global database (7%) [11], the UMD-DMD database (4%) [32], and the Leiden database (9%) [41].

The overall rate of consanguinity in KSA is 57.7%, ranging from 34 to 80.6% [51], with lower rates in Mecca

(North Western region) than in Riyadh (Central region) (44.1% versus 62.8%) [51]. This may account for the increased deletions in patients of Riyadh (21/27, 77.8%) when compared with those in our study [37]. During Muslim immigration from the Levant, Africa, in Ancient Islamic times, much intermarriage reinforced gene flow of the *DMD* gene to the Saudi people. This has likely influenced the prevalence of different Mendelian patterns, particularly X-linked types, exemplified by the consistency of data for *DMD* rearrangements between our study and a recent Spanish cohort study (46.1%, 131/284 for deletions and 56/284, 19.7% for duplications) [52].

It is noteworthy that some populations have inherent reproductive barriers that prevent interbreeding, which keeps them at native levels without merging (i.e., cryptic taxa). Other populations may lack inherent reproductive isolation. Therefore, admixture among different geographical populations might increase genetic variations and perhaps create new genotypic combinations within non-isolated (or non-native) populations [53]. Thus, genetic variations among Gulf Arabs and some Middle Eastern individuals (e.g., Barbarians in North Africa, Kurdish, Upper Egyptian) [54, 55] should be handled with caution, as increased consanguinity, extensive reproductive isolation, and admixture with native source populations (e.g., Black Africans, South Eastern Asians, Caucasians) have had substantial roles in gene flow or founder effects in these populations.

In our study, the analysis of *DMD* cases showed an apparent diagnostic delay, as 69.8% of our patients showed their first symptoms at an early age (1–3 years), but 44% of these patients were 9–12 years old at first clinical examination. Other countries have also reported long delays in diagnosis of the disease, with a mean delay between 1.6 and 2.5 years [56, 57]. In south China, the first symptoms occurred by 3 years of age, but the age at clinical evaluation was 6–8 years [36]. Numerous studies have advocated raising public awareness to identify early symptoms in DMD patients [47, 57, 58], as parents are usually the first to notice symptoms, which prompt them to visit a health professional. To further reduce diagnostic delay, creatine phosphokinase (CPK) testing should be emphasized in primary care and performed as a routine test in children's physical examinations.

Earlier clinical trials reported the safety and biochemical efficacy of intravenous or intramuscular administration of antisense oligonucleotides (20-30 mer) to bring hope to DMD patients with large deletions [59]. Therefore, inducing exon 51 skipping to restore the open reading frame is an attractive therapeutic strategy that can be achieved with splice-switching oligomers. After the US Food and Drug Administration (FDA) accelerated approval of AVI-4658/eteplirsen (Exondys 51; Sarepta Therapeutics Inc., Cambridge, MA, USA), targeting *DMD* exon 51 skipping, *eteplirsen* was approved and introduced in some countries [60–63]. Eteplirsen is useful for patients with amenable *DMD* deletions, ending at exon 50 and starting at exon 52 [64]. To date, eteplirsen has not been approved by the Saudi FDA (https://www.sfda.gov.sa). Hence, numerous efforts have used antisense oligomers to target exon skipping of exon 53 (SRP4053, PRO053), exon 45 (DS-514b, SRP4045), and exon 44 (PRO044) (https://www.clinicaltrials.gov/beta/home) [62, 65]. Based on our data for deletions, exon skipping could eventually apply to 81% (17/21) of DMD cases with large deletions. Among our Saudi patients with *DMD* gene deletions, the exons most frequently skipped were exons 51 (42.9%, 9/21), 53 (14.3%, 3/21), 44 (9.5%, 2/21), 45 (4.8%, 1/21), 43 (4.8%, 1/21), and 50 (4.8%, 1/21). Wein et al. have recently reported the efficiency of exon skipping in the *DMD* gene, with each duplicated exon expressing a wild-type, full-length mRNA [66]. For more than one duplicated exon, several antisense oligomers can be delivered as a cocktail of drugs to skip larger regions of the transcript. Thus, for duplications in exons 45–55, therapeutic skipping can be applied to more than 60% of all DMD patients [67].

Although the power is conventionally utilized for polygenic disorders, the power under different monogenic model of inheritance has not been systematically considered. This issue could be explained because of wide issues, for example, rate of background variation in disease-associated genes, mode of inheritance, extent of penetrance, and locus heterogeneity.

In contrast to dominant model of inheritance, the incomplete penetrance does not hold for the recessive model, and in consequence, much smaller sample sizes are needed under a recessive model, even in the presence of high locus heterogeneity [68]. According to our *priori* sample size estimations at the effect size "r" = 0.3 (medium effect), or "r" = 0.5 (strong effect), we would need 64 or 21 sample sizes, respectively, to ensure a power detection of 80%. Thus, post hoc analysis using our DMD sample size data in this study ($n = 45$ cases) could achieve the power of 66.7% ($r = 0.3$) and 98.5% ($r = 0.5$).

Pinning down the spectrum of mutations for *DMD* has been difficult because of poor replication of studies. First, when compared with our study, some studies have had populations with admixed ethnicities, conflicted outcomes, or small sample sizes, which lessen the strength of the overall results. Second, various molecular technologies have been utilized to examine *DMD* patients, resulting in a broad range of false-positive or false-negative results regarding rearrangements. Our study mainly used the *DMD* MLPA test, providing a cheap and straightforward DNA-based test that can screen for deletions and duplications and be performed in any

DNA laboratory. Third, insufficient communication between clinicians and geneticists, because of difficulty accessing hospitals of interest, may result in underdiagnosis of critical cases. However, precise coordination between clinicians and geneticists may help promote and improve the genetic diagnosis of dystrophinopathies and ameliorate potential therapies in these cases.

Conclusions

We detected nine previously undescribed exonic rearrangements within the *DMD* gene, including one unusual mixed rearrangement. MLPA or mPCR can be used to define the molecular characteristics of *DMD* rearrangements and hence the effects of the frame shifts on genotype-phenotype correlations in Saudi patients. This information will also be important for future gene therapy targeting exon skipping of the *DMD* gene. Our clinical characteristics revealed a diagnostic delay, suggesting the need for more public awareness about early symptoms of disease. However, CPK testing should also be performed as a routine test in children's hospitals and in primary care settings. In KSA, molecular testing of DMD patients should be covered by medical insurance, at least once in a lifetime. This single test could lead to genetic diagnosis of more patients. The large deletions and duplications we identified are predictive and intriguing, but the study needs to be replicated in different ethnic populations of the Middle East, as well as in other Saudi governorates. Though the sample size for this study might not have been large enough to explore the *DMD* mutational mechanisms, extensive sequencing analyses will be needed to discover the *DMD* breakpoints at the nucleotide level. Ongoing analyses of whole-exome sequences for Saudi patients with DMD are being carried out to identify the small breakpoints within the *DMD* gene.

Abbreviations
BMD: Becker muscular dystrophy; CPK: Creatine phosphokinase; DMD: Duchenne muscular dystrophy; FDA: Food and Drug Administration; IMD: Intermediate muscular dystrophy; KSA: Kingdom of Saudi Arabia; MLPA: Multiplex ligation-probe dependent amplification; ND: Not determined

Acknowledgements
The authors would like to thank the parents of the cases for their participation in this study. The authors also thank the Institute of Scientific Research at Umm Al-Qura University (Project #43309030) for financial support and the Faculty of Medicine, Cairo University-Giza, Egypt, for allowing ANE to assist the following up the clinical phenotypes of the *DMD* cases.

Funding
This work was funded through grants from the Institute of Scientific Research at Umm Al-Qura University (Project #43309030).

Authors' contributions
NAE and MTT designed the research; NAE, MTT, SJ, AD, EHJ, ANE, HK, and AK performed the research; NAE, MTT, and AHM analyzed the data; NAE, MTT, NB, KFA, and MR wrote the paper. Also, NAE and MTT initiated the grant funding through a contract with the Institute of Scientific Research at Umm Al-Qura University. All authors read and approved the final manuscript.

Consent for publication
Written informed consent was obtained from the parents of all study participants to publish the results.

Competing interests
The authors declare that they have no competing interests.

Author details
[1]Department of Medical Genetics, Medicine College, Umm Al-Qura University, P.O. Box 57543, Mecca 21955, Saudi Arabia. [2]Department of Molecular Genetics, Faculty of Medicine, Ain Shams University, Cairo 11566, Egypt. [3]Department of Medical Laboratory Technology, Faculty of Applied Medical Sciences, King Abdul-Aziz University, Jeddah, Saudi Arabia. [4]Department of Pediatrics, Al Hada Military Hospital, Al Hada, Saudi Arabia. [5]Department of Plan and Research, General Directorate of Health Affairs, Mecca Region, Ministry of Health, Mecca, Saudi Arabia. [6]Department of Pediatrics, Faculty of Medicine, Cairo University, Giza, Egypt.

References
1. Hoffman EP, Brown RH Jr, Kunkel LM. Dystrophin: the protein product of the Duchenne muscular dystrophy locus. Cell. 1987;51(6):919–28.
2. Koenig M, Monaco AP, Kunkel LM. The complete sequence of dystrophin predicts a rod-shaped cytoskeletal protein. Cell. 1988;53(2):219–28.
3. Emery AE. Population frequencies of inherited neuromuscular diseases—a world survey. Neuromuscul Disord. 1991;1(1):19–29.
4. Bushby K, Finkel R, Birnkrant DJ, Case LE, Clemens PR, Cripe L, Kaul A, Kinnett K, McDonald C, Pandya S, et al. Diagnosis and management of Duchenne muscular dystrophy, part 1: diagnosis, and pharmacological and psychosocial management. Lancet Neurol. 2010;9(1):77–93.
5. Jarmin S, Kymalainen H, Popplewell L, Dickson G. New developments in the use of gene therapy to treat Duchenne muscular dystrophy. Expert Opin Biol Ther. 2014;14(2):209–30.
6. Durbeej M, Campbell KP. Muscular dystrophies involving the dystrophin-glycoprotein complex: an overview of current mouse models. Curr Opin Genet Dev. 2002;12(3):349–61.
7. Monaco AP, Bertelson CJ, Liechti-Gallati S, Moser H, Kunkel LM. An explanation for the phenotypic differences between patients bearing partial deletions of the DMD locus. Genomics. 1988;2(1):90–5.
8. Hentze MW, Kulozik AE. A perfect message: RNA surveillance and nonsense-mediated decay. Cell. 1999;96(3):307–10.
9. Muntoni F, Torelli S, Ferlini A. Dystrophin and mutations: one gene, several proteins, multiple phenotypes. Lancet Neurol. 2003;2(12):731–40.
10. Flanigan KM, Dunn DM, von Niederhausern A, Howard MT, Mendell J, Connolly A, Saunders C, Modrcin A, Dasouki M, Comi GP, et al. DMD Trp3X nonsense mutation associated with a founder effect in North American families with mild Becker muscular dystrophy. Neuromuscul Disord. 2009; 19(11):743–8.
11. Bladen CL, Salgado D, Monges S, Foncuberta ME, Kekou K, Kosma K, Dawkins H, Lamont L, Roy AJ, Chamova T, et al. The TREAT-NMD DMD Global Database: analysis of more than 7,000 Duchenne muscular dystrophy mutations. Hum Mutat. 2015;36(4):395–402.
12. Chamberlain JS, Gibbs RA, Ranier JE, Nguyen PN, Caskey CT. Deletion screening of the Duchenne muscular dystrophy locus via multiplex DNA amplification. Nucleic Acids Res. 1988;16(23):11141–56.
13. Beggs AH, Koenig M, Boyce FM, Kunkel LM. Detection of 98% of DMD/BMD gene deletions by polymerase chain reaction. Hum Genet. 1990;86(1):45–8.
14. Forrest SM, Cross GS, Speer A, Gardner-Medwin D, Burn J, Davies KE. Preferential deletion of exons in Duchenne and Becker muscular dystrophies. Nature. 1987;329(6140):638–40.

15. Oudet C, Hanauer A, Clemens P, Caskey T, Mandel JL. Two hot spots of recombination in the DMD gene correlate with the deletion prone regions. Hum Mol Genet. 1992;1(8):599–603.

16. Den Dunnen JT, Grootscholten PM, Bakker E, Blonden LA, Ginjaar HB, Wapenaar MC, van Paassen HM, van Broeckhoven C, Pearson PL, van Ommen GJ. Topography of the Duchenne muscular dystrophy (DMD) gene: FIGE and cDNA analysis of 194 cases reveals 115 deletions and 13 duplications. Am J Hum Genet. 1989;45(6):835–47.

17. Hu XY, Ray PN, Murphy EG, Thompson MW, Worton RG. Duplicational mutation at the Duchenne muscular dystrophy locus: its frequency, distribution, origin, and phenotypegenotype correlation. Am J Hum Genet. 1990;46(4):682–95.

18. Lalic T, Vossen RH, Coffa J, Schouten JP, Guc-Scekic M, Radivojevic D, Djurisic M, Breuning MH, White SJ, den Dunnen JT. Deletion and duplication screening in the DMD gene using MLPA. Eur J Hum Genet. 2005;13(11):1231–4.

19. Elhawary NA, Shawky RM, Elsayed N. High-precision DNA microsatellite genotyping in Duchenne muscular dystrophy families using ion-pair reversed-phase high performance liquid chromatography. Clin Biochem. 2006;39(7):758–61.

20. Koenig M, Hoffman EP, Bertelson CJ, Monaco AP, Feener C, Kunkel LM. Complete cloning of the Duchenne muscular dystrophy (DMD) cDNA and preliminary genomic organization of the DMD gene in normal and affected individuals. Cell. 1987;50(3):509–17.

21. Ioannou P, Christopoulos G, Panayides K, Kleanthous M, Middleton L. Detection of Duchenne and Becker muscular dystrophy carriers by quantitative multiplex polymerase chain reaction analysis. Neurol. 1992;42(9):1783–90.

22. Kodaira M, Hiyama K, Karakawa T, Kameo H, Satoh C. Duplication detection in Japanese Duchenne muscular dystrophy patients and identification of carriers with partial gene deletions using pulsed-field gel electrophoresis. Hum Genet. 1993;92(3):237–43.

23. Yau SC, Bobrow M, Mathew CG, Abbs SJ. Accurate diagnosis of carriers of deletions and duplications in Duchenne/Becker muscular dystrophy by fluorescent dosage analysis. J Med Genet. 1996;33(7):550–8.

24. White S, Kalf M, Liu Q, Villerius M, Engelsma D, Kriek M, Vollebregt E, Bakker B, van Ommen GJ, Breuning MH, et al. Comprehensive detection of genomic duplications and deletions in the DMD gene, by use of multiplex amplifiable probe hybridization. Am J Hum Genet. 2002;71(2):365–74.

25. del Gaudio D, Yang Y, Boggs BA, Schmitt ES, Lee JA, Sahoo T, Pham HT, Wiszniewska J, Chinault AC, Beaudet AL, et al. Molecular diagnosis of Duchenne/Becker muscular dystrophy: enhanced detection of dystrophin gene rearrangements by oligonucleotide array-comparative genomic hybridization. Hum Mutat. 2008;29(9):1100–7.

26. Schouten JP, McElgunn CJ, Waaijer R, Zwijnenburg D, Diepvens F, Pals G. Relative quantification of 40 nucleic acid sequences by multiplex ligation-dependent probe amplification. Nucleic Acids Res. 2002;30(12):e57.

27. Schwartz M, Duno M. Improved molecular diagnosis of dystrophin gene mutations using the multiplex ligation-dependent probe amplification method. Genet Test. 2004;8(4):361–7.

28. Janssen B, Hartmann C, Scholz V, Jauch A, Zschocke J. MLPA analysis for the detection of deletions, duplications and complex rearrangements in the dystrophin gene: potential and pitfalls. Neurogenetics. 2005;6(1):29–35.

29. Chen C, Ma H, Zhang F, Chen L, Xing X, Wang S, Zhang X, Luo Y. Screening of Duchenne muscular dystrophy (DMD) mutations and investigating its mutational mechanism in Chinese patients. PLoS One. 2014;9(9):e108038.

30. Nobile C, Toffolatti L, Rizzi F, Simionati B, Nigro V, Cardazzo B, Patarnello T, Valle G, Danieli GA. Analysis of 22 deletion breakpoints in dystrophin intron 49. Hum Genet. 2002;110(5):418–21.

31. Cotton RG, Auerbach AD, Beckmann JS, Blumenfeld OO, Brookes AJ, Brown AF, Carrera P, Cox DW, Gottlieb B, Greenblatt MS, et al. Recommendations for locus-specific databases and their curation. Hum Mutat. 2008;29(1):2–5.

32. Tuffery-Giraud S, Beroud C, Leturcq F, Yaou RB, Hamroun D, Michel-Calemard L, Moizard MP, Bernard R, Cossee M, Boisseau P, et al. Genotype-phenotype analysis in 2,405 patients with a dystrophinopathy using the UMD-DMD database: a model of nationwide knowledgebase. Hum Mutat. 2009;30(6):934–45.

33. Mitsui J, Takahashi Y, Goto J, Tomiyama H, Ishikawa S, Yoshino H, Minami N, Smith DI, Lesage S, Aburatani H, et al. Mechanisms of genomic instabilities underlying two common fragile-site-associated loci, PARK2 and DMD, in germ cell and cancer cell lines. Am J Hum Genet. 2010;87(1):75–89.

34. Ankala A, Kohn JN, Hegde A, Meka A, Ephrem CL, Askree SH, Bhide S, Hegde MR. Aberrant firing of replication origins potentially explains intragenic nonrecurrent rearrangements within genes, including the human DMD gene. Genome Res. 2012;22(1):25–34.

35. Suh MR, Lee KA, Kim EY, Jung J, Choi WA, Kang SW. Multiplex ligation-dependent probe amplification in X-linked recessive muscular dystrophy in Korean subjects. Yonsei Med J. 2017;58(3):613–8.

36. Wang DN, Wang ZQ, Yan L, He J, Lin MT, Chen WJ, Wang N. Clinical and mutational characteristics of Duchenne muscular dystrophy patients based on a comprehensive database in South China. Neuromuscul Disord. 2017;27(8):715–22.

37. Al-Jumah M, Majumdar R, Al-Rajeh S, Chaves-Carballo E, Salih MM, Awada A, Al-Shahwan S, Al-Uthaim S. Deletion mutations in the dystrophin gene of Saudi patients with Duchenne and Becker muscular dystrophy. Saudi Med J. 2002;23(12):1478–82.

38. Tayeb MT. Deletion mutations in Duchenne muscular dystrophy (DMD) in Western Saudi children. Saudi J Biol Sci. 2010;17(3):237–40.

39. Elhawary NA, Nassir A, Saada H, Dannoun A, Qoqandi O, Alsharif A, Tayeb MT. Combined genetic biomarkers confer susceptibility to risk of urothelial bladder carcinoma in a Saudi population. Dis Markers. 2017;2017:1474560.

40. Aartsma-Rus A, Van Deutekom JC, Fokkema IF, Van Ommen GJ, Den Dunnen JT. Entries in the Leiden Duchenne muscular dystrophy mutation database: an overview of mutation types and paradoxical cases that confirm the reading-frame rule. Muscle Nerve. 2006;34(2):135–44.

41. White SJ, den Dunnen JT. Copy number variation in the genome; the human DMD gene as an example. Cytogenet Genome Res. 2006;115(3–4):240–6.

42. Kerr R, Robinson C, Essop FB, Krause A. Genetic testing for Duchenne/Becker muscular dystrophy in Johannesburg. South Africa S Afr Med J. 2013;103(12 Suppl 1):999–1004.

43. Manjunath M, Kiran P, Preethish-Kumar V, Nalini A, Singh RJ, Gayathri N. A comparative study of mPCR, MLPA, and muscle biopsy results in a cohort of children with Duchenne muscular dystrophy: a first study. Neurol India. 2015;63(1):58–62.

44. Okubo M, Minami N, Goto K, Goto Y, Noguchi S, Mitsuhashi S, Nishino I. Genetic diagnosis of Duchenne/Becker muscular dystrophy using next-generation sequencing: validation analysis of DMD mutations. J Hum Genet. 2016;61(6):483–9.

45. Liang WC, Wang CH, Chou PC, Chen WZ, Jong YJ. The natural history of the patients with Duchenne muscular dystrophy in Taiwan: a medical center experience. Pediatr Neonatol. 2017.

46. Ulgenalp A, Giray O, Bora E, Hizli T, Kurul S, Sagin-Saylam G, Karasoy H, Uran N, Dizdarer G, Tutuncuoglu S, et al. Deletion analysis and clinical correlations in patients with Xp21 linked muscular dystrophy. Turk J Pediatr. 2004;46(4):333–8.

47. Elhawary NA, Shawky RM, Hashem N. Frameshift deletion mechanisms in Egyptian Duchenne and Becker muscular dystrophy families. Mol Cells. 2004;18(2):141–9.

48. Nouri N, Fazel-Najafabadi E, Salehi M, Hosseinzadeh M, Behnam M, Ghazavi MR, Sedghi M. Evaluation of multiplex ligation-dependent probe amplification analysis versus multiplex polymerase chain reaction assays in the detection of dystrophin gene rearrangements in an Iranian population subset. Adv Biomed Res. 2014;3:72.

49. Sbiti A, El Kerch F, Sefiani A. Analysis of dystrophin gene deletions by multiplex PCR in Moroccan patients. J Biomed Biotechnol. 2002;2(3):158–60.

50. Madania A, Zarzour H, Jarjour RA, Ghoury I. Combination of conventional multiplex PCR and quantitative real-time PCR detects large rearrangements in the dystrophin gene in 59% of Syrian DMD/BMD patients. Clin Biochem. 2010;43(10–11):836–42.

51. el-Hazmi MA, al-Swailem AR, Warsy AS, al-Swailem AM, Sulaimani R, al-Meshari AA. Consanguinity among the Saudi Arabian population. J Med Genet. 1995;32(8):623–6.

52. Vieitez I, Gallano P, Gonzalez-Quereda L, Borrego S, Marcos I, Millan JM, Jairo T, Prior C, Molano J, Trujillo-Tiebas MJ, et al. Mutational spectrum of Duchenne muscular dystrophy in Spain: study of 284 cases. Neurologia. 2017;32(6):377–85.

53. Lavergne S, Molofsky J. Increased genetic variation and evolutionary potential drive the success of an invasive grass. Proc Natl Acad Sci U S A. 2007;104:3883–8.

54. Rund D, Cohen T, Filon D, Dowling CE, Warren TC, Barak I, Rachmilewitz E, Kazazian HH Jr, Oppenheim A. Evolution of a genetic disease in an ethnic

isolate: beta-thalassemia in the Jews of Kurdistan. Proc Natl Acad Sci U S A. 1991;88(1):310–4.

55. Jiffri EH, Bogari N, Zidan KH, Teama S, Elhawary NA. Molecular updating of β-thalassemia mutations in the upper Egyptian population. Hemoglobin. 2010;34(6):538–47.

56. Ciafaloni E, Fox DJ, Pandya S, Westfield CP, Puzhankara S, Romitti PA, Mathews KD, Miller TM, Matthews DJ, Miller LA, et al. Delayed diagnosis in duchenne muscular dystrophy: data from the Muscular Dystrophy Surveillance, Tracking, and Research Network (MD STARnet). J Pediatr. 2009; 155(3):380–5.

57. van Ruiten HJ, Straub V, Bushby K, Guglieri M. Improving recognition of Duchenne muscular dystrophy: a retrospective case note review. Arch Dis Child. 2014;99(12):1074–7.

58. Li X, Zhao L, Zhou S, Hu C, Shi Y, Shi W, Li H, Liu F, Wu B, Wang Y. A comprehensive database of Duchenne and Becker muscular dystrophy patients (0–18 years old) in East China. Orphanet J Rare Dis. 2015;10:5.

59. Cirak S, Arechavala-Gomeza V, Guglieri M, Feng L, Torelli S, Anthony K, Abbs S, Garralda ME, Bourke J, Wells DJ, et al. Exon skipping and dystrophin restoration in patients with Duchenne muscular dystrophy after systemic phosphorodiamidate morpholino oligomer treatment: an open-label, phase 2, dose-escalation study. Lancet. 2011;378(9791):595–605.

60. Mendell JR, Goemans N, Lowes LP, Alfano LN, Berry K, Shao J, Kaye EM, Mercuri E, Eteplirsen Study G, Telethon Foundation DMDIN. Longitudinal effect of eteplirsen versus historical control on ambulation in Duchenne muscular dystrophy. Ann Neurol. 2016;79(2):257–71.

61. Shimizu-Motohashi Y, Miyatake S, Komaki H, Takeda S, Aoki Y. Recent advances in innovative therapeutic approaches for Duchenne muscular dystrophy: from discovery to clinical trials. Am J Transl Res. 2016;8(6):2471–89.

62. Lee BL, Nam SH, Lee JH, Ki CS, Lee M, Lee J. Genetic analysis of dystrophin gene for affected male and female carriers with Duchenne/Becker muscular dystrophy in Korea. J Korean Med Sci. 2012;27(3):274–80.

63. Lim KR, Maruyama R, Yokota T. Eteplirsen in the treatment of Duchenne muscular dystrophy. Drug Des Devel Ther. 2017;11:533–45.

64. van Deutekom JC, van Ommen GJ. Advances in Duchenne muscular dystrophy gene therapy. Nat Rev Genet. 2003;4(10):774–83.

65. Mah JK. Current and emerging treatment strategies for Duchenne muscular dystrophy. Neuropsychiatr Dis Treat. 2016;12:1795–807.

66. Wein N, Vulin A, Findlay AR, Gumienny F, Huang N, Wilton SD, Flanigan KM. Efficient skipping of single exon duplications in DMD patient-derived cell lines using an antisense oligonucleotide approach. J Neuromuscul Dis. 2017; 4(3):199–207.

67. Aoki Y, Yokota T, Nagata T, Nakamura A, Tanihata J, Saito T, Duguez SM, Nagaraju K, Hoffman EP, Partridge T, et al. Bodywide skipping of exons 45-55 in dystrophic mdx52 mice by systemic antisense delivery. Proc Natl Acad Sci U S A. 2012;109(34):13763–8.

68. Guo MH, Dauber A, Lippincott MF, Chan YM, Salem RM, Hirschhorn JN. Determinants of power in gene-based burden testing for monogenic disorders. Am J Hum Genet. 2016;99(3):527–39.

18

The tale of histone modifications and its role in multiple sclerosis

Hui He, Zhiping Hu, Han Xiao, Fangfang Zhou and Binbin Yang[*]

Abstract

Epigenetics defines the persistent modifications of gene expression in a manner that does not involve the corresponding alterations in DNA sequences. It includes modifications of DNA nucleotides, nucleosomal remodeling, and post-translational modifications (PTMs). It is becoming evident that PTMs which act singly or in combination to form "histone codes" orchestrate the chromatin structure and dynamic functions. PTMs of histone tails have been demonstrated to influence numerous biological developments, as well as disease onset and progression. Multiple sclerosis (MS) is an autoimmune inflammatory demyelinating and neurodegenerative disease of the central nervous system, of which the precise pathophysiological mechanisms remain to be fully elucidated. There is a wealth of emerging evidence that epigenetic modifications may confer risk for MS, which provides new insights into MS. Histone PTMs, one of the key events that regulate gene activation, seem to play a prominent role in the epigenetic mechanism of MS. In this review, we summarize recent studies in our understanding of the epigenetic language encompassing histone, with special emphasis on histone acetylation and histone lysine methylation, two of the best characterized histone modifications. We also discuss how the current studies address histone acetylation and histone lysine methylation influencing pathophysiology of MS and how future studies could be designed to establish optimized therapeutic strategies for MS.

Keywords: Histone modifications, Multiple sclerosis, Immune-mediated injury, Myelin destruction, Neurodegeneration

Background

Epigenetic modifications is the ensemble of mechanisms of concurrent chromatin modification to modulate global patterns in gene expression and phenotype in a heritable manner, without affecting the DNA sequence itself, which can be classified into DNA modifications (methylation and hydroxymethylation) [1], (PTMs) [2], exchange of histone variants (e.g., H1, H3.3, H2A.Z, H2A.X) [3], and as non-coding RNA [4]. Unlike genes, which remain largely stable across a person's lifetime, the epigenome is highly dynamic. To get a better understanding of how this works, in 2008, the NIH invested in an exploration of the epigenome, launching its Roadmap Epigenomics Mapping Consortium. The project set out to produce a public resource of human epigenomic data that would help fuel basic biology and disease research.

Up to now, the most intensely studied epigenetic modification is DNA methylation; however, the most diverse modifications are on histone proteins. There are at least eight distinct types of modifications found on histones, including acetylation, methylation, phosphorylation [5], ubiquitylation [6], sumoylation [7], ADP ribosylation [8], deamination [9], and prolineisomerization [10]. Histone acetylation and histone methylation are among the most prevalent histone modifications. Researches in the last decades has greatly advanced our knowledge of not only histone modification but also modification of non-histone proteins, providing functional diversity of protein-protein interactions, as well as protein stability, localization and enzymatic activities. Given the complexity of the topic, in the current review, we will concentrate specifically on histone acetylation and histone lysine methylation, of which we now have the most information.

MS is a chronic debilitating disease that affects the brain and spinal cord. Familial clustering is one of

---

* Correspondence: yangbinbin@csu.edu.cn
Department of Neurology, 2nd Xiangya Hospital, Central South University, No 139, Renmin Road, Changsha, Hunan Province, China

important characteristics of MS, suggesting that a hereditary factor involved in determining the risk of MS [11]. However, twin studies showed that monozygotic twins are genetically identical, but a monozygotic twin whose co-twin afflicted with MS has only 25% risk of developing the disease [12]. This suggests that the disease phenotype results from genetic code itself, as well as the regulation of this code by other factors. Increasing evidence suggests that epigenetic modifications may hold the keys to explain the partial heritability of MS risk [13]. In addition, it is believed that epigenetic mechanisms mediate the response to many environmental influences including geographic location, month of birth, Epstein-Barr virus (EBV) infection [14], smoking [15], and latitude/vitamin D [16], which ultimately affect disease development. In this review, we propose a view of MS pathogenesis that specifically involves histone modulations.

Post-translational histone modifications

Histones are among the most highly conserved proteins that act as building blocks of the nucleosome, the fundamental structural and functional unit of chromatin. The nucleosome is an octamer, which is wrapped by 147 bp of DNA, consisting of two copies of four core histone (H) H2A, H2B, H3, and H4 around, tied together by linker histone H1 [17]. These five classes of histone proteins, bearing over 60 different residues, constitute the major protein components of the chromatin and provide a tight packing of the DNA. Meanwhile, the histones contain a flexible N-terminus, often named the "histone tail" [17], which can undergo various combinations of PTMs, dynamically allowing regulatory proteins access to the DNA to fine tune almost all chromatin-mediated processes including chromatin condensation, gene transcription, DNA damage repair, and DNA replication [18] (Fig. 1). Transcriptionally active and silent chromatin is characterized by distinct post-translational modifications on the histones or their combinations. H3K27ac and H3K4me1 are associated with active enhancers [19], and high levels of H3K4me3 and H3 and H4 acetylation are found at the promoters of active genes [20, 21]. The bodies of active genes are enriched in H3 and H4 acetylation [22], H3K79me3 [23], H2BK120u1, and a progressive shift from H3K36me1 to H3K36me3 between the promoters and the 3′ ends [24]. The methylation of H3K27 and H3K9 have emerged as hallmarks of repressive chromatin and are often found at silent gene loci. H3K27me3 are associated with the formation of facultative heterochromatin, whereas H3K9me2/3 has

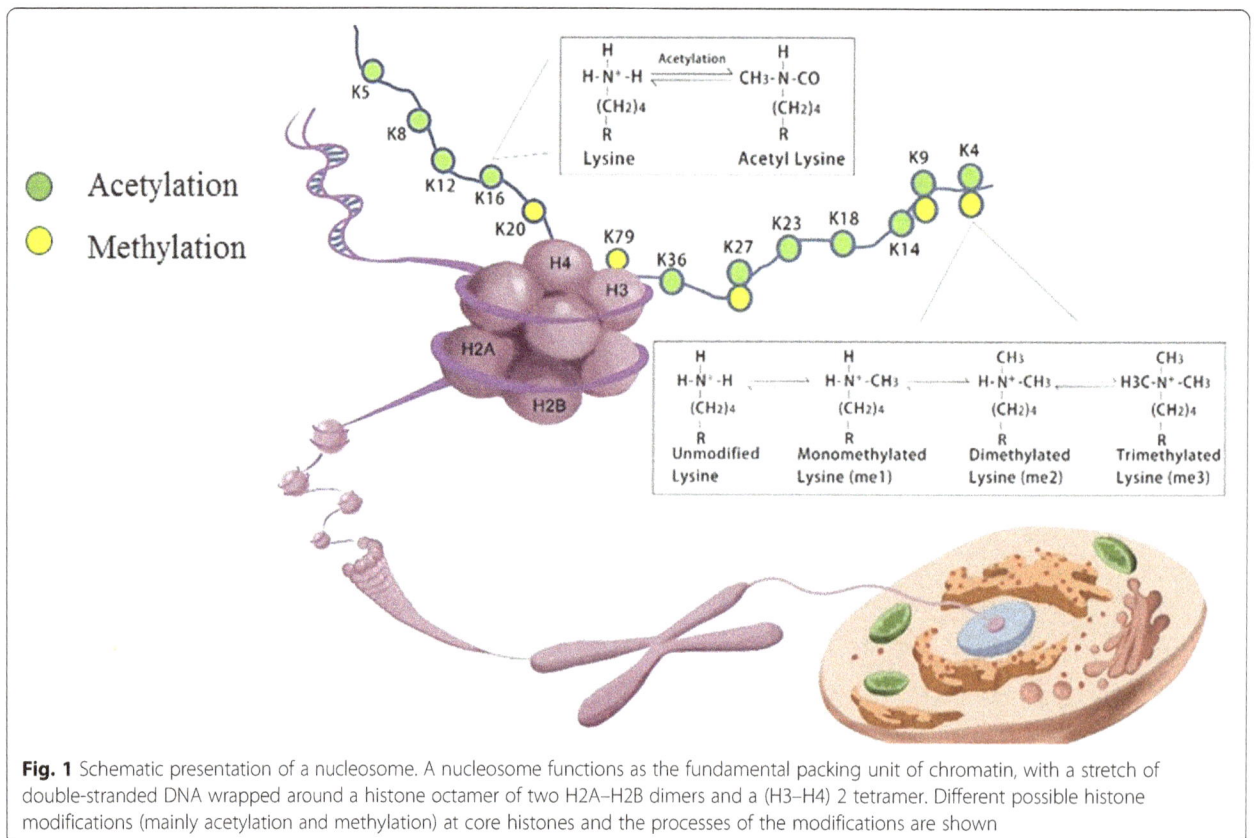

Fig. 1 Schematic presentation of a nucleosome. A nucleosome functions as the fundamental packing unit of chromatin, with a stretch of double-stranded DNA wrapped around a histone octamer of two H2A–H2B dimers and a (H3–H4) 2 tetramer. Different possible histone modifications (mainly acetylation and methylation) at core histones and the processes of the modifications are shown

important roles in the formation of constitutive hetero-chromatin [25]. H4K20m3 is a novel hallmark of peri-centric heterochromatin, whereas H4K20m1 regulates transcription both positively as well as negatively [26], suggesting that specific histone modifications can have dual functions. There are many combinations of modifications that are either occurring together or mutually exclusive, suggesting crosstalk between these marks. Combinations of PTMs, thus, may be associated with transcription in a manner that was not simply related to their individual effects. For example, Fischer et al. indicated that single-code histone acetylation, in particular H3 acetylation (H3ac), are better predictors of increased transcript levels than domains containing further modifications [27]. Single-code H3K4dimethylation (H3K4m2) or its combination with H3K4 tri-methylation (H3K4m3) showed no positive correlation with transcript levels [27]. It is interesting given that H3K4m3 is known to be associated with transcription-start sites of actively transcribed genes. The results from Fischer and his colleague suggested that H3K4me3 is actually not an optimal marker of active promoters and that the activating effect mainly results from its frequent colocalization with acetylations [27].

Histone proteins can undergo post-translational modifications by "writers" and "erasers," a set of enzymes responsible for the deposition and removal of the chemical modifications. Through different combinations and patterns of histone PTMs, they can form the "histone code" [28]. Then, how are these codes interpreted? There are several mechanisms that are not mutually exclusive. First, direct nucleosome-intrinsic effects, particularly by neutralization or addition of charge, PTMs weaken histone-DNA interaction and enable generation of a stably remodeled nucleosome with increased mobility [29]. Such conventional allosteric regulation usually relies on a highly specialized population of molecular interactions [30]. Second, in direct nucleosome-extrinsic effects, H4K16 has been demonstrated to be such a unique histone tail, the acetylation of which impedes the higher-order chromatin formation as a result of its modulation of internucleosomal contacts [31]. Third, the emerging effector-mediated paradigm posits that histone PTMs are "read" by protein modules termed as effectors, which translate them into patterns of gene activation or repression recruiting transcriptional or chromatin-remodeling machinery [30]. In the past decade, a wealth of conserved protein-interaction domains has been characterized as histone effectors, which recognize and bind histone PTMs specifically in a modification- and site-specific way. By covalent combinations of PTMs for binding, modified histone tails may function as integrating platforms for different chromatin-associated complexes, permitting them to receive inputs from upstream signaling cascades and transmit them to the downstream effectors [32].

Histone acetylation

Histone acetylation has been shown to be reversible. The N-terminal domains of histones bear a dozen of lysine residues subject to acetylation, with the exception of a lysine within the globular domain of H3K56, which was found to be acetylated in human by GCN5 [33]. This K residue is facing towards the major groove of the DNA within the nucleosome, so it is in good position to modulate nucleosome assembly by altering histone-DNA interactions when acetylated [34].

Readers of acetyl-lysines

The combinatorial effect of histone acetylation can be deciphered by two distinct, yet overlapping mechanisms—direct and effector-mediated readout mechanisms. In the direct mechanism, histone acetylation neutralizes the positive charge on lysine residues, thus destabilizing the DNA-histone interaction [35]. This results in an open, loosely packed chromatin structure (euchromatin) and consequently allows access for specific transcription factors and the general transcription machinery [31].

Alternatively, histone lysine acetylation marks may be interpreted indirectly via the intermediacy of effectors, which also generally serve to enhance transcriptional activation. The recognition of lysine residues is primarily initiated by bromodomains (BRD) [36]. In general, isolated BRD has been shown to bind to acetylated histones with relatively low affinity and relatively poor selectivity [37], yet, in the presence of multivalent binding, the specificity and affinity are frequently increased. For example, the tandem BRDs of human TATA-binding protein-associated factor-1 (TAF1) binds to multiple acetylated histone H4 peptides with increased affinity, each BRD engaging one acetyl-lysine mark in the same peptide [38]. In principle, the apposition of two BD modules rigidly confined in a relative orientation creates surfaces that are complementary to the spatial distributions of their substrates in chromatin. Therefore, the distances between discrete interactions become additional determinants in dictating specificity [38]. More recently, it has been demonstrated that two acetylated lysine residues might be simultaneously recognized by the same BRD module with significantly increased affinity. For example, a single binding pocket of BD1 of BRDT accommodates both acetyl-lysines of H4K5acK12ac and H4K8acK16ac peptides in a wider hydrophobic pocket, showing much stronger affinity than binding to either mark individually [39]. Moreover, the acetylated histone recognition by BD1 is complemented by a novel BRD-DNA interaction [40]. Simultaneous DNA and histone recognition enhances BRD's nucleosome binding affinity, specificity, and ability to localize to and compact acetylated chromatin [40].

Writers and erasers of acetylation

KATs, formally named as histone acetyltransferaces (HATs), can be generally classified into two categories based on subcellular localization. Type A KATs are located in the nucleus, involved in the acetylation of histones in chromatin, whereas type B KATs, predominantly cytoplasmic, acetylate newly translated histones to facilitate their transfer to the chromatin assembly factors [41]. In eukaryotes, the majority of canonical type A KATs has been grouped into three major families including p300/CBP, GCN5/PCAF, and MYST proteins [42] (Table 1). Two subfamilies of histone deacetylases (HDACs) have been identified in humans so far—Zn2+-dependent (classes I, II, and IV) and Zn2+-independent and NAD-dependent (class III). Generally speaking, class I HDACs are ubiquitously expressed and exhibit strongest enzymatic activity. Class II HDACs have sequence similarity to the yeast Hda1 protein which seems to be expressed in a more cell-specific manner [43]. They possess unique 14-3-3 binding sites at their N-termini. Following phosphorylation, the N-terminal regions recruit 14-3-3, with consequent export of the HDAC/14-3-3 complex from the nucleus to the cytoplasm [44, 45]. Thus, phosphorylation of class II HDACs provides a mechanism for coupling external signals to the genome. The class III HDACs, or sirtuins, display NAD+-dependent deacetylase activity and may specifically interact with and modify dozens of distinct substrates in various the biological processes.

Histone lysine methylation

Histone methylation occurs at lysine and arginine residues. In this review, we only focus on histone lysine methylation due to its prominence and its array of well-established roles in epigenetic gene control and chromatin domains organization. Histone lysine methylations have been found on a range of lysine residues in various histones, including K4, K9, K27, K36, and K79 residues in histone H3, K20 in histone H4, K59 in the globular domain of histone H4 [46], and K26 in histone H1B [47]. Instead of influencing the net charge of the histone tails, methylation of histone tails contributes to regulation of the transcriptional activity by functioning as a recognition template to recruit effector proteins to local chromatin [48]. Thus, histone lysine methylation can be associated with either activation or repression of transcription ultimately determined by the effectors. When compared with acetylation, histone lysine methylation is a relatively stable modification with a generally low turnover [49]. Moreover, methylation is controlled by histone methyltransferases (KMTs) and demethylases (KDMs) that possess strong substrate specificity (Table 2) (Table 3).

Readers of methylysines

Chromodomain is the founding member of "readers" of histone methyllysine [50], Besides the well-known methy-lysine-binding family of chromodomain, a large family of reader proteins including Tudor, MBT, PWWP, plant homeodomain (PHD) finger, Ankyrin repeats, and WD repeats make up the so-called Royal family [51, 52]. Three elements determine the strength and specificity of a particular methylated lysine reader. The foremost trait of the methyllysine readers is the presence of an aromatic cage structure in their binding to methyllysines, consisting two to four aromatic residues. The exact composition and size of the pocket make the readers selective in recognizing mono-, di-, or trimethylated state of lysine. Effectors for mono- and dimethylation tend to have a small

Table 1 Enzymatic mechanisms used for histone acetylation

Canonical members of KAT	Former name in human	Histone protein acetylated	Mechanism of catalysis
P300/CBP family			
KAT3			Hit-and-run
KAT3A	CBP	H2A, H2B	
KAT3B	P300	H2A, H2B	
GCN5 family			
KAT2			KAT/Ac-CoA/substrate ternary complex
KAT2A	GCN5	H3, H4,H2B	
KAT2B	PCAF	H3	
MYST family			
KAT5	Tip60	H4, H2AZ, H2AX	Ping-pong mechanism or ternary mechanism
KAT6			
KAT6A	MOZ/MYST3	H3	
KAT6B	MORF/MYST4		
KAT7	HBO1/MYST2	H4	
KAT8	MOF/MYST1	H4	

Table 2 Substrate specificity of KMTs and KDMs

	H3K4 Me1	H3K4 Me2	H3K4 Me3	H3K9 Me1	H3K9 Me2	H3K9 Me3	H3K27 Me1	H3K27 Me2	H3K27 Me3	H3K36 Me1	H3K36 Me2	H3K36 Me3	H3K79 Me1	H3K79 Me2	H3K79 Me3	H4K20 Me1	H4K20 Me2	H4K20 Me3
WRITERS	KMT2A KMT2B KMT2C KMT2D KMT2E KMT2F KMT2G KMT3C KMT3D (?) KMT3E KMT7	KMT2A KMT2B KMT2C KMT2D KMT2E KMT2F KMT2G KMT3D (?) KMT3E	KMT2A KMT2B KMT2C KMT2D KMT2E KMT2F KMT2G KMT3D (?) KMT3E	KMT1C KMT1D KMT8	KMT1C KMT1D	KMT1A KMT1B KMT1E KMT1F	KMT6A (?) KMT6B (?)	KMT1C KMT6A KMT6B	SMYD3 KMT6A KMT6B	KMT3B	KMT3B KMT3C	KMT3A	KMT4	KMT4	KMT4	KMT5A	KMT3B KMT5B KMT5C	KMT3B KMT5B KMT5C
ERASERS	KDM1A KDM5	KDM1A KDM5	KDM5	KDM1A KDM7B	KDM1A KDM3 KDM4 KDM7B	KDM2B KDM4	KDM7B	KDM6 KDM7B	KDM6	KDM2	KDM2 KDM4 KDM8	KDM4				KDM7A KDM7B	KDM7C	KDM7C

Table 3 Histone methyltransferases and demethyltransferases

Writers		
	KMT1	
SUV family	KMT1A	SUV39H1
	KMT1B	SUV39H2
	KMT1C	G9a
	KMT1D	GLP
	KMT1E	SETDB1
	KMT1F	SETDB2
	KMT2	
MLL family	KMT2A	MLL1
	KMT2B	MLL2
	KMT2C	MLL3
	KMT2D	MLL4
	KMT2E	MLL5
	KMT2F	SET1A
	KMT2G	SET1B
	KMT2H	ASH1
	KMT3	
NSD family	KMT3A	SETD2
	KMT3B	NSD1
	KMT3F	NSD3
	KMT3G	NSD2
SMYD family	KMT3C	SMYD2
	KMT3D	SMYD1
	KMT3E	SMYD3
	KMT4	DOT1L
	KMT5	
	KMT5A	SET8
	KMT5B	SUV420H1
	KMT5C	SUV420H2
	KMT6	
	KMT6A	EZH2
	KMT6B	EZH1
	KMT7	SET7/9
	KMT8	PRDM2/RIZ1
Erasers		
	KDM1	
	KDM1A	LSD1
	KDM1B	LSD2
	KDM2	
FBXL cluster	KDM2A	JHDM1A
	KDM2B	JHDM1B
	KDM3	
JMJD1 cluster	KDM3A	JMJD1A
	KDM3B	JMJD1B

Table 3 Histone methyltransferases and demethyltransferases *(Continued)*

	KDM3C	JMJD1C
	KDM4	
JMJD2 cluster	KDM4A	JMJD2A
	KDM4B	JMJD2B
	KDM4C	JMJD2C
	KDM4D	JMJD2D
	KDM5	
JARID1 Cluster	KDM5A	JARID1A
	KDM5B	JARID1B
	KDM5C	JARID1C
	KDM5D	JARID1D
	KDM6	
UTX/JMJD3 cluster	KDM6A	UTX/UTY
	KDM6B	JMJD3
	KDM7	
	KDM7A	JHDM1D
	KDM7B	JHDM1E
	KDM7C	JHDM1F
	KDM8	JMJD5

keyhole-like cavity, which leads to steric hindrance to limit accessibility of a higher methylation state [53]. In contrast, the binding pockets of effectors for di- and trimethylation are wider and more accessible, which may also lower the stringency in the discrimination preferences [53]. Typically two ways are involved in the recognition of methyl states. At some lysines, selective effector is recruited to a specific methylation state. For instance, Pdp1 binds to H4K20me1 to facilitate chromatin maturation, whereas 53BP1 in mammals and Crb2 in fission yeast selectively bind the H4K20me2, required for DNA damage checkpoint activation [54]. At other sites, methyl states only influence the binding affinity of the same histone-methyl-lysine-binding proteins. For example, Rpd3S preferentially binds K36me2 and K36me3, with K36me3 displaying the highest affinity. By contrast, the affinity of K36me1 to Rpd3S is much lower, similar to that of the unmodified ones [55]. Secondly, interaction with flanking sequence may impart an additional layer of specificity for a particular methylated lysine. Free histone peptides are usually unstructured in aqueous solution. On binding, they adopt a β-sheet conformation, with extensive contacts with the flanking sequence of the readers [56]. This pairing interaction not only contributes to the overall robustness but also provides structural basis for functional specificity [53]. At last, methyllysines are located close to the end of a histone peptide; upon binding, the histone termini can be buried snugly into a shallow pocket, which greatly facilitates the overall affinity [53].

Writers and erasers of histone lysine methylation

KMTs catalyze methylation of lysine residues with high site- and methyl-level specificity (Table 2). In the last decades, numerous KMTs have been identified and crystallized, which use *S*-adenosylmethionine (SAM) as a methyl group donor [57]. Except for KMT4/DOT1L, all known KMTs contain a conserved SET domain harboring the enzymatic activity [58]. Based on the similarities in the sequence within and around the catalytic SET domain, as well as homology to other protein modules and their domain architectures, SET-containing KMTs have traditionally been categorized into distinct subfamilies [59].

Histone lysine methylation was previously considered static and enzymatically irreversible until the first histone KDM—LSD1/KDM1A identified by Shi et al. [60], which changed our view of histone methylation regulation and ushered in the identification of numerous histone demethylases. Subsequent to the discovery of KDM1A, a new class of KDM enzymes which comprises the JmjC domain-containing protein was discovered. While KDM1A is unable to catalyze the dimethylation of trimethylated lysine residues owing to its requirement for imine formation for catalytic activity, the JmjC-driven demethylase have demethylation activity for mono-, di-, and trimethylated histone lysine residues. Indeed, most of the JmjC histone demethylases identified so far are capable of efficiently catalyzing demethylation of trimethylated lysines, and in most cases, they preferentially bind the trimethylated substrates [61, 62].

Histone modifications in MS

A core of pathogenetic functions common to both the immune and neurodegenerative processes of MS has been characterized by deregulation of MS-risk genes and resulting dysfunction of their encoding proteins [63]. Epigenetic transcription-regulating mechanisms in nucleated cells including cells of the CNS have been widely accepted. Therefore, MS-specific alterations in epigenetic regulation of chromatin may play a central role in gene expression and be essential for the initiation and development of MS. Among which, histone modification is an important epigenetic mechanism.

Histone modifications in MS susceptibility

Twin studies have established that susceptibility to MS is partly genetic. One family of major histocompatibility complex (MHC) genes, the human leukocyte antigen (HLA) alleles, has identified as a genetic determinant for MS [64]. In particular, carriage of HLA-DR/DQ serotype has been identified as a major MS risk allele. Notably, expression of HLA-DR has been shown to be suppressed by HDAC1 [65], which suggests that MS susceptibility loci have histone regulation links.

Histone modifications in autoimmunity-related mechanisms

The hallmark of MS and experimental autoimmune encephalomyelitis (EAE) is that myelin injury and axonal damage driven by an immune-mediated inflammatory response begins at disease onset. Autoreactive myelin-specific CD4+ T cells are believed to play a crucial pathogenic role [66]. Upon encountering myelin antigen, antigen-presenting cells (APCs) acquire a mature phenotype and migrate to lymph nodes where they present exogenous antigens to naïve CD4+ T cells. Naive CD4+ T cells may then differentiate into diverse functional subsets, including the T helper (Th) 1, Th2, Th17 cells, and Treg cells [67]. Once activated, CD4+ T cells are translocated into the CNS by crossing the brain-blood barrier (BBB) and then are reactivated by resident APCs (such as microglia) [68], which in turn initiate the recruitment of other inflammatory cells, resulting in demyelination and axon injury. While interferon-γ (IFN-γ)-associated Th1 and interleukin-17 (IL-17)-associated Th17 cells are considered to lead to disease progression and worsening of symptoms, IL-4-associated Th2 and transforming growth factor-β (TGF-β)-associated Treg have been indicated to associate with inflammation reduction and improvement of symptoms in MS patients [69].

It is widely accepted that the activation of CD4+ autoreactive T cells and their differentiation into a Th1 or Th17 phenotype are crucial events in the initial steps of MS, though many studies have shown that monocytes and monocyte-derived macrophages are also the primary cell types responsible for cellular pathology and tissue damage. In MS pathology, activated monocytes, which facilitate the migration of T cells across the blood-brain barrier (BBB), largely represent the inflammatory infiltrate [70]. Knowledge on the features of blood monocytes in MS, however, are little understood. Circulating monocytes, as an important source of cytokines, have been hypothesized to play a key role in regulating crucial immune functions. The M1/M2 paradigm is currently used to categorize the monocyte/macrophage functions [71], and M1/M2 macrophage balance polarization governs the fate of an organ in inflammation. Generally, M1 monocytes/macrophages are generally characterized by an IL-12hi, IL-23hi, tumor necrosis factor (TNF)-αhi, and IL-10lo phenotype, which produce abundant reactive oxygen species and shift the immune response towards a Th1 profile [72]. M2 monocytes/macrophages typically have IL-12lo, IL-23lo, TNF-αlo, and IL-10hi responses to stimulation, which are thought to drive Th2 responses [73].

HDACs have been shown to be closely tied to regulation of CD4+ T cells differentiation and various cytokines production through regulating the changes in chromatin structure which then influence gene expression. Correspondingly, HDAC inhibitors have also been demonstrated to elicit control over the immune response, which in turn suppress systemic and local inflammation [74]. Several recent studies have shown the potential for the use of HDAC inhibitor therapy to inhibit the proliferative response of CD4+ T cells and abrogated IFN-γ production [75]. A growing literature indicated that HDAC inhibitors inhibit the proinflammatory cytokine IL-2 expression, which is secreted by Th1 cells, and IL-2 mediated gene expression as well. Moreover, HDAC inhibitors reduce macrophage production of pro-demyelinating cytokines involved in T helper (Th) cell differentiation, including IL-12, IL-6, and TNF-α. Consequently, HDAC inhibitors cause a Th1 to Th2 dominance shift [76], and expanding Tregs, which by virtue of its immunosuppressive role, may help ameliorate MS.

Actually, dysregulated Th cell responses are not unique for MS pathology, but also a characteristic of a wide variety of several other inflammatory diseases, including inflammatory bowel disease, arthritis, diabetes, asthma, and allergies [77]. Therefore, compounds that inhibit HDACs, especially, class I, II, and IV enzymes, have been pursued as therapeutic agents for a wide range of inflammatory diseases. However, treating cells with HDAC inhibitors has also been shown to increase the expression of cytokines IL-10 [76], contributing to pro-humoral and protective role in EAE, which, in systemic lupus erythematosus (SLE) cells, actually downregulated expression of IL-10 and other anti-inflammatory cytokines [78]. The contrasting effects might reflect disease-specific effects of these compounds and further studies are needed.

It is suggested that chromatin remodeling, via histone lysine methylation, is mechanistically important in the acquisition of the M2-macrophage phenotype. Ishii et al. demonstrated that at the promoters of the M2 marker genes, H3K4me3 was significantly upregulated, whereas H3K27me2/3 was significantly decreased. Increased Jmjd3 contributes to the decrease of H3K27me2/3 marks and skews macrophages to an M2 phenotype [79]. Therefore, target gene regulation by histone Lysine methylation is a dynamic process that modulates inflammatory responses in the development of a variety of autoimmune diseases, including MS.

Recent studies demonstrated that KDM6 modulate immune functions by determining Th cell maturation and egress from the thymus [80], as well as CD4+ Th cell lineage differentiation [66], thereby significantly affecting immune responses in multiple biological systems. It is reported that Jmjd3 positively regulate the differentiation of Th17 cells, which play critical roles in proinflammatory reactions in autoimmue disorders, such as

rheumatoid arthritis and systemic lupus erythematosis [81]. Jmjd3-deficient mice were demonstrated to be resistant to the induction of EAE [66]. Correspondingly, H3K27 demethylase-specific inhibitor GSK-J4 markedly inhibited Th17 cell differentiation in vitro [66]. However, another independent research demonstrated that while Th1 and Th17 differentiation were not affected, 10 or 25 nM GSK-J4 significantly increased differentiation of anti-inflammatory Treg cells in vivo, which could partly explain the beneficial effects of GSK-J4 on EAE. GSK-J4 promoted Treg differentiation was proposed to be dependent on its direct effect on the maturation status of dendrite cells (DCs). DCs, the professional APC, being the key players in maintaining immune tolerance, now have gained increasing attention [82]. Specifically, H3K27me3 demethylase activity would skew DC differentiation towards a tolerogenic phenotype [83]. Accordingly, through altering the permissive H3K4me3/repressive H3K27me3 ratio at specific gene promoters, GSK-J4 induced a tolerogenic phenotype on DCs and subsequently inhibited the development of EAE [83].

Moreover, T cell anergy is thought to be a critical mechanism for preventing autoimmunity and failure of this tolerance mechanism causes MS [84]. The upregulated Sirt1 protein has been demonstrated to suppress T cell activation and lead to anergy induction in mice. Conversely, Sirt1 deficiency was reported to result in increased T cell activation and failed to maintain CD4+ T cell tolerance and increased susceptibility to EAE [85]. Mice with DC-specific deletion of SIRT1 showed remarkable resistance to EAE through enhanced IL-27 and IFN-β activation, two anti-inflammatory cytokines that negatively regulate Th17 cell differentiation [86]. These findings make the role of HDAC in MS quite controversial (Fig. 2).

Histone modifications in myelin destruction

Another cardinal feature of multiple sclerosis is the failure of remyelination caused by impaired differentiation of endogenous oligodendrocyte progenitor cells (OPCs). Unlike other neuronal lineages, in the oligodendrocyte lineage, high levels of histone acetylation are important in undifferentiated progenitor cells [87], which favor the expression of transcriptional repressors of myelin gene expression. Increased histone H3 acetylation in oligodendrocytes is associated with high levels of transcriptional inhibitors of oligodendrocyte differentiation which subsequently might lead to impaired remyelination in patients with MS [88]. Conversely, histone deacetylation enables expression of an oligodendrocyte transcriptional profile during developmental myelination, as well as remyelination [87]. While a large number of oligodendrocytes with deacetylated histone was observed in early MS lesions, a shift towards high levels of histone

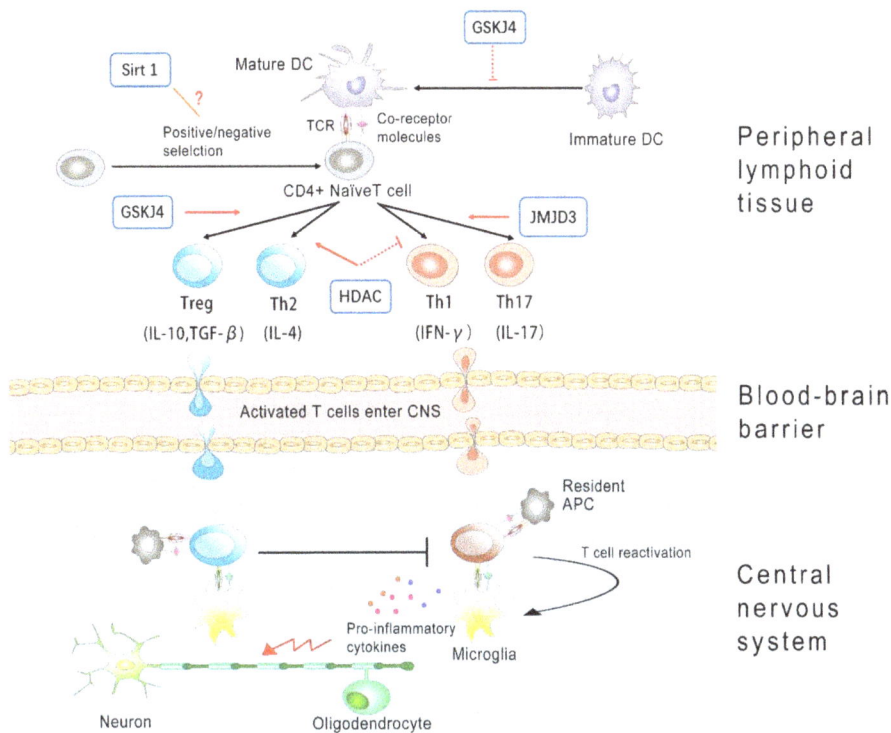

Fig. 2 A model of immune mechanism in MS. Cascade of events possibly underlying autoimmunity-related demyelination in MS and putative mechanisms of action of histone-modifying enzyme inhibitors are demonstrated

acetylation has been detected in oligodendrocyte lineage cells within normal-appearing white matter (NAWM) in the brain of patients with chronic MS [89]. The data suggested negative correlations between histone deacetylation efficiency and duration of disease.

Histone modifications in neurodegeneration

For decades, MS research has heavily focused on inflammatory white matter pathology. However, recent studies have discovered neurodegenerative components of the disease such as insidious axonal degeneration and neuronal atrophy, which seem to be the histopathological correlates of progressive clinical disability in MS patients [90]. Mitochondrial injury and subsequent energy failure are indicated as key factors in the induction of neurodegeneration. Betaine, a methyl donor, was found to be decreased in MS cortex, which was correlated with decreased H3K4me3 in neuronal (NeuN+) nuclei in MS cortex, in comparison to controls [91]. Mechanistic studies demonstrated that reduced methylation of H3K4me3 may result in the downregulation of oxidative phosphorylation genes and defects of respiratory chain enzymes in MS cortex [91]. A recent study showed that variant carriers of certain HDAC genes, including mitochondrial-related gene variants in SIRT4 and SIRT5, have been linked to more pronounced brain volume loss

(atrophy) during the clinical course of MS [92]. These results indicate that the histone modifications might be centrally linked with neurodegenerative processes in MS.

Potential treatment methods based on epigenetic mechanisms

Disturbance of transcriptional balance may promote dysregulation of immune system and neurodegeneration, both of which contribute to the clinical profile of MS. Animal model experiments support that deliberate epigenetic reprogramming for oligodendrocyte, immune cells, and neurons to perform properly may be a potential therapeutic strategy for MS.

There is a growing list of pharmacological agents that affect histone PTMs, among, which the most studied and used are histone deacetylase inhibitors (HDACi). For example, Camelo et al. showed that intraperitoneal administration of the HDACi, Trichostatin A (TSA) attenuated inflammation, reduced demyelination and axonal loss, and thus decreased disease severity in mice with spinal cord homogenate induced EAE [74]. The HDACi, vorinostat (SAHA), was shown to suppress DCs function and ameliorate EAE in C57BL/6 female mice [93]. VPA administration suppresses systemic and local inflammation to improve outcome of EAE in Lewis rats [94]. Likewise, curcumin, which inhibits the activity of

KATs, has been shown to ameliorate EAE through suppression of inflammatory cells infiltration in the spinal cord [95]. As previously mentioned, systemic administration with the epigenetic drug GSK-J4 prevented the development of EAE in mice [83]. Thus, the inhibitors of histone deacetylation or demethylation may be promising agents for MS treatment. However, systemic use of HDACis negatively affects the generation of new myelin since histone deacetylation is important for progenitor cell differentiation into myelin-forming oligodendrocytes [96] and is critical for remyelination efficiency in adults [88], as we reviewed previously. The potential detrimental consequence on myelin might counteract the beneficial effects, thus cautioning against the use of broad inhibitors of histone deacetylases in MS. Therefore, more targeted therapy that specifically epigenetically modifies certain pathogenic loci need to be developed. In the recent years, the CRISPR-dCas9 system is poised to become the most promising targetable epigenome-editing tools. The results of two recent seminal studies have strongly supported the capability of epigenome editing by a CRISPR-Cas9 to activate or silence transcription by targeting histone PTMs [97, 98]. Moreover, CRISPR-dCas9 epigenome-editing approach has been demonstrated to produce long-lasting changes in expression of targeted genes both in vitro and in vivo. Its simplicity and efficiency may facilitate the clinical application of this technology by avoiding repetitive or chronic administration. However, the research on CRISPR-mediated technology is still in its early stage, and it is important to continue to probe for its feasibility and safety for clinical purposes. An additional challenge for treating MS with these inhibitors is the lack of specificity, which would cause a relatively high risk of adverse effects. Correspondingly, successful epigenetic therapy would be the tissue specificity of the therapeutic effect. Receptor-coated nanoparticles or microvesicles as highly effective drug carriers pertaining to BBB may hold great promise in MS therapy. Several studies have recently demonstrated that treatment of mice with nanoparticles effectively decreased EAE progression [99]. Collectively, translational use of epigenetics might offer hope for a new class of therapeutics to treat MS and the development of targeted epigenetic therapies open new avenues for effective personalized treatment of patients with MS.

Conclusion

MS is the most prevalent autoimmune disease with highly variable clinical course and disease progression, in which the main common pathogenetic pathway involves an immune-mediated cascade [100]. Recently, huge steps have been made in the field of MS immunotherapy. Moreover, emerging evidence has shed light on the epigenetic

mechanisms contributing MS. Several epigenetic drugs which are in clinical trials or under investigation in human diseases have been proven to have immunomodulatory effects [101]. In addition, other expected changes also may occur in response to epigenetic treatment. In particular, histone PTMs in regulation of myelination and degeneration gene associated with MS and amelioration of EAE symptoms by drugs with PTM effects, such as HDAC inhibitors and KDM inhibitors, all emphasize the critical role of histone PTMs in the pathogenesis of MS. The amalgamation and crystallization of histone PTMs research and MS promises novel pleiotropic treatment strategies. However, given the potential for off-target potential to cause deleterious effects from HDAC and KDM inhibitors with broad activity, the endeavor to completely understand molecular mechanisms governing histone modifications and their precise molecular targets will hold the key to successfully translate the drug candidates to clinical practice.

Abbreviations

Ac-CoA: Acetyl coenzyme A; APC: Antigen-presenting cells; BBB: Brain-blood barrier; BRD: Bromodomain; DCs: Dendrite cells; EAE: Experimental autoimmune encephalomyelitis; EBV: Epstein-Barr virus; HATs: Histone acetyltransferaces; HDACs: Histone deacetylases; HLA: Human leukocyte antigen; IFN: Interferon; IL: Interleukin; MHC: Major histocompatibility complex; NAWM: Normal-appearing white matter; PHD: Plant homeodomain; PTMs: Post-translational modifications; TAF1: TATA-binding protein-associated factor-1; TGF: Transforming growth factor; Th: T helper

Acknowledgements

The authors would like to acknowledge Dr. Shiyu Chen for the artwork.

Authors' contributions

BY conceived and planned the review. ZH and BY drafted the manuscript. HH revised it critically for important intellectual content with support from FZ, HX, and BY. All authors contributed to the final manuscript.

Consent for publication

Not applicable.

Competing interests

The authors declare that they have no competing interests.

References

1. Laird PW. The power and the promise of DNA methylation markers. Nat Rev Cancer. 2003;3:253–66.
2. Jenuwein T, Allis CD. Translating the histone code. Science. 2001;293:1074–80.
3. Pusarla RH, Bhargava P. Histones in functional diversification: core histone variants. FEBS J. 2005;272:5149–68.
4. Mattick JS, Makunin I V. Non-coding RNA. Hum Mol Genet 2006; 15 Spec No 1:R17–R29.
5. Nowak SJ, Corces VG. Phosphorylation of histone H3: a balancing act between chromosome condensation and transcriptional activation. Trends Genet. 2004;20:214–20.

6. Li W, Nagaraja S, Delcuve GP, Hendzel MJ, Davie JR. Effects of histone acetylation, ubiquitination and variants on nucleosome stability. Biochem J. 1993;296:737–44.

7. Shiio Y, Eisenman RN. Histone sumoylation is associated with transcriptional repression. Proc Natl Acad Sci U S A. 2003;100:13225–30.

8. Boulikas T. DNA strand breaks alter histone ADP-ribosylation. Proc Natl Acad Sci U S A. 1989;86:3499–503.

9. Cuthbert GL, Daujat S, Snowden AW, Erdjument-Bromage H, Hagiwara T, Yamada M, et al. Histone deimination antagonizes arginine methylation. Cell. 2004;118:545–53.

10. Nelson CJ, Santos-Rosa H, Kouzarides T. Proline isomerization of histone H3 regulates lysine methylation and gene expression. Cell. 2006;126:905–16.

11. Sadovnick AD, Baird PA, Ward RH, Optiz JM, Reynolds JF. Multiple sclerosis: updated risks for relatives. Am J Med Genet. 1988;29:533–41.

12. Willer CJ, Dyment DA, Risch NJ, Sadovnick AD, Ebers GC, Canadian Collaborative Study Group. Twin concordance and sibling recurrence rates in multiple sclerosis. Proc Natl Acad Sci U S A. 2003;100:12877–82.

13. Huynh JL, Casaccia P. Epigenetic mechanisms in multiple sclerosis: implications for pathogenesis and treatment. Lancet Neurol. 2013;12:195–206.

14. Niller HH, Wolf H, Minarovits J. Epigenetic dysregulation of the host cell genome in Epstein-Barr virus-associated neoplasia. Semin Cancer Biol. 2009;19:158–64.

15. Wan ES, Qiu W, Baccarelli A, Carey VJ, Bacherman H, Rennard SI, et al. Cigarette smoking behaviors and time since quitting are associated with differential DNA methylation across the human genome. Hum Mol Genet. 2012;21:3073–82.

16. Pereira F, Barbáchano A, Singh PK, Campbell MJ, Muñoz A, Larriba MJ. Vitamin D has wide regulatory effects on histone demethylase genes. Cell Cycle. 2012;11:1081–9.

17. Luger K, Mäder W, Richmond RK, Sargent DF, Richmond TJ. Crystal structure of the nucleosome core particle at 2.8 A resolution. Nature. 1997;389:251–60.

18. Önder Ö, Sidoli S, Carroll M, Garcia BA. Progress in epigenetic histone modification analysis by mass spectrometry for clinical investigations. Expert Rev Proteomics. 2015;12:499–517.

19. Creyghton MP, Cheng AW, Welstead GG, Kooistra T, Carey BW, Steine EJ, et al. Histone H3K27ac separates active from poised enhancers and predicts developmental state. Proc Natl Acad Sci U S A. 2010;107:21931–6.

20. Deckert J, Struhl K. Histone acetylation at promoters is differentially affected by specific activators and repressors. Mol Cell Biol. 2001;21:2726–35.

21. Liang G, Lin JCY, Wei V, Yoo C, Cheng JC, Nguyen CT, et al. Distinct localization of histone H3 acetylation and H3-K4 methylation to the transcription start sites in the human genome. Proc Natl Acad Sci U S A. 2004;101:7357–62.

22. Myers FA, Evans DR, Clayton AL, Thorne AW, Crane-Robinson C. Targeted and extended acetylation of histones H4 and H3 at active and inactive genes in chicken embryo erythrocytes. J Biol Chem. 2001;276:20197–205.

23. Ng HH, Ciccone DN, Morshead KB, Oettinger MA, Struhl K. Lysine-79 of histone H3 is hypomethylated at silenced loci in yeast and mammalian cells: a potential mechanism for position-effect variegation. Proc Natl Acad Sci U S A. 2003;100:1820–5.

24. Pokholok DK, Harbison CT, Levine S, Cole M, Hannett NM, Tong IL, et al. Genome-wide map of nucleosome acetylation and methylation in yeast. Cell. 2005;122:517–27.

25. Zhang T, Cooper S, Brockdorff N. The interplay of histone modifications—writers that read. EMBO Rep. 2015;16:1467–81.

26. Nishioka K, Rice JC, Sarma K, Erdjument-Bromage H, Werner J, Wang Y, et al. PR-Set7 is a nucleosome-specific methyltransferase that modifies lysine 20 of histone H4 and is associated with silent chromatin. Mol Cell. 2002;9:1201–13.

27. Fischer JJ, Toedling J, Krueger T, Schueler M, Huber W, Sperling S. Combinatorial effects of four histone modifications in transcription and differentiation. Genomics. 2008;91:41–51.

28. Latham JA, Dent SY. Cross-regulation of histone modifications. Nat Struct Mol Biol. 2007;14:1017–24.

29. Cosgrove MS, Boeke JD, Wolberger C. Regulated nucleosome mobility and the histone code. Nat Struct Mol Biol. 2004;11:1037–43.

30. Seet BT, Dikic I, Zhou MM, Pawson T. Reading protein modifications with interaction domains. Nat Rev Mol Cell Biol. 2006;7:473–83.

31. Shogren-Knaak M, Ishii H, Sun JM, Pazin MJ, Davie JR, Peterson CL. Histone H4-K16 acetylation controls chromatin structure and protein interactions. Science. 2006;311:844–7.

32. Cheung P, Allis CD, Sassone-Corsi P. Signaling to chromatin through histone modifications. Cell. 2000;103:263–71.

33. Kenseth JR, Coldiron SJ. High-throughput characterization and quality control of small-molecule combinatorial libraries. Curr Opin Chem Biol. 2004;8:418–23.

34. Kouzarides T. Chromatin modifications and their function. Cell. 2007;128:693–705.

35. Tessarz P, Kouzarides T. Histone core modifications regulating nucleosome structure and dynamics. Nat Rev Mol Cell Biol. 2014;15:703–8.

36. Tamkun JW, Deuring R, Scott MP, Kissinger M, Pattatucci AM, Kaufman TC, et al. Brahma: a regulator of Drosophila homeotic genes structurally related to the yeast transcriptional activator SNF2SWI2. Cell. 1992;68:561–72.

37. Filippakopoulos P, Picaud S, Mangos M, Keates T, Lambert JP, Barsyte-Lovejoy D, et al. Histone recognition and large-scale structural analysis of the human bromodomain family. Cell. 2012;149:214–31.

38. Jacobson RH, Ladurner AG, King DS, Tjian R. Structure and function of a human TAFII250 double bromodomain module. Science. 2000;288:1422–5.

39. Morinière J, Rousseaux S, Steuerwald U, Soler-López M, Curtet S, Vitte AL, et al. Cooperative binding of two acetylation marks on a histone tail by a single bromodomain. Nature. 2009;461:664–8.

40. Miller TCR, Simon B, Rybin V, Grötsch H, Curtet S, Carlomagno T, et al. A bromodomain-DNA interaction facilitates acetylation-dependent bivalent nucleosome recognition by the BET protein BRDT. Nat Commun. 2016;7:13855.

41. Richman R, Chicoine LG, Collini MP, Cook RG, Allis CD. Micronuclei and the cytoplasm of growing Tetrahymena contain a histone acetylase activity which is highly specific for free histone H4. J Cell Biol. 1988;106:1017–26.

42. Hodawadekar SC, Marmorstein R. Chemistry of acetyl transfer by histone modifying enzymes: structure, mechanism and implications for effector design. Oncogene. 2007;26:5528–40.

43. Seto E, Yoshida M. Erasers of histone acetylation: the histone deacetylase enzymes. Cold Spring Harb Perspect Biol. 2014;6:a018713.

44. Vega RB, Harrison BC, Meadows E, Roberts CR, Papst PJ, Olson EN, et al. Protein kinases C and D mediate agonist-dependent cardiac hypertrophy through nuclear export of histone deacetylase 5. Mol Cell Biol. 2004;24:8374–85.

45. McKinsey TA, Zhang CL, Lu J, Olson EN. Signal-dependent nuclear export of a histone deacetylase regulates muscle differentiation. Nature. 2000;408:106–11.

46. Zhang L, Eugeni EE, Parthun MR, Freitas MA. Identification of novel histone post-translational modifications by peptide mass fingerprinting. Chromosoma. 2003;112:77–86.

47. Cai Y, Jin J, Swanson SK, Cole MD, Choi SH, Florens L, et al. Subunit composition and substrate specificity of a MOF-containing histone acetyltransferase distinct from the male-specific lethal (MSL) complex. J Biol Chem. 2010;285:4268–72.

48. Cloos PAC, Christensen J, Agger K, Helin K. Erasing the methyl mark: histone demethylases at the center of cellular differentiation and disease. Genes Dev. 2008;22:1115–40.

49. Trojer P, Reinberg D. Histone lysine demethylases and their impact on epigenetics. Cell. 2006;125:213–7.

50. Blus BJ, Wiggins K, Khorasanizadeh S. Epigenetic virtues of chromodomains. Crit Rev Biochem Mol Biol. 2011;46:507–26.

51. Kim J, Daniel J, Espejo A, Lake A, Krishna M, Xia L, et al. Tudor, MBT and chromo domains gauge the degree of lysine methylation. EMBO Rep. 2006;7:397–403.

52. Nameki N, Tochio N, Koshiba S, Inoue M, Yabuki T, Aoki M, et al. Solution structure of the PWWP domain of the hepatoma-derived growth factor family. Protein Sci. 2005;14:756–64.

53. Yun M, Wu J, Workman JL, Li B. Readers of histone modifications. Cell Res. 2011;21:564–78.

54. Greeson NT, Sengupta R, Arida AR, Jenuwein T, Sanders SL. Di-methyl H4 lysine 20 targets the checkpoint protein Crb2 to sites of DNA damage. J Biol Chem. 2008;283:33168–74.

55. Li B, Jackson J, Simon MD, Fleharty B, Gogol M, Seidel C, et al. Histone H3 lysine 36 dimethylation (H3K36me2) is sufficient to recruit the Rpd3s histone deacetylase complex and to repress spurious transcription. J Biol Chem. 2009;284:7970–6.

56. Klein BJ, Lalonde ME, Côté J, Yang XJ, Kutateladze TG. Crosstalk between epigenetic readers regulates the MOZ/MORF HAT complexes. Epigenetics. 2014;9:186–93.

57. Dillon SC, Zhang X, Trievel RC, Cheng X. The SET-domain protein superfamily: protein lysine methyltransferases. Genome Biol. 2005;6:227.

58. Van Leeuwen F, Gafken PR, Gottschling DE. Dot1p modulates silencing in yeast by methylation of the nucleosome core. Cell. 2002;109:745–56.

59. Cheng X, Collins RE, Zhang X. Structural and sequence motifs of protein (histone) methylation enzymes. Annu Rev Biophys Biomol Struct. 2005;34:267–94.

60. Shi Y, Sawada J, Sui G, Affar EB, Whetstine JR, Lan F, et al. Coordinated histone modifications mediated by a CtBP co-repressor complex. Nature. 2003;422:735–8.

61. Whetstine JR, Nottke A, Lan F, Huarte M, Smolikov S, Chen Z, et al. Reversal of histone lysine trimethylation by the JMJD2 family of histone demethylases. Cell. 2006;125:467–81.

62. Tsukada Y, Fang J, Erdjument-Bromage H, Warren ME, Borchers CH, Tempst P, et al. Histone demethylation by a family of JmjC domain-containing proteins. Nature. 2006;439:811–6.

63. Van Den Elsen PJ, Van Eggermond MCJA, Puentes F, Van Der Valk P, Baker D, Amor S. The epigenetics of multiple sclerosis and other related disorders. Mult Scler Relat Disord. 2014;3:163–75.

64. Hillert J. Human leukocyte antigen studies in multiple sclerosis. Ann Neurol. 1994;36(Suppl):S15–7.

65. Gray SG, Dangond F. Rationale for the use of histone deacetylase inhibitors as a dual therapeutic modality in multiple sclerosis. Epigenetics. 2006;1:67–75.

66. Liu Z, Cao W, Xu L, Chen X, Zhan Y, Yang Q, et al. The histone H3 lysine-27 demethylase Jmjd3 plays a critical role in specific regulation of Th17 cell differentiation. J Mol Cell Biol. 2015;7:505–16.

67. Zhu J, Paul WE. CD4 T cells: fates, functions, and faults. Blood. 2008;112:1557–69.

68. Furtado GC, Marcondes MCG, Latkowski J-A, Tsai J, Wensky A, Lafaille JJ. Swift entry of myelin-specific T lymphocytes into the central nervous system in spontaneous autoimmune encephalomyelitis. J Immunol. 2008;181:4648–55.

69. Seder RA, Ahmed R. Similarities and differences in CD4+ and CD8+ effector and memory T cell generation. Nat Immunol. 2003;4:835–42.

70. Larochelle C, Alvarez JI, Prat A. How do immune cells overcome the blood-brain barrier in multiple sclerosis? FEBS Lett. 2011;585:3770–80.

71. Gordon S. Alternative activation of macrophages. Nat Rev Immunol. 2003;3:23–35.

72. Gordon S, Taylor PR. Monocyte and macrophage heterogeneity. Nat Rev Immunol. 2005;5:953–64.

73. Sica A, Mantovani A. Macrophage plasticity and polarization: in vivo veritas. J Clin Invest. 2012;12:787–95.

74. Camelo S, Iglesias AH, Hwang D, Due B, Ryu H, Smith K, et al. Transcriptional therapy with the histone deacetylase inhibitor trichostatin A ameliorates experimental autoimmune encephalomyelitis. J Neuroimmunol. 2005;164:10–21.

75. Su R-C, Becker AB, Kozyrskyj AL, Hayglass KT. Epigenetic regulation of established human type 1 versus type 2 cytokine responses. J Allergy Clin Immunol. 2008;121:57–63.e3.

76. Säemann MD, G a B, Osterreicher CH, Burtscher H, Parolini O, Diakos C, et al. Anti-inflammatory effects of sodium butyrate on human monocytes: potent inhibition of IL-12 and up-regulation of IL-10 production. FASEB J. 2000;14:2380–2.

77. Antignano F, Zaph C. Regulation of CD4 T-cell differentiation and inflammation by repressive histone methylation. Immunol Cell Biol. 2015;93:245–52.

78. Mishra N, Reilly CM, Brown DR, Ruiz P, Gilkeson GS. Histone deacetylase inhibitors modulate renal disease in the MRL-lpr/lpr mouse. J Clin Invest. 2003;111:539–52.

79. Satoh T, Takeuchi O, Vandenbon A, Yasuda K, Tanaka Y, Kumagai Y, et al. The Jmjd3-Irf4 axis regulates M2 macrophage polarization and host responses against helminth infection. Nat Immunol. 2010;11:936–44.

80. Manna S, Kim JK, Baugé C, Cam M, Zhao Y, Shetty J, et al. Histone H3 lysine 27 demethylases Jmjd3 and Utx are required for T-cell differentiation. Nat Commun. 2015;6:8152.

81. Singh RP, Hasan S, Sharma S, Nagra S, Yamaguchi DT, Wong DTW, et al. Th17 cells in inflammation and autoimmunity. Autoimmun Rev. 2014;13:1174–81.

82. Kushwah R, Hu J. Dendritic cell apoptosis: regulation of tolerance versus immunity. J Immunol. 2010;185:795–802.

83. Doñas C, Carrasco M, Fritz M, Prado C, Tejón G, Osorio-Barrios F, et al. The histone demethylase inhibitor GSK-J4 limits inflammation through the induction of a tolerogenic phenotype on DCs. J Autoimmun. 2016;75:105–17.

84. Waldner H, Collins M, Kuchroo VK. Activation of antigen-presenting cells by microbial products breaks self tolerance and induces autoimmune disease. J Clin Invest. 2004;113:990–7.

85. Zhang J, Lee SM, Shannon S, Gao B, Chen W, Chen A, et al. The type III histone deacetylase Sirt1 is essential for maintenance of T cell tolerance in mice. J Clin Invest. 2009;119:3048–58.

86. Yang H, Lee SM, Gao B, Zhang J, Fang D. Histone deacetylase sirtuin 1 deacetylates IRF1 protein and programs dendritic cells to control Th17

87. protein differentiation during autoimmune inflammation. J Biol Chem. 2013;288:37256–66.

87. Marin-Husstege M, Muggironi M, Liu A, Casaccia-Bonnefil P. Histone deacetylase activity is necessary for oligodendrocyte lineage progression. J Neurosci. 2002;22:10333–45.

88. Shen S, Sandoval J, Swiss VA, Li J, Dupree J, Franklin RJM, et al. Age-dependent epigenetic control of differentiation inhibitors is critical for remyelination efficiency. Nat Neurosci. 2008;11:1024–34.

89. Pedre X, Mastronardi F, Bruck W, Lopez-Rodas G, Kuhlmann T, Casaccia P. Changed histone acetylation patterns in normal-appearing white matter and early multiple sclerosis lesions. J Neurosci. 2011;31:3435–45.

90. Mahad DH, Trapp BD, Lassmann H. Pathological mechanisms in progressive multiple sclerosis. Lancet Neurol. 2015;14:183–93.

91. Singhal NK, Li S, Arning E, Alkhayer K, Clements R, Sarcyk Z, et al. Changes in methionine metabolism and histone H3 trimethylation are linked to mitochondrial defects in multiple sclerosis. J Neurosci. 2015;35:15170–86.

92. Inkster B, Strijbis EMM, Vounou M, Kappos L, Radue EW, Matthews PM, et al. Histone deacetylase gene variants predict brain volume changes in multiple sclerosis. Neurobiol Aging. 2013;34:238–47.

93. Ge Z, Da Y, Xue Z, Zhang K, Zhuang H, Peng M, et al. Vorinostat, a histone deacetylase inhibitor, suppresses dendritic cell function and ameliorates experimental autoimmune encephalomyelitis. Exp Neurol. 2013;241:56–66.

94. Zhang Z, Zhang ZY, Wu Y, Schluesener HJ. Valproic acid ameliorates inflammation in experimental autoimmune encephalomyelitis rats. Neuroscience. 2012;221:140–50.

95. Xie L, Li XK, Funeshima-Fuji N, Kimura H, Matsumoto Y, Isaka Y, et al. Amelioration of experimental autoimmune encephalomyelitis by curcumin treatment through inhibition of IL-17 production. Int Immunopharmacol. 2009;9:575–81.

96. Shen S, Li J, Casaccia-Bonnefil P. Histone modifications affect timing of oligodendrocyte progenitor differentiation in the developing rat brain. J Cell Biol. 2005;169:577–89.

97. Hilton IB, D'Ippolito AM, Vockley CM, Thakore PI, Crawford GE, Reddy TE, et al. Epigenome editing by a CRISPR-Cas9-based acetyltransferase activates genes from promoters and enhancers. Nat Biotechnol. 2015;33:510–7.

98. Thakore PI, D'Ippolito AM, Song L, Safi A, Shivakumar NK, Kabadi AM, et al. Highly specific epigenome editing by CRISPR-Cas9 repressors for silencing of distal regulatory elements. Nat Methods. 2015;12:1143–9.

99. Ghalamfarsa G, Hojjat-Farsangi M, Mohammadnia-Afrouzi M, Anvari E, Farhadi S, Yousefi M, et al. Application of nanomedicine for crossing the blood–brain barrier: theranostic opportunities in multiple sclerosis. J Immunotoxicol. 2016;13:603–19.

100. O'Brien K, Gran B, Rostami A. T-cell based immunotherapy in experimental autoimmune encephalomyelitis and multiple sclerosis. Immunotherapy. 2010;2:99–115.

101. Dunn J, Rao S. Epigenetics and immunotherapy: the current state of play. Mol Immunol. 2017;87:227–39.

19

A short guide to long non-coding RNA gene nomenclature

Mathew W Wright

Abstract

The HUGO Gene Nomenclature Committee (HGNC) is the only organisation authorised to assign standardised nomenclature to human genes. Of the 38,000 approved gene symbols in our database (www.genenames.org), the majority represent protein-coding (pc) genes; however, we also name pseudogenes, phenotypic loci, some genomic features, and to date have named more than 8,500 human non-protein coding RNA (ncRNA) genes and ncRNA pseudogenes. We have already established unique names for most of the small ncRNA genes by working with experts for each class. Small ncRNAs can be defined into their respective classes by their shared homology and common function. In contrast, long non-coding RNA (lncRNA) genes represent a disparate set of loci related only by their size, more than 200 bases in length, share no conserved sequence homology, and have variable functions. As with pc genes, wherever possible, lncRNAs are named based on the known function of their product; a short guide is presented herein to help authors when developing novel gene symbols for lncRNAs with characterised function. Researchers must contact the HGNC with their suggestions prior to publication, to check whether the proposed gene symbol can be approved. Although thousands of lncRNAs have been predicted in the human genome, for the vast majority their function remains unresolved. lncRNA genes with no known function are named based on their genomic context. Working with lncRNA researchers, the HGNC aims to provide unique and, wherever possible, meaningful gene symbols to all lncRNA genes.

Keywords: Long non-coding RNA, Nomenclature, ncRNA, lncRNA

Introduction

Since its inception in the 1970s, the HUGO Gene Nomenclature Committee (HGNC) [1] has kept apace with the discovery and characterisation of new human genes, providing each gene with a unique symbol and name and thus aiding effective scientific communication. By the time the initial sequence of the Human Genome was published in 2001 [2], the HGNC database (www.genenames.org) [3] contained more than 13,000 approved gene names, mostly for protein-coding genes with only around 200 non-coding RNA (ncRNA) gene names. With the burgeoning research and interest in ncRNAs over the last decade, the number of ncRNA loci with gene names has vastly expanded to more than 8,500 currently; about 2,000 of these represent long non-coding RNA (lncRNA) genes. Whereas classes of small ncRNAs can be defined by their shared homology and common function [4], in contrast, lncRNA genes are a disparate set of loci related only by their size (more than 200 bases in length), are non-homologous, and have variable functions [5]. Their discovery has been further complicated because they are expressed at very low levels, sometimes only at specific developmental stages, and in specific tissues [6]. Large-scale transcriptomic analyses, such as RNA-Seq, have now revealed thousands of putative long non-coding RNAs [7]; these present unique nomenclature challenges, especially because for the vast majority, the function of the resultant transcript(s) remains unknown. Below, we present a brief guide to the nomenclature of lncRNA genes and provide examples of some of the genes named to date.

lncRNA gene naming guidelines

The HGNC endeavours to approve symbols and names that have been used in publications, but this is not always possible. To ensure their symbol can be approved authors must contact the HGNC prior to publication to agree the nomenclature for each novel lncRNA gene. When

Correspondence: hgnc@genenames.org
HUGO Gene Nomenclature Committee (HGNC), EMBL-EBI, Wellcome Trust Genome Campus, Hinxton, Cambridge CB10 1SD, UK

creating a new lncRNA gene name, there are a number of factors that should be taken into account:

Each approved gene symbol must be unique

This is the paramount nomenclature rule and cannot be broken. Uniqueness enables unambiguous communication and this utility of approved gene nomenclature ensures that everyone knows they are speaking about the same gene. If an author publishes a lncRNA name that is already in use for another locus, then the HGNC will have to assign an alternative symbol. For instance, a novel lncRNA required to keep epidermal cells in an undifferentiated state was published as *ANCR* [8] but this could not be approved since this was already in use for the 'Angelman syndrome chromosome region'; so, in agreement with the authors, it was approved as *DANCR* for 'differentiation antagonizing non-protein coding RNA'.

Symbols are short-form representations of the descriptive gene name

Each lncRNA is assigned a gene symbol that is an abbreviation or acronym of a descriptive name. For example, the symbol *BANCR* is an abbreviation of the full name 'BRAF-activated non-protein coding RNA'. Gene symbols are the primary descriptors used in communications about genes and their brevity makes them user friendly.

Symbols should only contain Latin letters and Arabic numerals

Gene symbols should only contain Latin letters and Arabic numerals, e.g. *NEAT1* (nuclear paraspeckle assembly transcript 1). Punctuation is not used and will generally be removed or replaced by a letter or number. The use of hyphens is limited to specific exceptions, such as genes named as antisense to protein-coding genes (discussed later), e.g. *BACE1-AS* (BACE1 antisense RNA).

Human gene symbols are all uppercase

By long-established convention, all human gene symbols are written in uppercase letters. This distinguishes them from rodent genes where only the first letter is uppercase and the rest lowercase. For instance the mouse gene *Hotair* is the ortholog of the human *HOTAIR* (HOX transcript antisense RNA) gene.

Symbols should not contain any reference to species

Symbols should not contain any reference to species, for example 'H/h' for human. The use of 'human' in gene names should also be avoided because approved human gene names are transferred across to homologous genes in other species, where 'human' would be potentially confusing and misleading.

Symbols should not spell out commonly used words

Whilst authors might be tempted to use commonly used words for gene symbols because they are easily recognized and pronounced, they should be avoided because they generate unnecessary confusion and make searching for information about a gene much more difficult. A good example of this is *AIRN*, which was first published as AIR [9]. A search with 'AIR' in PubMed returns more than 220,000 unrelated hits, whereas a search with the approved symbol '*AIRN*' returns only the 10 publications specific to this gene. Other examples include EGO [10], since approved as *EGOT* (eosinophil granule ontogeny transcript), and PANDA [11] now *PANDAR* (promoter of CDKN1A antisense DNA damage activated RNA).

If possible, names should be based on function

Genes are preferentially named based on the function of the gene product. Examples include the well-known '*XIST*' which is short for 'X (inactive)-specific transcript' because the transcript is involved in transcriptionally silencing one

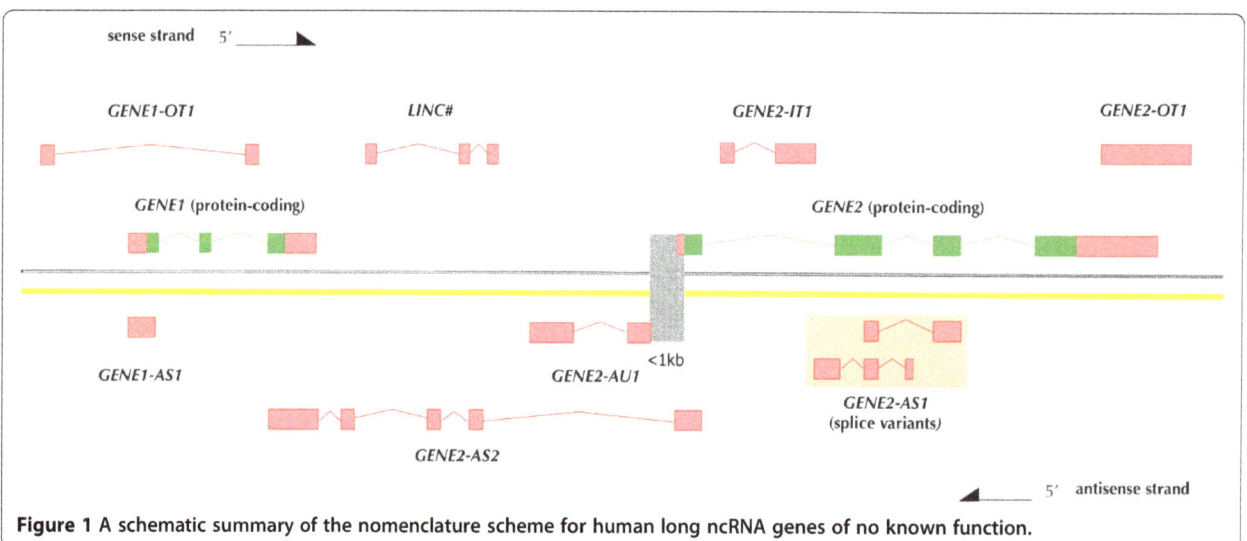

Figure 1 A schematic summary of the nomenclature scheme for human long ncRNA genes of no known function.

of the pair of X chromosomes, and more recently '*TINCR*' [12] which stands for 'tissue differentiation-inducing non-protein coding RNA' because the product is required for epidermal tissue differentiation. If possible, the name of a gene should be based on the normal function of the gene product and not a mutant phenotype. Gene names should be concise and not attempt to represent all known information about a gene. The following are a few other things to consider in gene symbols and names:

- Must not be offensive or pejorative
- Must not be used to acknowledge individuals or places
- Should not reference names of mythical, fictional, or historical figures
- Should not be whimsical or impart no meaningful information about the gene

Functional transcribed pseudogenes should retain their pseudogene name

A small number of transcribed pseudogenes have now been shown to be functional, e.g. *PTENP1* regulates levels of *PTEN* by binding to *PTEN*-targeting miRNA [13]. Transcribed pseudogenes with published function will retain their pseudogene nomenclature and not be renamed based on function; however, '(functional)' is added to the end of the gene name so that these genes can be found in a search, e.g. the full name of *PTENP1* is 'phosphatase and tensin homolog pseudogene 1 (functional)'.

Naming genes with no known function

LncRNA genes with no known function are named pragmatically based on their genomic context. A schematic of the naming protocol is presented in Figure 1. This figure demonstrates how gene nomenclature can be applied in these instances but should not be used independently by researchers to generate lncRNA gene names with potentially different numbering to the approved HGNC names. If there is a proximal pc gene then the lncRNA genes are given a gene symbol beginning with the pc symbol and assigned a suffix according to whether they are: antisense (AS) e.g. *BACE1-AS*; intronic (IT) e.g. *SPRY4-IT1*; or overlapping (OT) e.g. *SOX2-OT*. Long intergenic lncRNAs (lincRNAs) that lie between pc gene loci are named with a common root symbol (*LINC*, 'long intergenic non-coding RNA') and an iterated, numerical suffix. The HGNC naming schema is consistent with the lncRNA categories annotated by GENCODE: antisense RNAs, sense intronic, sense overlapping, and lincRNA [14]. A new locus category is under consideration for lncRNAs that lie in a head-to-head orientation with a pc gene and hence putatively share a bidirectional promoter; the HGNC proposes naming these as antisense upstream (AU), e.g. *GENE2-AU1*. It should be noted that the HGNC does not approve names for splice variants so the two variant transcripts

opposite *GENE2* in Figure 1 are named as one lncRNA gene (*GENE2-AS1*). Also if an lncRNA gene encodes transcripts that span more than one protein-coding gene, then the first protein-coding gene from the 5′ end of the lncRNA is used to name it, e.g. *GENE2-AS2* in Figure 1. This naming schema is applicable to most lncRNA genes but some lncRNA genes within gene dense regions may not fit into these discrete categories and require individual assessment by the HGNC (Additional file 1: Figure S1 shows the HGNC decision tree for naming lncRNAs with no known function).

Conclusions

Working together with the lncRNA community, the HGNC aims to provide informative names for all lncRNA genes in the human genome. The simple guidelines stated in this paper are intended to guide researchers, but the only way to approve a new lncRNA gene symbol is to contact the HGNC. For further information on lncRNA nomenclature please see the HGNC lncRNA webpage: *www.genenames. org/rna/LNCRNA* and email us at hgnc@genenames.org.

Competing interests
The authors declare that they have no competing interests.

Acknowledgements
Thank you to the lncRNA research community for their invaluable input over the years. Special thanks to fellow HGNC team members Dr Elspeth Bruford and Dr Ruth Seal for their helpful discussions and continuing camaraderie. This work was supported by the Wellcome Trust (099129/Z/12/Z) and the National Human Genome Research Institute (P41 HG03345).

References
1. Povey S, Lovering R, Bruford E, Wright M, Lush M, Wain H: **The HUGO Gene Nomenclature Committee (HGNC).** *Hum Genet* 2001, **109**(6):678–680.
2. Lander ES, Linton LM, Birren B, Nusbaum C, Zody MC, Baldwin J, Devon K, Dewar K, Doyle M, FitzHugh W, Funke R, Gage D, Harris K, Heaford A, Howland J, Kann L, Lehoczky J, LeVine R, McEwan P, McKernan K, Meldrim J, Mesirov JP, Miranda C, Morris W, Naylor J, Raymond C, Rosetti M, Santos R, Sheridan A, Sougnez C, *et al*: **Initial sequencing and analysis of the human genome.** *Nature* 2001, **409**(6822):860–921.
3. Gray KA, Daugherty LC, Gordon SM, Seal RL, Wright MW, Bruford EA: **Genenames.org: the HGNC resources in 2013.** *Nucleic Acids Res* 2013, **41**(Database issue):D545–D552.
4. Wright MW, Bruford EA: **Naming 'junk': human non-protein coding RNA (ncRNA) gene nomenclature.** *Hum Genom* 2011, **5**(2):90–98.
5. Kung JT, Colognori D, Lee JT: **Long noncoding RNAs: past, present, and future.** *Genetics* 2013, **193**(3):651–669.
6. Clark MB, Mattick JS: **Long noncoding RNAs in cell biology.** *Semin Cell Dev Biol* 2011, **22**(4):366–376.
7. Derrien T, Johnson R, Bussotti G, Tanzer A, Djebali S, Tilgner H, Guernec G, Martin D, Merkel A, Knowles DG, Lagarde J, Veeravalli L, Ruan X, Ruan Y, Lassmann T, Carninci P, Brown JB, Lipovich L, Gonzalez JM, Thomas M, Davis CA, Shiekhattar R, Gingeras TR, Hubbard TJ, Notredame C, Harrow J, Guigo R: **The GENCODE v7 catalog of human long noncoding RNAs: analysis of their gene structure, evolution, and expression.** *Genome Res* 2012, **22**(9):1775–1789.
8. Kretz M, Webster DE, Flockhart RJ, Lee CS, Zehnder A, Lopez-Pajares V, Qu K, Zheng GX, Chow J, Kim GE, Rinn JL, Chang HY, Siprashvili Z, Khavari PA: **Suppression of progenitor differentiation requires the long noncoding RNA ANCR.** *Genes Dev* 2012, **26**(4):338–343.
9. Lyle R, Watanabe D, Te Vruchte D, Lerchner W, Smrzka OW, Wutz A, Schageman J, Hahner L, Davies C, Barlow DP: **The imprinted antisense RNA**

at the Igf2r locus overlaps but does not imprint Mas1. *Nat Genet* 2000, **25**(1):19–21.

10. Wagner LA, Christensen CJ, Dunn DM, Spangrude GJ, Georgelas A, Kelley L, Esplin MS, Weiss RB, Gleich GJ: **EGO, a novel, noncoding RNA gene, regulates eosinophil granule protein transcript expression.** *Blood* 2007, **109**(12):5191–5198.

11. Hung T, Wang Y, Lin MF, Koegel AK, Kotake Y, Grant GD, Horlings HM, Shah N, Umbricht C, Wang P, Wang Y, Kong B, Langerod A, Borresen-Dale AL, Kim SK, van de Vijver M, Sukumar S, Whitfield ML, Kellis M, Xiong Y, Chang HY: **Extensive and coordinated transcription of noncoding RNAs within cell-cycle promoters.** *Nat Genet* 2011, **43**(7):621–629.

12. Kretz M, Siprashvili Z, Chu C, Webster DE, Zehnder A, Qu K, Lee CS, Flockhart RJ, Groff AF, Chow J, Johnston D, Kim GE, Spitale RC, Flynn RA, Zheng GX, Aiyer S, Raj A, Rinn JL, Chang HY, Khavari PA: **Control of somatic tissue differentiation by the long non-coding RNA TINCR.** *Nature* 2013, **493**(7431):231–235.

13. Poliseno L, Salmena L, Zhang J, Carver B, Haveman WJ, Pandolfi PP: **A coding-independent function of gene and pseudogene mRNAs regulates tumour biology.** *Nature* 2010, **465**(7301):1033–1038.

14. Harrow J, Frankish A, Gonzalez JM, Tapanari E, Diekhans M, Kokocinski F, Aken BL, Barrell D, Zadissa A, Searle S, Barnes I, Bignell A, Boychenko V, Hunt T, Kay M, Mukherjee G, Rajan J, Despacio-Reyes G, Saunders G, Steward C, Harte R, Lin M, Howald C, Tanzer A, Derrien T, Chrast J, Walters N, Balasubramanian S, Pei B, Tress M, *et al*: **GENCODE: the reference human genome annotation for The ENCODE Project.** *Genome Res* 2012, **22**(9):1760–1774.

20

16th Carbonyl Metabolism Meeting: from enzymology to genomics

Edmund Maser

Abstract

The 16th International Meeting on the Enzymology and Molecular Biology of Carbonyl Metabolism, Castle of Ploen (Schleswig-Holstein, Germany), July 10–15, 2012, covered all aspects of NAD(P)-dependent oxido-reductases that are involved in the general metabolism of xenobiotic and physiological carbonyl compounds. Starting 30 years ago with enzyme purification, structure elucidation and enzyme kinetics, the Carbonyl Society members have meanwhile established internationally recognized enzyme nomenclature systems and now consider aspects of enzyme genomics and enzyme evolution along with their roles in diseases. The 16th international meeting included lectures from international speakers from all over the world.

Keywords: Carbonyl metabolism, Alcohol dehydrogenase (ADH), Aldehyde dehydrogenase (ALDH), Medium-chain dehydrogenase (MDR), Short-chain dehydrogenase/reductase (SDR), Aldo-keto reductase (AKR)

Background

The International Meeting on the Enzymology and Molecular Biology of Carbonyl Metabolism was established 30 years ago by Prof. Dr. Henry Weiner, Purdue University, West Lafayette, IN, USA, and was later co-organized by his wife Esther. Starting in 1982 in Bern, Switzerland, the meeting was held every other year, alternately in USA and elsewhere in the world. This has been the recurring plan until today. At these meetings, talks and posters are presented on oxido-reductases that use a carbonyl as a substrate: alcohol dehydrogenases (ADH), aldehyde dehydrogenases (ALDH), short-chain dehydrogenases/reductases (SDR) and aldo-keto reductases (AKR). Previously, topics ranged from biochemical enzymology to gene regulation and the function of the enzyme in the cell. Important aspects of molecular toxicology were introduced in parallel, such as the deleterious effects of lipid-derived reactive aldehydes and their metabolic detoxification. Later, important aspects on enzyme genomics and enzyme evolution were included and now cover eukaryotic as well as prokaryotic organisms. This is why this meeting brings so many interested and engaged

scientists from various disciplines together and has such a great tradition and tremendous success.

The aim of the meeting is not only to provide timely updates on all aspects within the field of carbonyl metabolism to the senior scientists and to attract new investigators, but also to educate and train younger scientists and to provide them an opportunity to discuss with senior scientists in a relaxed atmosphere. Excellent personal interactions during the meeting have resulted in a number of international collaborations that emerged among the participants and were very important for the advancement of science in the field of carbonyl metabolism.

Each of the previous 15 meetings was a unique experience for the participants. Moreover, the half-day social excursions have become one of the most anticipated events at each meeting.

The scientific papers of each meeting were published in the form of a book under the editorship of Henry Weiner and the help of three co-editors, who ensured a review process that meets the criteria of international peer-reviewed journals. Later, in 2001, 2003, 2009 and 2011, the manuscripts of every volume appeared as a special issue of the international journal *Chemico-Biological Interactions*. The articles represent a permanent record of what was presented during the conferences and provide a view on the continuous

Correspondence: maser@toxi.uni-kiel.de

Institute of Toxicology and Pharmacology for Natural Scientists, University Medical School, Schleswig-Holstein, Campus Kiel, Brunswiker Str. 10, Kiel D-24105, Germany

scientific progress achieved on the carbonyl metabolizing enzyme super-families within the last 30 years.

The 16th meeting on the enzymology and molecular biology of carbonyl metabolism

The 16th meeting was no exception. Sixteen countries were represented, with 53% of participants coming from Europe, 32% from North America and 15% from Asia and Australia. This year's program was put together by Hans-Joerg Martin and Edmund Maser. Over 70 talks and posters provided the participants with a wide variety of presentations dealing with enzymology, molecular biology and metabolic aspects of carbonyl metabolizing oxido-reductases. Much new information was presented, including three dimensional structures of enzymes not previously reported, new aspects of gene regulation, and metabolism and enzyme mechanisms, as well as enzyme genomics and evolution along with sequence alignments. In addition, there were nomenclature updates for the different enzyme super-families. Importantly, there was an increased emphasis and information pertaining to the emerging physiological/pathological roles and significance of the carbonyl-metabolizing enzymes. Specifically, several new milestone findings were associated to human diseases, and the development of respective medical treatment strategies was discussed. The level of the science presented was top notch, a fact that was recognized by a speaker from the pharmaceutical industry. The full program can be seen on the homepage of the meeting: http://www.carbonyl.toxi.uni-kiel.de/.

A new feature was the competition for the best poster presentation by pre-doctoral and post-doctoral researchers. Members of the Poster Award Committee reviewed the posters and selected two finalists who shared the 'Henry and Esther Weiner Award 2012' which was named after the founder and his wife of the carbonyl metabolism study group.

Details of the scientific programme

The conference was kicked off by Edmund Maser (Kiel University Medical School, Germany) who first welcomed all the participants and then gave a brief overview on the 30 years' history of the Carbonyl Metabolism meetings. He also summarized several recent hot spots in carbonyl metabolism research.

ALDH session

Vasilis Vasiliou (University of Colorado, USA) opened the scientific section on the super-family of ALDH with an introduction into ALDH evolution. He gave an update on the enzymes' systematic nomenclature and summarized recent findings on the physiological and pathophysiological role of these enzymes in humans and other organisms.

Lilian Gonzalez-Segura (Universidad Nacional Autónoma de México, México) continued by discussing the structural, phylogenetic and functional evidence for the existence of specific potassium-binding sites in ALDHs. Based on her results, she concluded that most of the ALDH enzymes possess intra-subunit sites, few have inter-subunit sites, and only the ALDH9s from *Pseudomonas* spp. have central-cavity sites.

Roger Holmes (Griffith University, Australia) focused on the comparative genomics and proteomics of ALDH2 and ALDH1B1, which are both mitochondrial enzymes that metabolize acetaldehyde and other biological aldehydes in the human body. He proposed that a dominant negative heterodimerization of ALDH2*2 subunits with ALDH1B1 may partially explain a lack of compensation by ALDH1B1 in ALDH2*2 individuals and that the *ALDH1B1* gene originated in early vertebrates from a retro-transposition of the *ALDH2* gene.

Tom Hurley (Indiana University School of Medicine, USA) gave a perspective on the discovery of selective and non-selective inhibitors of ALDH1A1 and ALDH3A1 to overcome resistance towards cyclophosphamide upon chemotherapy. He and his group have identified new compounds which will be further developed in different cell line studies.

Sergey Krupenko (Medical University of South Carolina, USA) reported on an unusual mode of coenzyme binding in the carboxy-terminal domain of ALDH1L1 which is a natural fusion of three unrelated genes and catalyzes the two-step conversion of 10-formyltetrahydrofolate to tetrahydrofolate and CO_2 with $NADP^+$ as cofactor. A variety of amino acid substitutions led him to conclude that Glu673 restricts the affinity for the co-factor, whereas Cys707 acts as the sensor of the co-factor redox state.

In her talk, Rosario Muñoz-Clares (Universidad Nacional Autónoma de México, México) focused on the structural and functional aspects of plant ALDH10 enzymes and their role in glycine betaine (GB) production by oxidation of betaine aldehyde (BAL). Biochemical and phylogenetic analyses support the existence of two kinds of ALDH10 isozymes: those with low-BAL affinity, present in most plants, and those with high-BAL affinity, only present in GB-accumulator plants.

Naim Stiti (University of Bonn, Germany) discussed the role of ALDHs in abiotic stress and the detoxification of reactive aldehydes derived from lipid peroxidation in *Arabidopsis thaliana*. He found that the enzyme activities were redox-dependent and that thiol groups of redox-sensitive cysteines at the surface of the protein subunits are critical to dimerization and inactivation.

Vasilis Vasiliou (University of Colorado, USA) introduced a novel concept of an ALDH being involved in gout, a common form of inflammatory arthritis that results from hyperuricemia. He discussed the structural determinants of ALDH16A1 during evolution and postulated that ALDH16A1 may interact with several other proteins associated with uric-acid formation, diabetes, vesicular transport and protein degradation.

François Talfournier (CNRS-Université de Lorraine, France) focused on key roles played by the structural dynamics associated with the co-factor binding in the catalytic mechanism of ALDHs. Structural analyses together with kinetic data from amino acid substitutions at critical residues led him to hypothesize that the co-factor binding mode is in part responsible for the different kinetic behaviour of hydrolytic and CoA-dependent ALDHs.

Shih-Jiun Yin (National Defense Medical Center, Taiwan) presented his work on the expression pattern, activities and protein contents of ADHs and ALDHs in human liver. He emphasized that functional allelic variations at the *ADH1B* and *ALDH2* gene loci affect the development of alcoholism and may be involved in the pathogenesis of alcohol-related liver diseases.

ADH session

Hans Jörnvall (Karolinska Institutet Stockholm, Sweden) opened the scientific section on the ADH and highlighted the origin and evolution of the super-families of medium-chain alcohol dehydrogenases/reductases (MDR) and SDR. He concluded that ADHs have a common origin, with at least three separate emergences, the SDRs, the metal-free MDRs and the Zn-dependent MDRs, from the common ancestor.

Hector Riveros-Rosas (Universidad Nacional Autónoma, México) continued with evolutionary insights on the metabolic role developed by ADHs in animals. He claimed that the presence of ADHs in vertebrates has not been a consequence of chronic ethanol exposure and that their participation in ethanol metabolism can be considered incidental, and not adaptive.

Bryce Plapp (University of Iowa, USA) described the biochemical basis for yeast strains that are fitter for growth in the presence of allyl alcohol. He and his coworkers did site-directed mutagenesis at critical sites in ADH1 and determined steady-state kinetic constants. They found that the fitter yeasts are 'bradytrophs' (slow growing) under fermentative conditions because the ADHs have decreased catalytic efficiency and produce less of the toxic acrolein.

Jan-Olov Höög (Karolinska Institutet Stockholm, Sweden) addressed the yet unknown function of mammalian ADH5, which so far could not be isolated as native protein. Computational methods including sequence alignments, homology modelling and molecular dynamics implied that the protein does not fold properly, is not stable, is regulated on the mRNA level, or might even be a pseudogene.

Stephanie MacAllister (University of Toronto, Canada) presented her research comparing the toxicity mechanisms of acrolein and chloroacetaldehyde (CAA) which are derived from the metabolism of the anti-cancer drugs ifosfamide and cyclophosphamide. By using a variety of model systems, they found that acrolein was significantly more toxic than CAA, and that thiol groups, such as *N*-acetylcysteine and cysteine, were the most effective protective agents against both acrolein and CAA biomarker toxicities.

David Kopecny (Palacky University, Czech Republic) presented a structure-functional study on *S*-nitrosoglutathione reductase from tomato. This enzyme is also known as *S*-(hydroxymethyl)glutathione dehydrogenase, which belongs to the ADH3 family branch of the ADH superfamily.

AKR/SDR session

Edmund Maser (Kiel University Medical School, Germany) opened the scientific session on the AKR and SDR superfamilies with a brief introduction into their history. He then gave a perspective on important events during the evolution of these two super-families and highlighted the association of important AKR and SDR to disease including diabetes, hypertension, the metabolic syndrome, osteoporosis and cancer.

Bengt Persson (Linköping University, Sweden) gave a talk on the nomenclature of the SDR and explained the sub-division of this large super-family using bio-informatics methods. In his opinion, identification of new SDR families is slowing down, even though the number of SDR enzymes increases continuously, indicating that the classification system works and that we will know all SDR families in a not-too-distant future.

Natalia Kedishvili (University of Alabama, Birmingham, USA) focused on the role of the SDR16C family of proteins in retinoic acid homeostasis, including SDR16C4 (retinol dehydrogenase 10), which is indispensable for retinoic acid biosynthesis during vertebrate embryogenesis, SDR16C1, SDR16C5 and SDR16C6. She concluded that the SDR16C family is essential in the regulation of retinoic acid biosynthesis during embryogenesis and possibly in adulthood.

Xavier Parés (Universitat Autònoma de Barcelona, Spain) first reviewed the role of cytosolic MDR and membrane-bound SDR in retinoid metabolism. Then, he reported on the identification of cytosolic AKR1C3 and AKR1B10 with retinaldehyde reductase activity, the latter having a possible relevance in cancer. He suggested that retinoid analogues could be a good starting point for searching AKR1B10-selective inhibitors.

Jaume Farrés (Universitat Autònoma de Barcelona, Spain) continued with his data on AKR1C3 as a highly efficient 9-cis-retinaldehyde reductase which is elevated in different cancer types and is also involved in chemotherapeutic-drug resistance. Farrés could show that the proliferative action of AKR1C3 involves the retinoic acid signalling pathway and that this is in part due to the retinaldehyde reductase activity of AKR1C3.

Petra Haberzettl (University of Louisville, USA) reported on the interesting observation that aldose reductase (AKR1B1) protects against age-dependent insulin resistance in type 2 diabetes, which she concluded might be due to the fact that AKR1B1 metabolizes excess glucose via the polyol pathway.

Chi-Ching Hwang (Kaohsiung Medical University, Taiwan) explained the catalytic role of the flexible substrate binding loop in 3α-hydroxysteroid dehydrogenase/carbonyl reductase (3α-HSD/CR) from *Comamonas testosteroni*, which catalyzes the oxidation of androsterone to androstandione.

Adrian Lapthorn (University of Glasgow, UK) presented his results on the structure and function of the aldo-keto reductase AKR14A1 from *Escherichia coli* which shares strong structural similarities with tetrameric voltage-gated potassium channel β-subunit AKR6A2.

Trevor Penning (University of Pennsylvania, USA) described his groundbreaking work to surmount castration-resistant prostate cancer with AKR1C3 inhibitors. He and co-workers have developed three major classes of inhibitors based on non-steroidal anti-inflammatory drug analogues in which compounds were identified to have nano-molar potency for selective AKR1C3 inhibition.

Susanne Weber (Helmholtz Zentrum München, Germany) described a novel human gene locus, *AKR1B15*, which clusters with other members of the AKR1B subfamily, *AKR1B1* and *AKR1B10*, on chromosome 7. She found highest levels of *AKR1B15* mRNA in placenta and testis, and intermediate levels in prostate, ovary and skeletal muscle.

Paul-Georg Germann (Takeda Pharmaceutical, Europe) gave a perspective from industry on the translation from biochemical research into medicinal drugs. He suggested that the biomedical knowledge that is produced in basic research in many research institutions, including universities, should be exploited together with the industry to develop new therapeutic strategies.

Tea Lanišnik Rižner (University of Ljubljana, Slovenia) discussed the involvement of 17-ketosteroid reductases, AKR1C3 and 17β-HSD type 1, in estrogen biosynthesis in endometrial cancer. Based on her data, she concluded that in endometrial cancer, estrogen is formed from estrogen-sulphate and not from androstendione, suggesting that the sulphatase pathway has more importance for possible anti-estrogenic strategies than the aromatase pathway.

Mark Petrash (University of Colorado, USA) introduced an interesting concept where clues from Ayurveda led to a potential diabetes therapy directed at aldose reductase AKR1B1. Based on the observation that the Indian gooseberry (*Emblica officinalis*, amla) is used in traditional Indian medicine to minimize diabetic complications, and that these effects were consistent with AKR1B1 inhibition, he and co-workers identified AKR1B1 inhibitors in amla extracts.

Aruni Bhatnagar (University of Louisville, USA) addressed the function of aldose reductase (AKR1B1) in cardiac autophagy and reported on studies with wild-type and AKR1B1-null mice. He concluded that AKR1B1 prevents the activation of autophagic responses either by metabolizing glucose and/or removing aldehydes generated by oxidative changes that precede autophagy.

Oleg Barski (University of Louisville, USA) presented his provocative theory that aldehyde quenchers such as carnosine could serve as potential therapies for the prevention and treatment of atherosclerosis and inflammation. He and colleagues found that carnosine rapidly reacted with oxidized lipid-derived reactive aldehydes and that carnosine feeding inhibits atherogenesis by facilitating aldehyde removal and inhibiting ER stress.

Conclusion and future perspectives

One of the strengths of this meeting was the laid-back, congenial atmosphere, which fostered the interaction between delegates. This not only created an environment where younger researchers could meet senior scientists, but also further facilitated the exchange of ideas and the start of collaborations. It was particularly gratifying to see so much new, unpublished data being presented. In summary, this year's Carbonyl Meeting continued the high standard of excellence set by previous conferences.

Election of an executive committee and proposal to found a society

Attendance remained strong throughout the meeting, including the last session, which featured a lively debate on the future of the 'International Meeting on he Enzymology and Molecular Biology of Carbonyl Metabolism' (led by Edmund Maser). Several attendees addressed their concerns over budgetary and organisational issues, which finally resulted in the proposal to found a society. Seven senior scientists were nominated for the Executive Committee. It was agreed to find a name for the Society (together with an acronym) and select a president (chairperson), president-elect, treasurer and four additional persons within 2 years, i.e. before the meeting in 2014 which will be held in Philadelphia, USA.

Social program

The highlights of the social program were a boat excursion on Lake Ploen on a Ploen cruiser followed by a guided walk through Princes' Island and a concert presented by the famous 'Shanty Chorus' during the gala dinner.

Publication

Scientific contributions will appear as full peer-reviewed papers in a Special Issue of the international journal *Chemico-Biological Interactions*. The managing guest editor (Bryce Plapp) and four associate guest editors (Natasha Kedishvili, Tea Lanišnik-Rižner, Peter O'Brien and Edmund Maser) will work in the review process, which is planned to be completed in December 2012, such that the Special Issue of *Chemico-Biological Interactions* can appear in early 2013.

Competing interests

The author declares that he has no competing interests.

Acknowledgements

The meeting was financially supported by a grant from the German Research Council (DFG) and a donation by Takeda, for which the organizers express their appreciation. The organization of the meeting reflects the precise and careful work of the staff of the Institute of Toxicology and Pharmacology for Natural Scientists, University Medical School Schleswig-Holstein, Kiel, Germany. Mrs. Nicole Newiger, Mrs. Elli Kubach, Mr. Chris Steffens, Dr. Hans-Joerg Martin and Prof. Dr. Edmund Maser should be especially recognized. This report is dedicated to the memory of Prof. Dr. Henry Weiner.

21

Identification of functional DNA variants in the constitutive promoter region of *MDM2*

Marie-Eve Lalonde[1], Manon Ouimet[1], Mathieu Larivière[1], Ekaterini A Kritikou[1] and Daniel Sinnett[1,2]*

Abstract

Although mutations in the oncoprotein murine double minute 2 (MDM2) are rare, *MDM2* gene overexpression has been observed in several human tumors. Given that even modest changes in MDM2 levels might influence the p53 tumor suppressor signaling pathway, we postulated that sequence variation in the promoter region of *MDM2* could lead to disregulated expression and variation in gene dosage. Two promoters have been reported for *MDM2*; an internal promoter (P2), which is located near the end of intron 1 and is p53-responsive, and an upstream constitutive promoter (P1), which is p53-independent. Both promoter regions contain DNA variants that could influence the expression levels of *MDM2*, including the well-studied single nucleotide polymorphism (SNP) SNP309, which is located in the promoter P2; i.e., upstream of exon 2. In this report, we screened the promoter P1 for DNA variants and assessed the functional impact of the corresponding SNPs. Using the dbSNP database and genotyping validation in individuals of European descent, we identified three common SNPs (−1494 G > A; indel 40 bp; and −182 C > G). Three major promoter haplotypes were inferred by using these three promoter SNPs together with rs2279744 (SNP309). Following subcloning into a gene reporter system, we found that two of the haplotypes significantly influenced *MDM2* promoter activity in a haplotype-specific manner. Site-directed mutagenesis experiments indicated that the 40 bp insertion/deletion variation is causing the observed allelic promoter activity. This study suggests that part of the variability in the *MDM2* expression levels could be explained by allelic p53-independent P1 promoter activity.

Keywords: MDM2, SNP, Promoter analysis, Functional validation, Site-directed mutagenesis

Introduction

The p53 tumor suppressor has a key role in orchestrating cellular responses to various types of stresses, including DNA damage and oncogene activation with apoptosis, cell-cycle arrest, senescence, DNA repair, cell metabolism, or autophagy [1,2]. Malfunction and mutations of *p53* have been found in most human cancers, leading to a deregulated p53 activity that allows cells to proliferate and survive [3]. The activity of p53 is regulated by many proteins, and one of the most extensively studied regulators of p53 is the murine double minute 2 (MDM2) oncoprotein. MDM2 can regulate p53 activity in different ways and even modest modifications of *MDM2* levels can affect the p53 pathway [4]. Firstly, MDM2 directly binds to the p53 transactivation domain,

thus inhibiting its transcriptional activity. Secondly, MDM2 promotes ubiquitin-dependent proteasomal degradation of p53 by functioning as an E3 ubiquitin ligase [5,6]. Finally, MDM2 shuttles p53 out of the nucleus to the cytoplasm of the cell, promoting the degradation of p53. Importantly, MDM2 forms a negative-feedback loop in regulating p53 activity, in which p53 induces transcription of *MDM2*, and, in turn, the MDM2 protein inhibits p53 activity (reviewed by Momand et al. [7]).

Although mutations in *MDM2* are rare, *MDM2* overexpression is observed in a number of human tumors due to various mechanisms including gene amplification [8-10] and increased transcription [11,12]. *MDM2* overexpression predisposed transgenic mice to spontaneous tumor formation [13] and therefore, overexpression of *MDM2* may substitute for inactivating mutations in p53 [9]. Because MDM2 is an important negative regulator of p53 activity, overexpression of *MDM2* can result in

* Correspondence: daniel.sinnett@umontreal.ca
[1]Division of Hematology-Oncology, Research Center, Sainte-Justine Hospital, 3175 Chemin de la Cote-Sainte-Catherine, Montreal H3T 1C5, Canada
[2]Department of Pediatrics, Faculty of Medicine, University of Montreal, Montreal QC H3T 1C5, Canada

the inhibition of p53-mediated-transcriptional activation, thereby promoting human carcinogenesis.

Functional sequence variants in promoter regions can lead to variable gene expression levels [14,15]; single nucleotide polymorphisms (SNPs) in promoters of genes implicated in DNA-damage responses and apoptosis could have an impact in an individual's susceptibility to develop cancer [16-21]. Because MDM2 is a key component of the p53-mediated DNA-damage response, promoter SNPs in this gene might influence this highly regulated pathway by modifying cellular MDM2 protein levels [22]. The *MDM2* gene has a basal promoter (P1) and an alternative promoter (P2) starting in the intron 1 [23]. The promoter P2 contains a p53-responsive element and has been shown to regulate *MDM2* levels in stressed cells, whereas the promoter P1 functions mainly in a non-stressed environment [23,24]. The rs2279744 (SNP309) in the intronic *p53*-responsive promoter of the *MDM2* gene has been shown to increase the affinity of the transcriptional activator Sp1, resulting in higher levels of *MDM2* mRNA and protein. This SNP has been shown to attenuate apoptotic activity and accelerate tumor formation [22,25-27]. Several studies have reported associations between rs2279744 and the risk of different types of cancer [28-30]; however, this association has not always been confirmed [31-33]. In an attempt to obtain a more complete view of the *MDM2* promoters, we determined the SNP content and the haplotype structure of the constitutive P1 promoter. Here, we show that distinct P1 promoter haplotypes can influence the p53-independent promoter activity in an allele-specific manner.

Methods

SNP discovery in *MDM2* proximal promoter region

The initial search for promoter SNPs (pSNPs) in *MDM2* proximal promoter defined as 2.0 kb upstream of the transcription start site was done using the dbSNP database (build 128) [34]. Seven SNPs were selected for genotyping in a panel of 91 individuals of Western European descent. The Institutional Review Board approved the research protocol and informed consent was obtained from all participants. The corresponding promoter region was amplified in one polymerase chain reaction (PCR) fragment in a 50μL reaction volume, using the following conditions: 20 pmole of 5′AAAGCCCAAATTTCCTTGCT3′ (forward) and 5′CTCCATCTTTCCGACACACA3′ (reverse) primers, 2 mM MgCl$_2$, 0.2 mM deoxynucleoside triphosphates (dNTPs), 1× Fast Start Taq DNA polymerase buffer and GC rich buffer, 2U Fast Start Taq DNA polymerase (Roche Diagnostics, Laval, Canada) and 15 ng of genomic DNA. The PCR program was 95°C for 3 min; 10 cycles with a denaturation at 95°C for 15 s; annealing at 55–50°C (each cycle decreases by 0.5°C) for 20 s and

elongation at 68°C for 2 min; followed by 25 cycles at 50°C for annealing. The amplicons were dot-blotted in duplicate on a nylon membrane and were hybridized with allele-specific oligonucleotides (ASOs) as previously described [35]. Oligonucleotide probes specific for each promoter SNP were used for ASO analysis and are available upon request. A 40 bp insertion/deletion (indel) polymorphism was genotyped by amplification of a 260 bp fragment containing the indel region followed by electrophoresis of the resulting amplicons on a 3% agarose gel to detect one (homozygous) or two bands (heterozygous). PCR conditions were as follows: 20 pmole of 5′TTTCCTTTCTGGTAGGCTGG3′ (forward) and 5′CACCTACTTTCCCACAGAGA3′ (reverse) primers, 1.5 mM MgCl$_2$, 0.2 mM dNTPs, 1× Fast Start Taq DNA polymerase buffer and GC rich buffer, 1U Fast Start Taq DNA polymerase (Roche Diagnostics, Laval, Canada), and 15 ng of genomic DNA. The PCR program was 95°C for 3 min; 32 cycles with a denaturation at 95°C for 30 s; annealing at 52°C for 30 s; and elongation at 72°C for 20 s. Hardy-Weinberg equilibrium was tested with a χ^2 test for goodness of fit. Haplotypes were generated by PHASE software (version 2; University of Washington, Seattle, WA, USA) [36].

Gene reporter assays and site-directed mutagenesis
Constructs

The two major promoter haplotypes (approximately 2.0 kb region) were amplified from genomic DNA of known homozygous individuals and cloned individually in the promoterless pGL3basic Firefly luciferase vector (Promega Corp., Fitchburg, WI, USA) using the Gateway Technology (Invitrogen Corporation, Carlsbad, CA, USA). Specific mutations were introduced by site-directed mutagenesis (Quickchange multi site-directed mutagenesis kit, Stratagene from Agilent Technologies, Santa Clara, CA, USA) according to the manufacturer's instructions. Clones chosen for transfection were sequenced to confirm the presence of the SNPs and then purified using the Qiagen plasmid mini kit (Qiagen Company, Toronto, Canada) prior to transfection.

Transfection

The resulting constructs were used to transiently transfect three cell lines (HeLa, HepG2, and JEG3) using lipofectamine reagent according to the manufacturer's protocol (Invitrogen). Constructs (99 ng) and SV40-driven (1 ng) *Renilla* luciferase cytomegalovirus (CMV) immediate early enhancer/promoter region (pRL-CMV) (ratio 100:1) were co-transfected to control transfection efficiency. The pGL3basic promoterless plasmid (Promega) was used as a negative control and the pGL3SV40 plasmid (Promega) was used as a positive control. The transfected cells were plated in 96-well plates with approximately 6×10^4 cells per well. The cells were harvested 24 h following transfection, and luciferase reporter gene activity was measured with dual-luciferase reporter assay system (Promega) in a SpectraMax 190 luminometer according to the manufacturer's protocol (Molecular Devices, LLC, Sunnyvale, CA, USA). Firefly luciferase activities of the allelic constructs were normalized using the *Renilla* luciferase pRL-CMV activity. The results were expressed as the ratio of Firefly luciferase activity divided by the pRL-CMV internal control activity and expressed as relative luciferase (means ± standard deviation) of four replicates. Three independent experiments were carried out for each cell line. Statistical analyses were performed using unpaired Student's t test to determine p values. Global p value is calculated with Fischer's inverse Chi-squared test [37].

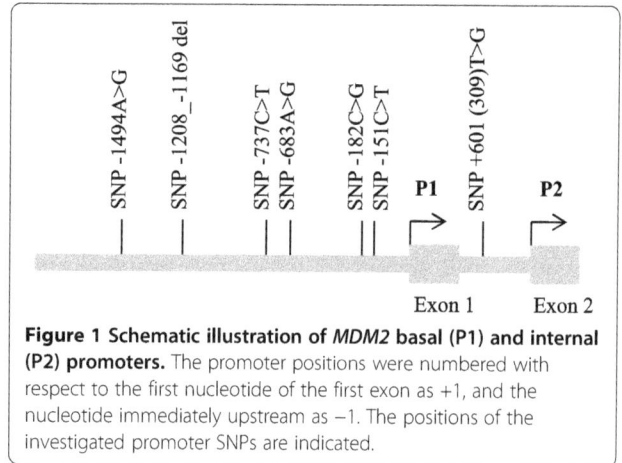

Figure 1 Schematic illustration of *MDM2* basal (P1) and internal (P2) promoters. The promoter positions were numbered with respect to the first nucleotide of the first exon as +1, and the nucleotide immediately upstream as −1. The positions of the investigated promoter SNPs are indicated.

In silico predictions of putative TFBS
MatInspector program from Genomatix Software GmbH (Bayerstrasse, Munich, Germany, www.genomatix.de) was used to determine the presence of putative binding sites for known transcription factors. The predicted gain or loss of putative transcription factor binding site (TFBS) due to a given SNP was determined by the optimized matrix threshold as defined in the MatInspector program.

Results
The search for SNPs in the constitutive P1 promoter of *MDM2* led to the identification of eight pSNPs, including a 40 bp indel (see Table 1). In addition to these pSNPs, we included the well-studied rs2279744 located in the P2 promoter for haplotype analysis (see Figure 1 for a schematic representation of *MDM2* promoters). By genotyping a panel of 91 unrelated Western Europeans, we found four pSNPs (−1494 G > A, rs1144944; indel 40 bp, rs3730485;

−182 C > G, rs937282; and SNP309/601 T > G, rs2279744) to be polymorphic. For rs2279744 the observed minor allele frequency of 35% was similar to the one previously reported for Caucasians [38]. Among the five non-polymorphic SNPs, both −1166 T > G (rs2904506) and −1164 C > G (rs3930427) are located in the 40 bp indel sequence thus creating in some individuals a near identical (except for 2 bps) tandem duplication. Therefore, individuals carrying the deletion behave like they have different alleles at these two positions. Because single variants might not be sufficient to capture the genetic variability relative to a given phenotype, we constructed haplotypes using all four polymorphic pSNPs. Based on these data, we estimated haplotype phase and the corresponding frequencies (Table 2). The three most common promoter haplotypes (1A, 1B, and 2) represented 92.3% of the observed haplotypes in Europeans. Haplotypes 1A and 1B differ at rs2279744, whereas haplotype 2 differs at all four positions (Figure 2a). To evaluate the extent of linkage disequilibrium between the SNPs studied, we measured D'

Table 1 List of the SNPs found in dbSNP database for *MDM2* promoter

Rs number[a]	Position[b]	SNPs ID[c]	MAF[d]
rs1144944	g.67,486,752 G > A	−1494	25%
rs3730485	g.67,487,038_67,487,077del	40 bp indel (−1208 to −1169)	37%
rs2904506	g.67,487,080 T > G	−1166	-
rs3930427	g.67,487,082 C > G	−1164	-
rs3730486	g.67,487,509 C > T	−737	0%
rs3730487	g.67,487,563A > G	−683	0%
rs937282	g.67,488,064 C > G	−182	50%
rs3730491	g.67,488,095 C > T	−151	0%
rs2279744[e]	g.67,488,847 T > G	+601	35%

[a]From dbSNP build 128.
[b]NCBI Build 36.1.
[c]Position relative to the transcription start site (based on reference sequence mRNA).
[d]Minor allele frequencies (MAF) were calculated with 91 unrelated European individuals in this study.
[e]For comparison purposes, we included SNP309 (+601 in our nomenclature) associated with the internal promoter P2.

Table 2 Most frequent *MDM2* promoter haplotypes

Haplotype	−1494 G > A	40 bp deletion	−182 C > G	+309 T > G	Frequency[a]
1A	A	No deletion	C	G	36.8%
1B	A	No deletion	C	T	17.0%
2	G	Deletion	G	T	38.5%

[a]Frequencies calculated with genotyping results of the 91 unrelated European individuals.

and R^2; these values between rs1144944 and rs937282 are 0.977 and 0.934, respectively, and 0.968 and 0.471 between these two SNPs and rs2279744. This indicates that rs2279744 (SNP309) is tightly linked with the P1 promoter's variants.

To assess the functional impact of the major promoter haplotypes 1 and 2, we subcloned the promoter haplotypes in the promoterless pGL3 basic Firefly luciferase reporter vector and we carried out transient transfection experiments for each haplotype-specific constructs in three cell lines (Figure 2b). Because these constructs contain only the proximal P1 promoter (rs2279744 was not included), we could not test differential promoter activities between haplotypes 1A and 1B. Significant

differences were found between H1 and H2 (Figure 2a), with the promoter haplotype H1 having stronger promoter activity in all cell lines tested (Figure 2b). The relative luciferase activity driven by H1 was up to 2.3-fold higher than the luciferase levels driven by H2, indicating variable haplotype-specific expression levels of *MDM2*. The 309 G allele was only present in 1.0% of individuals carrying haplotypes other than H1A (data not shown); therefore, European individuals carrying the allele G of this SNP are more likely to have the high P1 promoter activity haplotype because of the linkage disequilibrium.

Using *in silico* predictive tools, none of these SNPs seem to affect the putative binding of known transcription factors. However, the 40 bp indel contains several

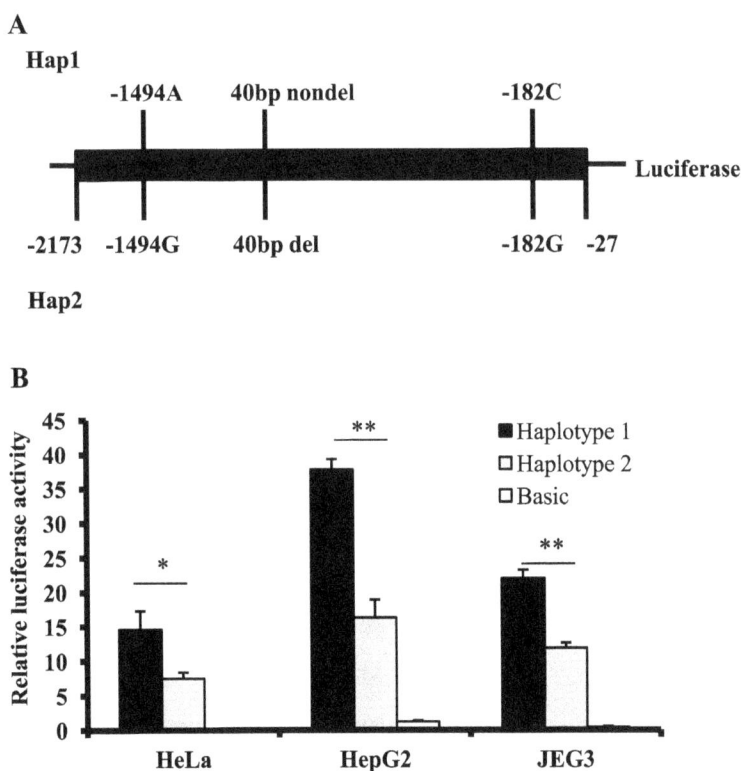

Figure 2 Gene reporter assays to evaluate the functional impact of the most frequent *MDM2* basal promoter haplotypes. (A) Schematic representation of constructs tested for luciferase gene reporter assays in pGL3 basic vector. Haplotypes H1A and H1B are identical when excluding the position +309 T > G (see Table 2). **(B)** Relative luciferase activity of *MDM2* promoter haplotypes was measured following transient transfection in HeLa, HepG2 and JEG3 cells. The empty promoterless pGL3 basic vector was used as negative control. Results are expressed in a ratio of Firefly/*Renilla* activity multiplied by 100. Promoter haplotype H2 was used as reference against which relative expression was compared. Haplotype H1 showed significantly higher expression levels across all three cell lines. The *p* values are calculated from four replicates with unpaired student's *t* test. Significant differences are marked with an asterisk (*$p < 3 \times 10^{-3}$; **$p < 8 \times 10^{-6}$).

predicted transcription factor binding sites (data not shown). In an attempt to identify the *cis*-acting elements responsible for the observed changes in *MDM2* P1 promoter activity, we modified the allele combination in both haplotypes using site-directed mutagenesis (Figure 3). None of the allele combinations in the context of the 40 bp insertion (defining H1) significantly affected the promoter activity of the corresponding H1-derived haplotypes. In the context of the 40 bp deletion (defining H2), the −1494A > G variant (rs1144944) does not affect the H2-derived promoter activity. However, the introduction of allele −182 C (instead of allele G) completely abrogated the promoter activity when combined with the 40 bp deletion compared to the H1-derived construct. This indicates the role of the 40 bp indel variation in the observed allelic promoter activity and the presence of a putative *cis*-acting element at position −182. Taken together, these results support the functional impact of *MDM2* promoter haplotypes on the promoter activity.

Discussion

In more than half the tumors with a fault in the p53 pathway, TP53 itself is not mutated but the p53 pathway is abrogated. Mechanisms that result in this abrogation include increased expression of MDM2 [7] and deletion or epigenetic inactivation of the p53-positive regulator and MDM2 inhibitor ADP-ribosylating factor [39,40]. MDM2 might influence cancer risk through its interaction with other key cancer genes with various functions [41-44]. The MDM2 oncogene is overexpressed in various human cancers and its expression correlates with the phenotypes of high-grade, late-stage, and resistant tumors [45,46]. MDM2 has an important role in cancer development, mostly through inactivation of the p53 pathway [46]. By contrast, the p53-independent MDM2-mediated tumorigenesis is less understood.

At the promoter level, regulation of *MDM2* expression is complex involving two promoters, P1 and P2, which govern transcripts with different translational potentials. In this report, we characterized the two major haplotypes that correspond to the upstream p53-independent constitutive P1 promoter. Unlike the p53-responsive P2 promoter, the P1 promoter lacks an identified TATA box and p53-responsive element [23,47]. We showed that the constitutive expression levels of *MDM2* might at least be partially regulated by distinct promoter SNPs, particularly the 40 bp deletion and the corresponding promoter haplotypes (see Results section). Previous work has shown a

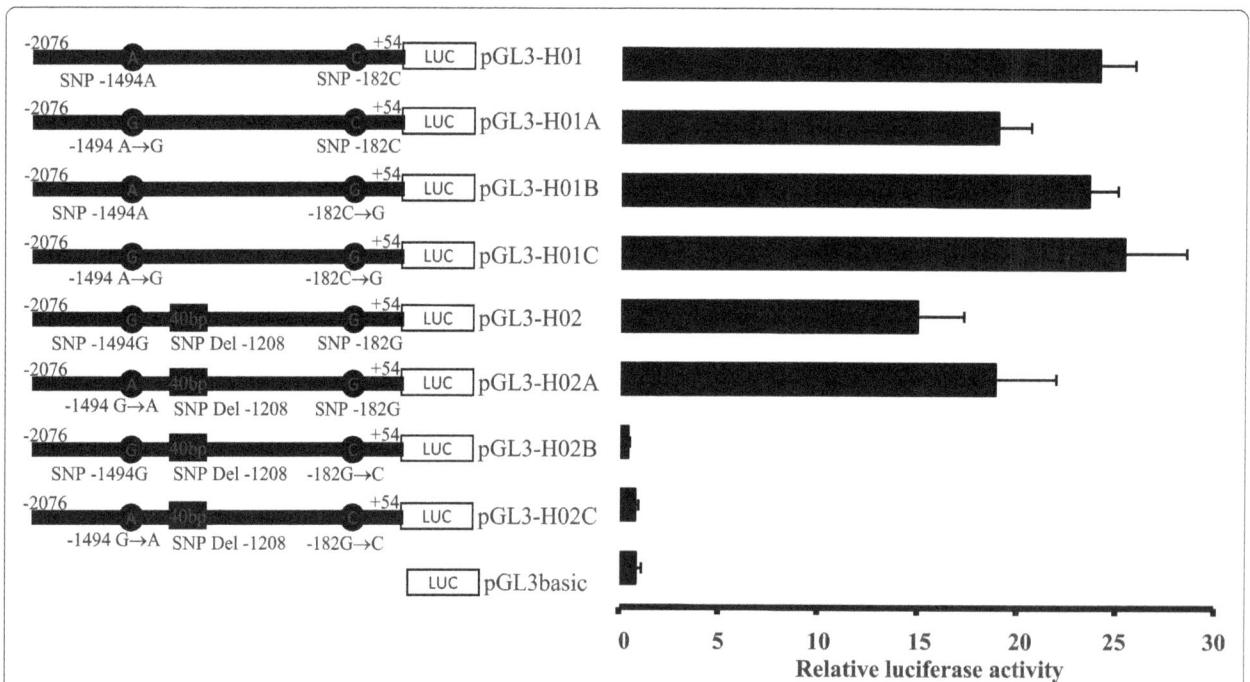

Figure 3 Functional analyses of *MDM2* promoter haplotype H1- and H2-derived mutations in HeLa cells. H1- and H2-derived constructs carrying mutations introduced by site-directed mutagenesis and tested for luciferase gene reporter assays in pGL3 basic vector (left panel). Relative luciferase activity of the H1- and H2-derived promoter haplotypes was measured following transient transfection in HeLa cells (right panel). The empty promoterless pGL3 basic vector was used as negative control. Results are expressed in a ratio of Firefly/*Renilla* activity multiplied by 100. Promoter haplotype H2 was used as reference against which relative expression was compared. The *p* values are calculated from four replicates with unpaired student's *t* test. The *p* value between H1 and H2 is 0.0008.

correlation of rs937282 with allelic differences in promoter activity, with the allele –182 G having high promoter activity [48]. However, in our hands, the –182 G allele was associated with the low-activity P1 promoter haplotype. This discrepancy could be explained by the fact that the extended promoter P1 haplotype was not determined in their study. The latter is particularly relevant when considering the observed impact of the SNP-182 C > G alleles in the context of the presence/absence of the 40 bp deletion.

Most previous studies have been focused on the impact of rs2279744 (SNP309 (T > G)), which is located in the p53-dependent promoter P2. *In vitro* studies have shown that the allele SNP309G increased the affinity of Sp1 transcription activator for a putative binding site and increase the steady-state levels of *MDM2*, which in turn reduced the basal p53 levels [26,49]. Although many studies have attempted to assess the association between rs2279744 and different cancer types, the data remains controversial [27,38,50]. A clear association between rs2279744 and cancer risk was reported in Asians but not in Europeans and in Africans in a meta-analysis [38]. The explanation for this observation is unclear but could be explained by genetic heterogeneity because the observed SNP309 frequencies are variable in different populations ranging from approximately 50% in Asians to 33% and 10% in Caucasians and Africans, respectively [38]. A recent meta-analysis indicates that MDM2 SNP309 serves as a tumor susceptibility marker [51]. Finally, the transcription factor influenced by rs2279744 might be cell-type specific so that this variant does not affect *MDM2* expression in certain tissues [22].

These conflicting rather than conclusive results might be explained by several reasons, including linkage disequilibrium between SNP309 and another, yet unknown, functional SNP in *MDM2*. This linkage disequilibrium could also contribute to cancer associations with SNP309 suggesting that haplotype constructions of *MDM2* pSNPs would add force to these association studies. In this report, we showed that SNP309G was associated with the high P1 promoter activity haplotype. We believe that looking at the impact of haplotypes rather than individual SNPs on promoter activity is a more suitable approach because it takes into account the putative interaction between SNPs. In conclusion, this study revealed differential constitutive P1 promoter activities, at least *in vitro*. This observation implies that individuals who carry distinct p53-independent P1 promoter haplotypes might have a modified risk for cancer development. Association studies in large patient cohorts will helps us to further determine the importance of these haplotypes in cancer.

Competing interests
The authors declare that they have no competing interests.

Authors' contributions
MEL carried out most molecular genetics experiments and drafted the manuscript. MO and ML participated in some molecular studies. MEL, MO, EK, and DS contributed to the interpretation of the data. MEL, DS and EK conceived the study, and participated in its design and coordination. All authors read and approved the final manuscript.

Acknowledgments
This study was supported by research funds provided by the Canadian Institutes of Health Research as well as Genome Quebec/Canada. MEL is the recipient of a Natural Sciences and Engineering Research Council (NSERC) Canada Graduate's scholarship. DS holds the François-Karl Viau Chair in Pediatric Oncogenomics and is a scholar of the Fonds de la Recherche en Santé du Québec (FRSQ).

References
1. Hu W, Feng Z, Ma L, Wagner J, Rice JJ, Stolovitsky G, Levine AJ: A single nucleotide polymorphism in the MDM2 gene disrupts the oscillation of p53 and MDM2 levels in cells. *Cancer Res* 2007, **67**:2757–2765.
2. Harris SL, Levine AJ: The p53 pathway: positive and negative feedback loops. *Oncogene* 2005, **24**:2899–2908.
3. Bennett WP, Hussain SP, Vahakangas KH, Khan MA, Shields PG, Harris CC: Molecular epidemiology of human cancer risk: gene-environment interactions and p53 mutation spectrum in human lung cancer. *J Pathol* 1999, **187**:8–18.
4. Bond GL, Hu W, Levine AJ: MDM2 is a central node in the p53 pathway: 12 years and counting. *Curr Cancer Drug Targets* 2005, **5**:3–8.
5. Haupt Y, Maya R, Kazaz A, Oren M: Mdm2 promotes the rapid degradation of p53. *Nature* 1997, **387**:296–299.
6. Kubbutat MH, Jones SN, Vousden KH: Regulation of p53 stability by Mdm2. *Nature* 1997, **387**:299–303.
7. Momand J, Zambetti GP, Olson DC, George D, Levine AJ: The mdm-2 oncogene product forms a complex with the p53 protein and inhibits p53-mediated transactivation. *Cell* 1992, **69**:1237–1245.
8. Momand J, Jung D, Wilczynski S, Niland J: The MDM2 gene amplification database. *Nucleic Acids Res* 1998, **26**:3453–3459.
9. Oliner JD, Kinzler KW, Meltzer PS, George DL, Vogelstein B: Amplification of a gene encoding a p53-associated protein in human sarcomas. *Nature* 1992, **358**:80–83.
10. Meddeb M, Valent A, Danglot G, Nguyen VC, Duverger A, Fouquet F, Terrier-Lacombe MJ, Oberlin O, Bernheim A: MDM2 amplification in a primary alveolar rhabdomyosarcoma displaying a t(2;13)(q35;q14). *Cytogenet Cell Genet* 1996, **73**:325–330.
11. Bueso-Ramos CE, Yang Y, deLeon E, McCown P, Stass SA, Albitar M: The human MDM-2 oncogene is overexpressed in leukemias. *Blood* 1993, **82**:2617–2623.
12. Watanabe T, Hotta T, Ichikawa A, Kinoshita T, Nagai H, Uchida T, Murate T, Saito H: The MDM2 oncogene overexpression in chronic lymphocytic leukemia and low-grade lymphoma of B-cell origin. *Blood* 1994, **84**:3158–3165.
13. Jones SN, Hancock AR, Vogel H, Donehower LA, Bradley A: Overexpression of Mdm2 in mice reveals a p53-independent role for Mdm2 in tumorigenesis. *Proc Nat Acad Sci U S A* 1998, **95**:15608–15612.
14. Knight JC: Regulatory polymorphisms underlying complex disease traits. *J Mol Med (Berlin, Germany)* 2005, **83**:97–109.
15. Pastinen T, Sladek R, Gurd S, Sammak A, Gel B, Lepage P, Lavergne K, Villeneuve A, Gaudin T, Brandstrom H, Beck A, Verner A, Kingsley J, Harmsen E, Labuda D, Morgan K, Vohl M-C, Naumova AK, Sinnett D, Hudson TJ: A survey of genetic and epigenetic variation affecting human gene expression. *Physiol Genomics* 2004, **16**:184–193.
16. Park JY, Park JM, Jang JS, Choi JE, Kim KM, Cha SI, Kim CH, Kang YM, Lee WK, Kam S, Park RW, Kim IS, Lee JT, Jung TH: Caspase 9 promoter polymorphisms and risk of primary lung cancer. *Hum Mol Genet* 2006, **15**:1963–1971.
17. Harris SL, Gil G, Robins H, Hu W, Hirshfield K, Bond E, Bond G, Levine AJ: Detection of functional single-nucleotide polymorphisms that affect apoptosis. *Proc Nat Acad Sci U S A* 2005, **102**:16297–16302.

18. Ho SY, Wang YJ, Chen HL, Chen CH, Chang CJ, Wang PJ, Chen HH, Guo HR: Increased risk of developing hepatocellular carcinoma associated with carriage of the TNF2 allele of the −308 tumor necrosis factor-alpha promoter gene. *Cancer Causes Control* 2004, **15**:657–663.

19. Wagner K, Hemminki K, Grzybowska E, Klaes R, Burwinkel B, Bugert P, Schmutzler RK, Wappenschmidt B, Butkiewicz D, Pamula J, Pekala W, Försti A: Polymorphisms in genes involved in GH1 release and their association with breast cancer risk. *Carcinogenesis* 2006, **27**:1867–1875.

20. Ma X, Ruan G, Wang Y, Li Q, Zhu P, Qin YZ, Li JL, Liu YR, Ma D, Zhao H: Two single-nucleotide polymorphisms with linkage disequilibrium in the human programmed cell death 5 gene 5′ regulatory region affect promoter activity and the susceptibility of chronic myelogenous leukemia in Chinese population. *Clin Cancer Res* 2005, **11**:8592–8599.

21. Moshynska O, Sankaran K, Saxena A: Molecular detection of the G(−248)A BAX promoter nucleotide change in B cell chronic lymphocytic leukaemia. *Mol Pathol* 2003, **56**:205–209.

22. Bond GL, Hu W, Levine A: A single nucleotide polymorphism in the MDM2 gene: from a molecular and cellular explanation to clinical effect. *Cancer Res* 2005, **65**:5481–5484.

23. Zauberman A, Flusberg D, Haupt Y, Barak Y, Oren M: A functional p53-responsive intronic promoter is contained within the human mdm2 gene. *Nucleic Acids Res* 1995, **23**:2584–2592.

24. Ries S, Biederer C, Woods D, Shifman O, Shirasawa S, Sasazuki T, McMahon M, Oren M, McCormick F: Opposing effects of Ras on p53: transcriptional activation of mdm2 and induction of p19ARF. *Cell* 2000, **103**:321–330.

25. Bond GL, Hirshfield KM, Kirchhoff T, Alexe G, Bond EE, Robins H, Bartel F, Taubert H, Wuerl P, Hait W, Toppmeyer D, Offit K, Levine AJ: MDM2 SNP309 accelerates tumor formation in a gender-specific and hormone-dependent manner. *Cancer Res* 2006, **66**:5104–5110.

26. Bond GL, Hu W, Bond EE, Robins H, Lutzker SG, Arva NC, Bargonetti J, Bartel F, Taubert H, Wuerl P, Onel K, Yip L, Hwang SJ, Strong LC, Lozano G, Levine AJ: A single nucleotide polymorphism in the MDM2 promoter attenuates the p53 tumor suppressor pathway and accelerates tumor formation in humans. *Cell* 2004, **119**:591–602.

27. Bond GL, Menin C, Bertorelle R, Alhopuro P, Aaltonen LA, Levine AJ: MDM2 SNP309 accelerates colorectal tumour formation in women. *J Med Genet* 2006, **43**:950–952.

28. Bond GL, Levine AJ: A single nucleotide polymorphism in the p53 pathway interacts with gender, environmental stresses and tumor genetics to influence cancer in humans. *Oncogene* 2007, **26**:1317–1323.

29. Dharel N, Kato N, Muroyama R, Moriyama M, Shao RX, Kawabe T, Omata M: MDM2 promoter SNP309 is associated with the risk of hepatocellular carcinoma in patients with chronic hepatitis C. *Clin Cancer Res* 2006, **12**:4867–4871.

30. Ohmiya N, Taguchi A, Mabuchi N, Itoh A, Hirooka Y, Niwa Y, Goto H: MDM2 promoter polymorphism is associated with both an increased susceptibility to gastric carcinoma and poor prognosis. *J Clin Oncol* 2006, **24**:4434–4440.

31. Wilkening S, Hemminki K, Rudnai P, Gurzau E, *et al*: No association between MDM2 SNP309 promoter polymorphism and basal cell carcinoma of the skin. *Br J Dermatol* 2007, **157**:375–377.

32. Petenkaya A, Bozkurt B, Akilli-Ozturk O, Kaya HS, Gur-Dedeoglu B, Yulug IG: Lack of association between the MDM2-SNP309 polymorphism and breast cancer risk. *Anticancer Res* 2006, **26**:4975–4977.

33. Pine SR, Mechanic LE, Bowman ED, Welsh JA, Chanock SC, Shields PG, Harris CC: MDM2 SNP309 and SNP354 are not associated with lung cancer risk. *Cancer Epidemiol Biomarkers Prev* 2006, **15**:1559–1561.

34. Sherry ST, Ward MH, Kholodov M, Baker J, Phan L, Smigielski EM, Sirotkin K: dbSNP: the NCBI database of genetic variation. *Nucleic Acids Res* 2001, **29**:308–311.

35. Labuda D, Krajinovic M, Richer C, Skoll A, Sinnett H, Yotova V, Sinnett D: Rapid detection of CYP1A1, CYP2D6, and NAT variants by multiplex polymerase chain reaction and allele-specific oligonucleotide assay. *Anal Biochem* 1999, **275**:84–92.

36. Stephens M, Smith NJ, Donnelly P: A new statistical method for haplotype reconstruction from population data. *Am J Hum Genet* 2001, **68**:978–989.

37. Idelman G, Taylor JG, Tongbai R, Chen RA, Haggerty CM, Bilke S, Chanock SJ, Gardner K: Functional profiling of uncommon VCAM1 promoter polymorphisms prevalent in African American populations. *Hum Mutat* 2007, **28**:824–829.

38. Hu Z, Jin G, Wang L, Chen F, Wang X, Shen H: MDM2 promoter polymorphism snp309 contributes to tumor susceptibility: evidence from 21 case–control studies. *Cancer Epidemiol Biomarkers Prev* 2007, **16**:2717–2723.

39. Esteller M, Cordon-Cardo C, Corn PG, Meltzer SJ, Pohar KS, Watkins DN, Capella G, Peinado MA, Matias-Guiu X, Prat J, Baylin SB, Herman JG: p14ARF silencing by promoter hypermethylation mediates abnormal intracellular localization of MDM2. *Cancer Res* 2001, **61**:2816–2821.

40. Sherr CJ, Weber JD: The ARF/p53 pathway. *Curr Opin Genet Dev* 2000, **10**:94–99.

41. Dobbelstein M, Wienzek S, Konig C, Roth J: Inactivation of the p53-homologue p73 by the mdm2-oncoprotein. *Oncogene* 1999, **18**:2101–2106.

42. Xiao ZX, Chen J, Levine AJ, Modjtahedi N, Xing J, Sellers WR, Livingston DM: Interaction between the retinoblastoma protein and the oncoprotein MDM2. *Nature* 1995, **375**:694–698.

43. Zhang Y, Xiong Y, Yarbrough WG: ARF promotes MDM2 degradation and stabilizes p53: ARF-INK4a locus deletion impairs both the Rb and p53 tumor suppression pathways. *Cell* 1998, **92**:725–734.

44. Zhang Z, Zhang R: p53-independent activities of MDM2 and their relevance to cancer therapy. *Curr Cancer Drug Targets* 2005, **5**:9–20.

45. Dickens MP, Fitzgerald R, Fischer PM: Small-molecule inhibitors of MDM2 as new anticancer therapeutics. *Semin Cancer Biol* 2010, **20**:10–18.

46. Brown CJ, Lain S, Verma CS, Fersht AR, Lane DP: Awakening guardian angels: drugging the p53 pathway. *Nat Rev Cancer* 2009, **9**:862–873.

47. Barak Y, Gottlieb E, Juven-Gershon T, Oren M: Regulation of mdm2 expression by p53: alternative promoters produce transcripts with nonidentical translation potential. *Genes Dev* 1994, **8**:1739–1749.

48. Wang M, Zhang Z, Zhu H, Fu G, Wang S, Wu D, Zhou J, Wei Q, Zhang Z: A novel functional polymorphism C1797G in the MDM2 promoter is associated with risk of bladder cancer in a Chinese population. *Clin Cancer Res* 2008, **14**:3633–3640.

49. Arva NC, Gopen TR, Talbott KE, Campbell LE, Chicas A, White DE, Bond GL, Levine AJ, Bargonetti J: A chromatin-associated and transcriptionally inactive p53-Mdm2 complex occurs in mdm2 SNP309 homozygous cells. *J Biol Chem* 2005, **280**:26776–26787.

50. Wilkening S, Bermejo JL, Hemminki K: MDM2 SNP309 and cancer risk: a combined analysis. *Carcinogenesis* 2007, **28**:2262–2267.

51. Wan Y, Wu W, Yin Z, Guan P, Zhou B: MDM2 SNP309, gene-gene interaction, and tumor susceptibility: an updated meta-analysis. *BMC Cancer* 2011, **11**:208.

22

Conservation of the three-dimensional structure in non-homologous or unrelated proteins

Konstantinos Sousounis[1], Carl E Haney[1], Jin Cao[2], Bharath Sunchu[3] and Panagiotis A Tsonis[1*]

Abstract

In this review, we examine examples of conservation of protein structural motifs in unrelated or non-homologous proteins. For this, we have selected three DNA-binding motifs: the histone fold, the helix-turn-helix motif, and the zinc finger, as well as the globin-like fold. We show that indeed similar structures exist in unrelated proteins, strengthening the concept that three-dimensional conservation might be more important than the primary amino acid sequence.

Keywords: 3D protein structure, Conserved motifs, Unrelated proteins

Introduction

When the human genome was sequenced (as well as that of other mammals), it was estimated that there are approximately 25,000 genes encoding for proteins [1,2]. After being synthesized, proteins assume their three-dimensional structure by a specific arrangement of beta strands, alpha helices, turns, or loops. In many cases, a combination of these structural features creates certain motifs, exerting a particular function (i.e., DNA binding) that is quite conserved in proteins from virtually all organisms. Interestingly, the number of these motifs is much smaller than the number of genes. However, it has also been noted that some structural motifs show significant robustness even though no significant homology exists among them at the primary amino acid sequence. It seems that evolutionary constraints have limited the ability of proteins to become vastly different. Moreover, it has been shown that protein structures are three to ten times more conserved than the amino acid sequence [3]. Thus, a particular motif, i.e., a zinc-binding domain of very similar or virtually identical structure, can be found in many different proteins, which could also be unrelated to each other when function is concerned. Thus, it seems that evolution does favor conservation of structural motifs in proteins.

The purpose of this tutorial/review is to illustrate this diversity that exists in the function of structurally conserved protein motifs. For this reason, protein folds with low homology in amino acid sequence and high structural similarity were used. The analysis for the obvious reason of space is not exhaustive and is focused on four specific protein structural folds: We have selected to present data with three different DNA-binding domains: the histone fold, the helix-turn-helix motif (HTH), and the zinc finger, as well the globin-like fold, part of an important protein in oxygen binding and transport. These four folds were chosen because they are ubiquitous in many different organisms and are well represented in many different proteins.

For our comparisons, an intensive search of the Vector Alignment Search Tool (VAST) [4,5], an algorithm to determine three-dimensional (3D) structure similarities according to geometric criteria, was done. A protein family was identified using a representative protein and, using VAST and the Molecular Modeling Database [6], dissimilar structure proteins were identified and annotated followed by root mean square deviation (RMSD) determination. The structures were then downloaded into Cn3D ('see in 3D') [7] for viewing the sequence alignment. The above are part of Entrez [8,9]. These structures were then aligned in PyMOL [10] for 3D viewing. The files for the PyMOL structures provided have been downloaded from the Protein Data Bank (PDB) [11]. The lower the RMSD means better structural alignment. Lower identity means that the two proteins do not share the same amino acids in the corresponding structural alignment. Though, depending on how big are the structures that we are comparing, the RMSD and sequence identity may vary. Small domains may

---

* Correspondence: ptsonis1@udayton.edu
[1]Department of Biology, University of Dayton, Dayton, OH 45469-2320, USA
Full list of author information is available at the end of the article

contain always certain amino acids increasing the identity. On the other hand, big proteins may not align well and may increase the RMSD. For the present analysis, we chose to set the limits as follows: RMSD to be lower than 3.5 and amino acid identity to be lower than 25% in order to conclude that this pair of proteins has similar structures but dissimilar sequences.

Globin-like fold

Globin-like fold is an all-alpha protein fold normally consisting of six alpha helices [12]. The number of helices can be altered in different families of globin-like proteins. These helices are not randomly distributed in the protein, but they are oriented following standard helix-helix packing rules in order to form a globular structure. Globin-like fold is mostly known from hemoglobins (Figure 1) and myoglobins which play an important role in transferring oxygen to all the tissues of an organism with the help of heme groups which can bind oxygen reversibly. The heme-binding proteins are part of the actual family of globins [12].

The globin family was the first example that showed structural conservation even in different organisms [14-17] and led scientists' pursuit to prove that 3D structures of proteins are more conserved than their sequences. It turned out that globin-like folds exist in many proteins with different functions. Hemoglobins and myoglobins play a role in oxygen transport; cyanoglobins [18] bind to oxygen to help in cellular processes; phycocyanins and phycoerythrins [19,20] play a role in absorbing light; cytokines and immuno-globins [21,22] play a role in the immune system; and fibronectin [23] is part of the extracellular matrix. Natural selection kept the 3D structure of this fold intact [24] while utilizing it for different functions to meet other required organismal requirements.

We have compared pairs of functionally different proteins or proteins from organisms that diversified long ago.

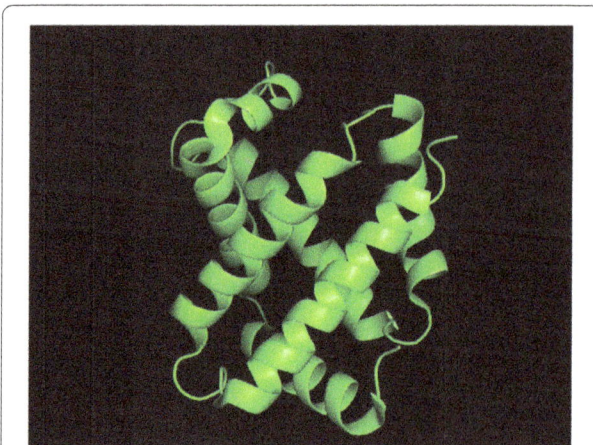

Figure 1 Human hemoglobin (PDB: 2DN2; chain A) [13].

Figure 2 shows the 3D structural conservation despite low sequence similarity. structure is conserved in a monomeric hemoglobin of a trematode (PDB: 1H97) compared to a hemoglobin which is part of a large protein (3.6 million Da) from an annelid (PDB: 2GTL). In this case, the single hemoglobin from a trematode can bind and transport oxygen. However, it is structurally relevant to hemoglobins that are part of a 3.6 million-Da protein, an erythrocruorin, which serves the same purpose but has more advantages such as resistance to oxidation and other cooperative binding properties [25,26]. Both proteins are part of the globin-like superfamily [12].

In the next example, structural conservation of a plant hemoglobin (PDB: 2GNW), which may play a role in binding free molecules that cause oxidation, and a globin-coupled sensor (PDB: 2W31), which plays a role in adapting the organism in the presence of oxygen via transmitted signals to a transmembrane protein, can be seen [27,28]. This example demonstrates how a globin-like fold has been used for different kinds of responses from scavenging hazardous active molecules to sense external stimuli and cooperate with other proteins to get the appropriate response. Both proteins are part of the globin-like superfamily [12].

Nitric oxide detoxification in *M. tuberculosis* occurs with the help of a truncated hemoglobin protein (PDB: 2GLN). Its structure is similar to an extracellular giant hemoglobin from an annelid (PDB: 2ZS1) that plays a role in binding oxygen [29,30].

Certain organisms absorb light through pigments. Allophycocyanin is a pigment and its structure is part of the phycobilisome family [12]. This structure (PDB: 1KN1) is similar to a protein that plays a role in regulating the sigma (s) factor during transcription (PDB: 2BNL) and belongs to the Rsbr_N superfamily (VAST) [31,32]. This is an example of using the globin-like fold as a building block to make a larger structure like the N-domain of the rsbr to serve a different role.

The last example is from two organisms that evolved separately for many millions of years: a neuroglobin (PDB: 1OJ6) from *Homo sapiens* and a protoglobin (PDB: 2VEB) from archaea. The role of globin-like proteins in archaea is not yet fully determined. It is proposed to play a role in metabolism of the strictly anaerobic *M. acetivorans* and to be the building block of globin-coupled sensors. The structure is similar to the neuroglobin from humans which play an important role in regulating oxygen transport in neural tissues [33,34].

Histone fold

This motif is most commonly associated with histones but can also be found in a multitude of proteins such as DNA-binding transcription initiation factors which are functionally conserved in archaea and eukaryotes [35].

PDB # - Function	RMSD - % Identity	Images from PyMOL and Cn3D	References
Example 1 - 2GTL_Chain A: *Lumbricus terrestris* (annelide) hemoglobin part of a 3.6million Dalton protein. Transports oxygen. - 1H97_Chain A: *Paramphistomum epiclitum* (trematode) monomeric hemoglobin. High affinity to oxygen.	RMSD: 2.3 Identity: 12.1%		 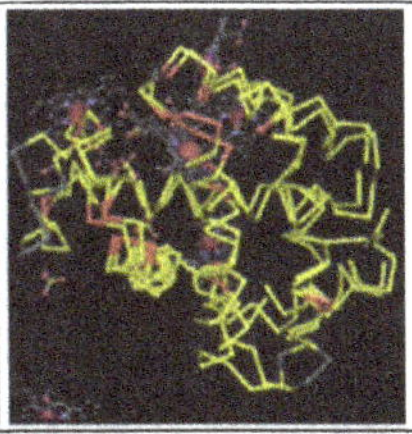 24, 25
Example 2 - 2GNW_Chain B: Found in plants. Its role is not yet determined. *Oryza sativa.* - 2W31_Chain A: detects oxygen and transmits signal. *Geobacter sulfurreducens.*	RMSD: 3.2 Identity: 13.4%		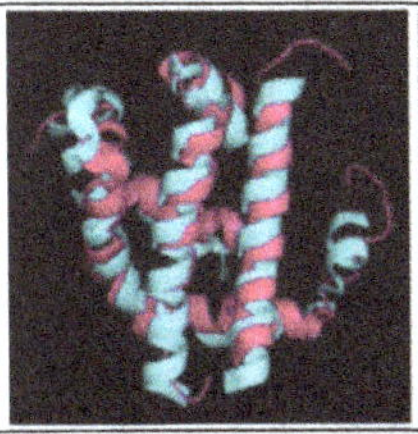 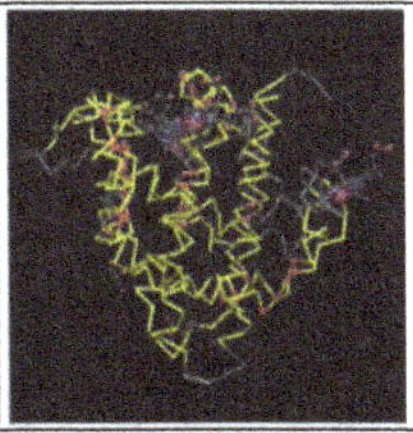 27, 26
Example 3 - 2GLN_Chain A: nitric oxide scavenging. *Mycobacterium tuberculosis.* - 2ZS1_Chain A: extracellular giant Hb. Cooperative oxygen binding via inorganic cations. *Oligobrachia mashikoi.*	RMSD: 2.4 Identity: 6.7%		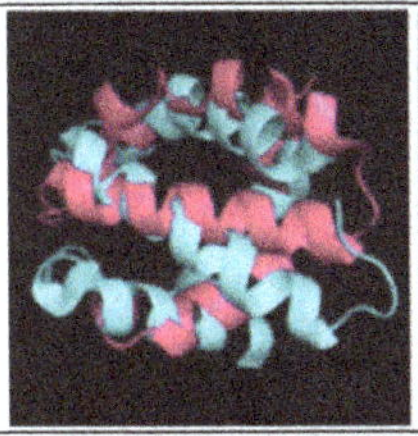  28, 29,
Example 4 - 1KN1_Chain A: allophycocyanin, absorbs light, part of phycobilisomes and phycobilisome structural family. *Pyropia yezoensis.* - 2BNL_Chain C: Non heme, regulates s factor after environmental stress. *Bacillus subtilis*	RMSD: 2.9 Identity:11.4%		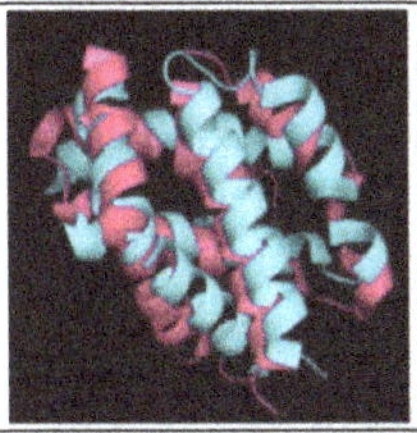  30, 31
Example 5 - 2VEB_Chain A: Found in archae, role is not yet determined. *Methanosarcina acetivorans.* - 1OJ6_Chain A: A neuroglobin found in human brain. Binds to oxygen. *Homo sapiens.*	RMSD: 2.9 Identity: 12.7%		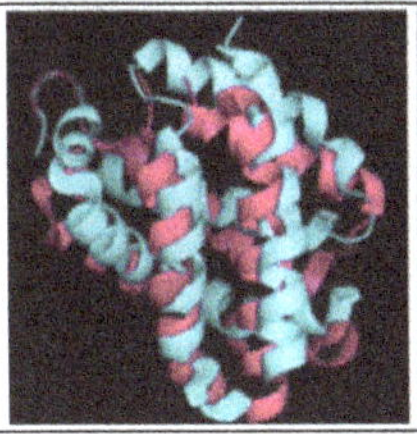 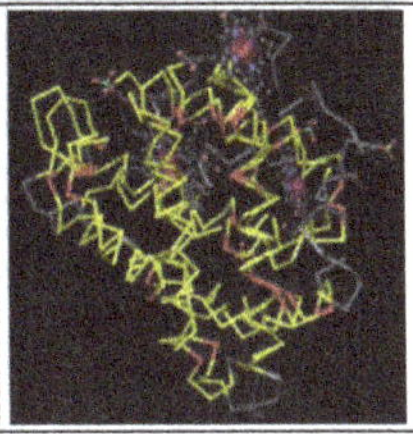 32, 33

Figure 2 Comparison of structure and sequence similarity of sample globin-like fold proteins according to PDB number. First column: PDB number and a brief description of the protein. Second column: RMSD and amino acid sequence identity as defined by VAST. Third column: Left is the alignment of the two proteins taken by PyMOL. In the structure representation, the first protein is in *pink*, and the second, in *cyan*. Right is the alignment of the two proteins taken by Cn3D. In the sequence representation, *red* indicates the same amino acid, whereas *yellow* indicates differing amino acids. Fourth column: references.

Because of this functional conservation in archaea and eukaryotes, the histone fold is thought to be an ancient motif [36]. Interestingly, the pure functionality of the histone fold is not found in eubacteria [37]. As seen in Figure 3, the basic structure of the histone fold comprises a central alpha helix flanked on each side by two smaller helices.

Due to the hydrophobic nature of the histone fold, it is only stable within histone fold-to-histone fold dimers. Eukaryotic histones, for example, dimerize specifically with H2A dimerizing with H2B and H3 dimerizing with H4, thereby creating the basis of the histone octamer. Archaea histones appear to have less specificity in dimerizing to a specific partner but, through dimerization, utilize the histone fold to produce a similar histone structure [38,39].

Since the function of histones and the histone fold are shared by archaea and eukarya, it is thought to have been derived from an early thermophile which initially utilized the histone fold to maintain the integrity of DNA under thermal stress. This increased integrity would have also brought about the added benefit of genome compaction which would have required a mechanism to unwind and transcribe those genes and thus the appearance of proteins such as TATA box-binding proteins and transcription initiation factors which also

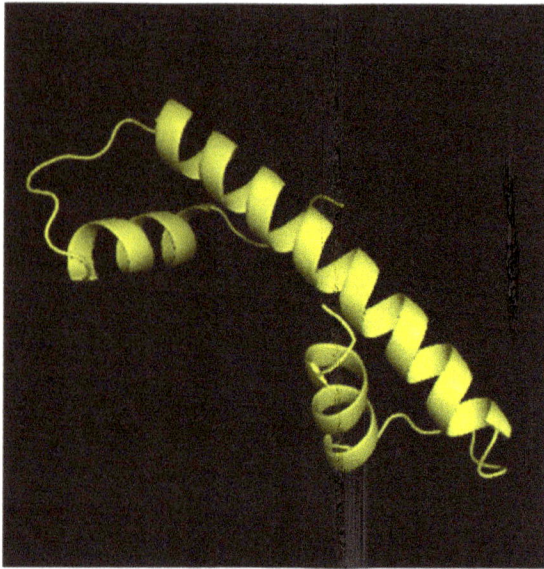

Figure 3 The typical histone fold. It consists of one central helix flanked on each side by a shorter alpha helix (PDB: 1HTA) [38].

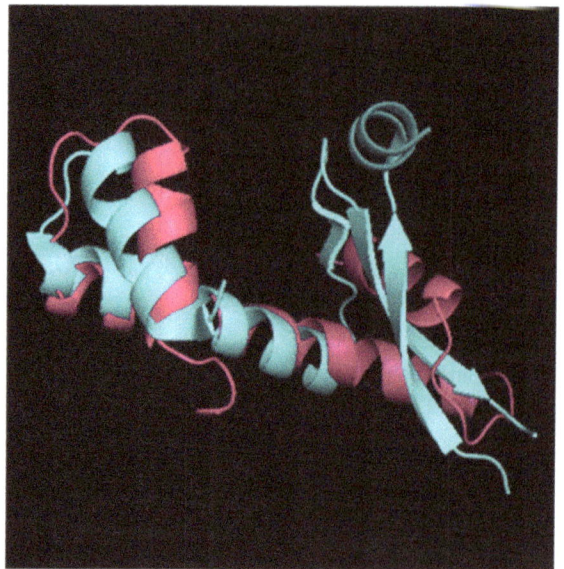

Figure 4 Comparison of the histone fold (PDB: 1HTA) [38] to eubacteria HU protein (PDB: 1MUL) [46]. *Hot pink*: histone fold, *cyan*: eubacteria HU protein. Notice the similarity in the helix-turn-helix and the size difference in the central helix.

utilize the histone fold and are functionally conserved in both eukaryotic and archaea organisms [35,40,41]. Since the packaging and protection of DNA is paramount along with the ability to transcribe DNA when needed, the numerous essential interactions have caused the histone fold to be conserved [42].

From a molecular point of view, the histone fold is thought to have evolved from the helix-strand-helix (HSH) motif where duplication caused two helices to merge, forming a larger central helix [36,43]. Alva et al. demonstrated how this could occur by shortening the HSH strand which led to a 3D swap and caused the dimerization of two HSH motifs. This dimerization recovered the interactions between the HSH motifs due to the strand shortening and thereby causing the histone fold [43].

As mentioned previously, eubacteria do not appear to contain the histone fold motif. They do, however, contain proteins which have histone-like proteins. The most ubiquitous of these proteins is the HU protein (H for histone-like and U from the U93 strain of *Escherichia coli*, in which it was identified from). HU proteins are essential in maintaining the nucleoid structure and are involved in all DNA-dependent functions [44]. Interestingly, the HSH-type motif is found in HU proteins of eubacteria which also have histone-like functionality [42,45]. Looking at the structures of HU and the histone fold (Figure 4), one can easily identify similarities in the HSH with respect to the histone fold, thereby showing how the functionality of DNA binding has been conserved through different but similar means.

Interestingly, some proteins have evolved a method to overcome the need of the dimerization of different proteins through a double histone fold. A double histone fold is essentially two histone folds occurring in a single peptide chain which can 'dimerize' with itself [47]. As seen in Figure 5, a great structural similarity between the H2A/H2B two-protein dimer has a great structural similarity to the single-protein Son of sevenless (Sos) protein [48,49]. With the double histone fold being so 'economical' by not needing to dimerize with another protein, it is not surprising that it was recently found in a virus where it is hypothesized to aid in the packaging and organization of DNA inside the capsid [50].

Due to the multiple interactions required of the histone fold, the selective pressures limit a large differentiation in sequence identity. For example, H3 and H4 histones are among the most highly conserved proteins in terms of sequence and length due to their specific interactions with DNA. H2A and H2B have regions which show greater variability but show great specificity to dimerizing with each other. Despite the conservation of the histone fold in the histone structure, these four core eukaryotic histones have little sequence similarity (15–20%) with one another [42]. Interestingly, even proteins such as the histone H2A/H2B and the cytoplasmic hSos [50] (Figure 5, example 4) which show strong structural similarity but do not seem to function as histones or DNA-binding factors still do not stray far from this sequence identity. This sequence similarity is seen in organisms which are obviously so evolutionary distant as archaea and eukaryotes [51,52]

PDB # - Function	RMSD - % identity	Images from PyMOL and Cn3D	References
Example 1 - 1BH8_Chain B: TATA binding protein associated factor (TAF)II28 in *Homo sapiens* - 1A7W: Histone HMfB from archaea *Methanothermus fervidus*	RMSD: 1.0 Identity: 22.6%		37, 69
Example 2 - 3R45_Chain A: Histone H3-like centromeric protein A in *Homo sapiens* - 1B6W: Histone HmfA from archaea *Methanothermus fervidus*	RMSD: 1.1 Identity: 24.6%		37, 70
Example 3 - 3AFA_Chain E: Histone 3 in *Homo sapiens* - 1H3O_Chain D: Transcription Initiation Factor (TFIID) in *Homo sapiens*	RMSD: 1.0 Identity: 12.3%		71, 72
Example 4 - 2JSS: Yeast (*Saccharomyces cerevesiae*) Histone H2A/H2B - 1Q9C: Double histone fold of *Homo sapiens* Son of sevenless (hSos)	RMSD:1.9 Identity: 22.7%		73, 74

Figure 5 Comparison of structure and sequence similarity of sample histone fold proteins according to PDB number. First column: PDB number and a brief description of the protein. Second column: RMSD and amino acid sequence identity as defined by VAST. Third column: Left is the alignment of the two proteins taken by PyMOL. In the structure representation, the first protein is in *pink*, and the second, in *cyan*. Right is the alignment of the two proteins taken by Cn3D. In the sequence representation, *red* indicates the same amino acid, whereas *yellow* indicates differing amino acids. Fourth column: references.

(Figure 5, examples 1 and 2). This may be due to the hydrophobic residue interactions required in all six helices of a histone fold dimer [39].

Helix-turn-helix motif

HTH motif consists of an α-helix, a turn, and a second α-helix which is often called the 'recognition' helix as the part of the HTH motif that fits into the DNA major groove. There are several positions significant to keep the HTH structure rather than to specify contacts with the DNA, while the amino acid residues in other positions are usually varied to determine the specificity of DNA-protein interactions [53]. This motif is found in many DNA-binding domains and transcriptional factors such as homeotic proteins. This sequence, which is conserved in many organisms for related proteins, was used to discover a large number of DNA-binding proteins [54]

Winged helix-turn-helix (wHTH, Figure 6) shares the same original ancestor as that of HTH in evolutionary history; it is also a DNA-binding domain which binds to

specific DNA sequences. The wHTH is formed by a three-helix bundle (α1, α2, α3) and a three- or four-

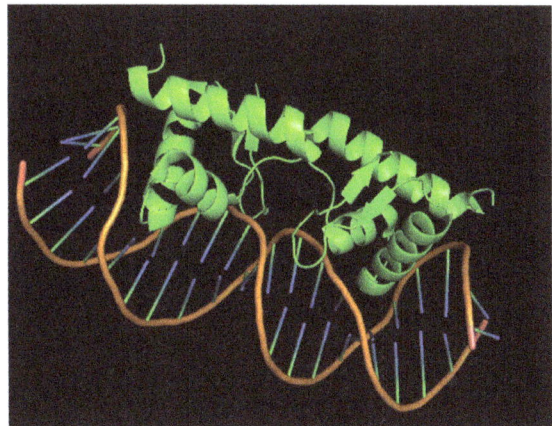

Figure 6 A typical winged helix-turn-helix structure (PDB: 3JSO) [55].

strand beta-sheet. The α2 and α3 helices are similar to those of the HTH motif except that wHTH has beta-sheet wings on the ends of HTH parts. Many repressor DNA-binding domains like LexA, arginine, Rex, ArsR, and MarR form a wHTH structure.

Figure 7 shows five examples of HTH comparisons of different proteins. All of them show high structural similarity and low sequence identity. In addition, the examples compare HTH motifs from different organisms that do different functions.

An ancestral archaea homolog of the N-terminal of the transcription factor II E subunit a (PDB: 1Q1H) [56] folds as a wHTH. This domain has a groove which is negatively charged. Thus, it cannot bind to negatively charged DNA as in vitro experiments show. Though, it promotes interactions with other proteins. This domain has structural similarities with a catabolite gene activator protein (PDB: 1RUN) [57], a protein that is known to bind DNA. This example clearly illustrates that natural selection chose structures to have different roles than the dominant ones. Cro repressor from the λ phage (PDB: 1D1L) [58] forms a dimer by two antiparallel b-strands in order to bind to DNA. This protein has structural similarities with the bacterial Fis protein (PDB: 3JRH) [59] which binds to DNA with no sequence specificity.

Transcriptional regulators can be triggered to function by different signals from the environment. Signals that

PDB # - Function	RMSD - % Identity	Images from PyMOL and Cn3D	References
Example 1 - 1Q1H_Chain A: The alpha subunit of transcription factor TFIIE homolog from archae *Sulfolobus solfataricus*. - 1RUN_Chain A: DNA binding domain of *Escherichia coli* regulatory proteins which belongs to catabolite activator protein family.	RMSD: 1.0 Identity: 22.7%	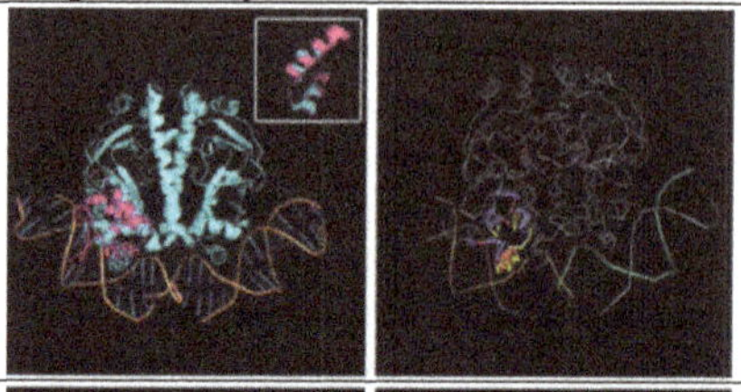	47, 48
Example 2 - 1D1L_Chain A: Cro repressor of *Enterobacteria phage lambda*. - 3JRH_Chain B: *Escherichia coli* protein from Fis family which binds to unspecific DNA.	RMSD: 2.0 Identity: 21.4%	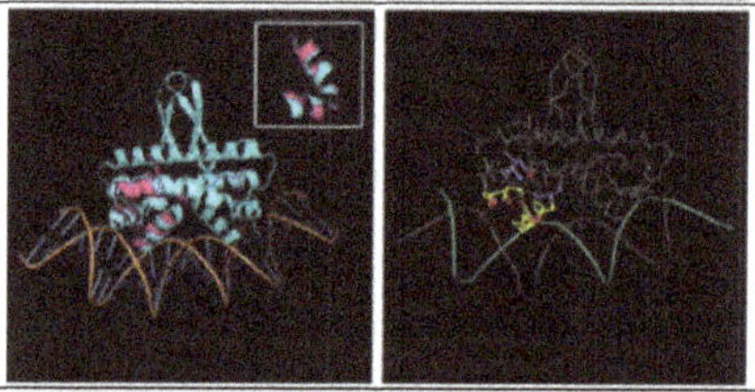	50, 49
Example 3 - 1R1T_Chain A: Negatively allogistic regulated DNA binding of SmtB protein in presence of metals. *Synechococcus elongatus*. - 1ZLK_Chain A: Hypoxia – induced DosR protein. *Mycobacterium tuberculosis*.	RMSD: 1.5 Identity: 20.5%	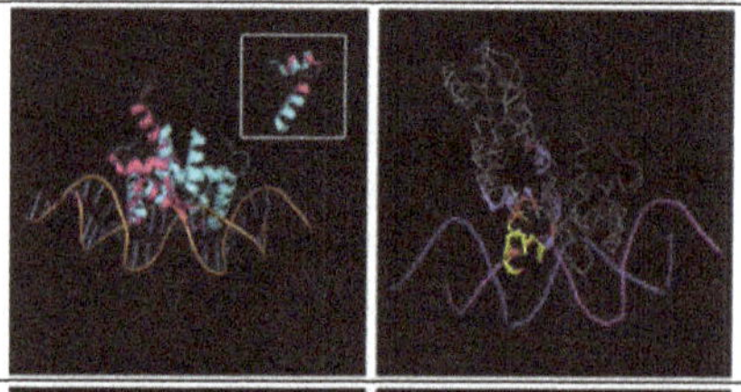	51, 52
Example 4 - 1RES_Chain A: DNA-binding domains of *Escherichia coli* γδ resolvase. - 1Z9C_Chain F: Chimeric winged helix-turn-helix (wHTH) DNA-binding domain of OhrR-ohrA complex. *Bacillus subtilis*.	RMSD: 1.4 Identity: 10.7%	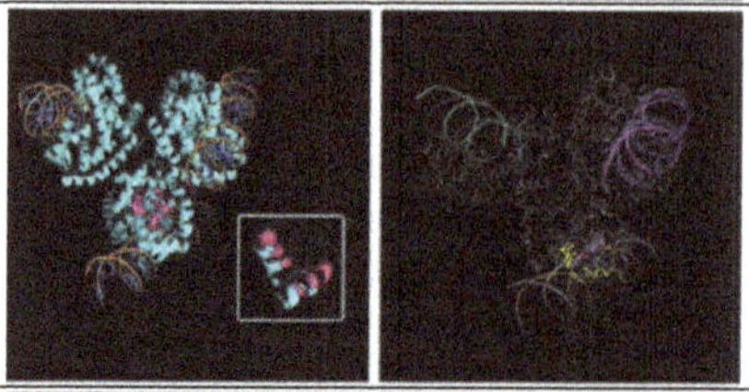	53, 54
Example 5 - 3OIO_Chain A: Bacterial regulatory helix-turn-helix proteins of AraC family from *Chromobacterium violaceum*. - 1XS9_Chain A: MarA *Escherichia coli* regulatory helix-turn-helix protein that binds DNA as a monomer.	RMSD: 2.2 Identity: 20.7%	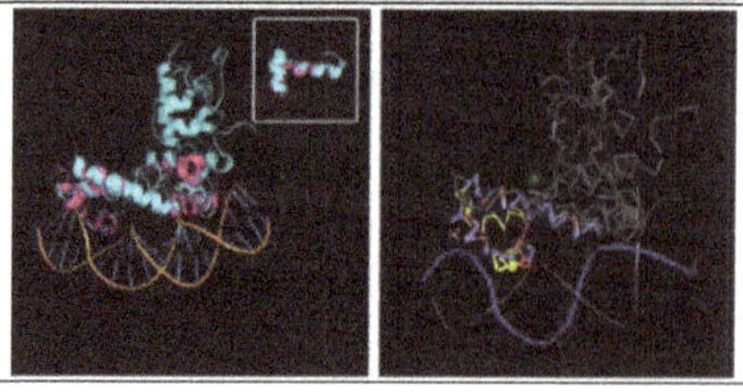	55 Chang et al., deposited in PDB, not published

Figure 7 Comparison of structure and sequence similarity of sample helix-turn-helix motif proteins according to PDB number. First column: PDB number and a brief description of the protein. Second column: RMSD and amino acid sequence identity as defined by VAST. Third column: Left is the alignment of the two proteins taken by PyMOL. In the structure representation, the first protein is in *pink*, and the second, in *cyan*. Right is the alignment of the two proteins taken by Cn3D. In the sequence representation, *red* indicates the same amino acid, whereas *yellow* indicates differing amino acids. Fourth column: references.

are not related with signal transduction cascades, which involve primarily phosphorylation or dephosphorylation of proteins, can involve smaller molecules like metals or oxygen. This is the case for SmtB (1R1T) [60], a cyanobacterial repressor protein that has reduced affinity for DNA in the presence of metals. The HTH motif of this repressor is structurally similar to the HTH motif of the bacterial DosR protein (PDB: 1ZLK) [61] which prolongs survival when the organism is left without oxygen.

OhrR is a bacterial protein (PDB: 1Z9C) [62] that has a HTH motif composed of eukaryotic-like wHTH, prokaryotic HTH motifs, and other helices. This protein is induced to function by oxidation of certain residues. This chimeric HTH motif is structurally similar to the HTH motif of a DNA-binding domain of a γδ resolvase in *E. coli* (PDB: 1RES) [63].

Finally, the HTH motif from a bacterial transcriptional regulator, AraC-type (PDB: 3OIO), is structurally similar to that of the transcriptional activator MarA (PDB: 1XS9) [64] which is associated with the RNA polymerase and binds to DNA as a monomer.

Zinc finger motif

Zinc (Zn) fingers (see Figure 8) are small structural motifs whose structure is stabilized by a zinc ion, and they are the most common DNA- or RNA-binding motif in different proteins. There are different structural types of Zn fingers and are present in proteins that perform a broad array of functions such as replication and repair, transcription and translation, metabolism and signaling, cell proliferation, and apoptosis [65]. Zn fingers occupy 3% of the genes in the human genome [66]. The major part of structural stability of Zn fingers is provided by zinc coordination and by the conserved hydrophobic core that flanks the Zn binding site. There are a relatively small number of conserved residues present in Zn fingers [67].

Classical Cys2-His2 (C2H2) Zn fingers have about 30 amino acids in which 25 of the 30 amino acid residues form a loop around the central Zn ion and the 5 other amino acids form the linkers between the consecutive Zn fingers. It consists of two secondary structural units: The first one is an antiparallel beta-sheet, which contains the loop formed by the two cysteines, and the second one is an alpha helix containing the His-His. These two structural units are held together by the zinc atom. The Zn ion tetrahedrally coordinates to the conserved pairs of cysteines and histidines, and this coordination is vital for the maintenance of the overall structure of the Zn finger. The majority of the 30 amino acids are polar and basic residues which are important in nucleic acid binding. In addition to the conserved cysteines and histidines which are vital for the formation of the Zn finger fold, there are other conserved amino acids, notably Tyr, Phe, and Leu, which form a hydrophobic structural core of the folded structure [66].

In the example shown in Figure 9, each pair of the compared Zn fingers have less sequence similarity, sometimes bind to different types of molecules, may have different functions, may belong to different species, but exhibit a great structural overlap. This supports the notion that only few small numbers of conserved residues are required for the maintenance of the overall structure of the zinc finger.

Example 1 in Figure 9 shows two DNA-binding proteins: a DNA-binding domain (DBD) from the GAGA factor (PDB: 1YUJ) [69] and one of the zinc finger domains from zinc finger protein 692 (PDB: 2D9H), which belong to *D. melanogaster* and *H. sapiens*, respectively. The DBD of the GAGA factor uses only one zinc finger in contrast to other zinc finger proteins which commonly use more than two in order to have a good affinity for the DNA. They show a great structural similarity despite low sequence identity.

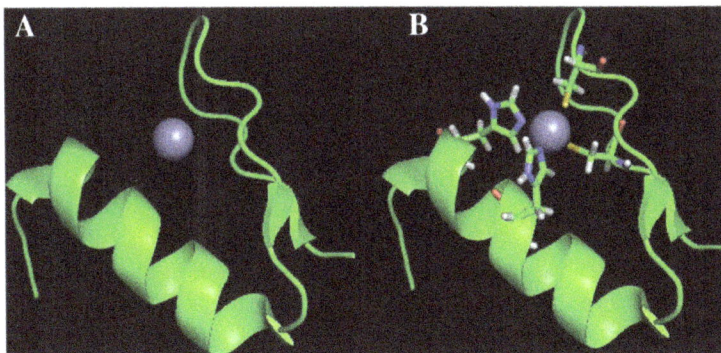

Figure 8 Structure of C2H2 zinc finger of transcription factor IIIA of *Xenopus laevis* (PDB: 2HGH, [68]). (**A**) Cartoon representation with zinc as a ball. (**B**) Includes the two cysteines and two histidines that interact with the zinc as sticks.

PDB # - Function	RMSD - % Identity	Images from PyMOL and Cn3D	References
Example 1 - 1YUJ_Chain A: structure of DNA binding domain of the GAGA factor(DNA binding protein) *Drosophila melanogaster* - 2D9H_Chain A: forth and fifth zf-C2H2 domains of zinc finger protein 692(DNA binding protein). *Homo sapiens*	RMSD: 0.8 Identity:16.0%		[59] Zhang et al., deposited in PDB, not published
Example 2 - 3ODC_Chain B: Human zinc finger 2. Binds to DNA. *Homo sapiens.* - 1MHZ_Chain D: hydroxylase component of methane monooxygenase. No evidence that binds to DNA. *Methylosinus trichosporium.*	RMSD: 1.5 Identity: 3%	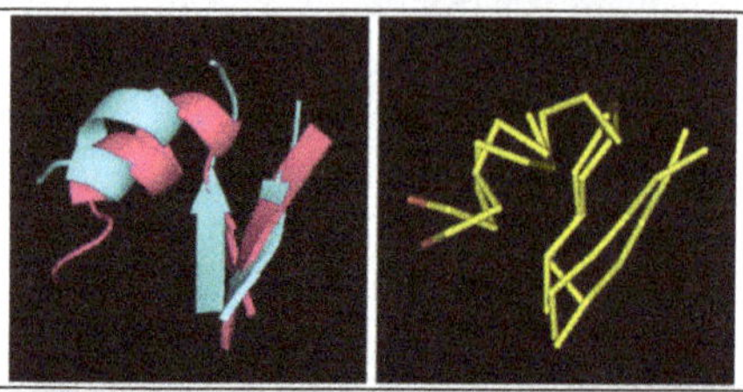	60, 61
Example 3 - 1MHZ_Chain D: hydroxylase component of methane monooxygenase *Methylosinus trichosporium* - 2INC_Chain A: Native Toluene/o-xylene Monooxygenase Hydroxylase *Pseudomonas stutzeri*	RMSD: 2.4 Identity: 21.0%	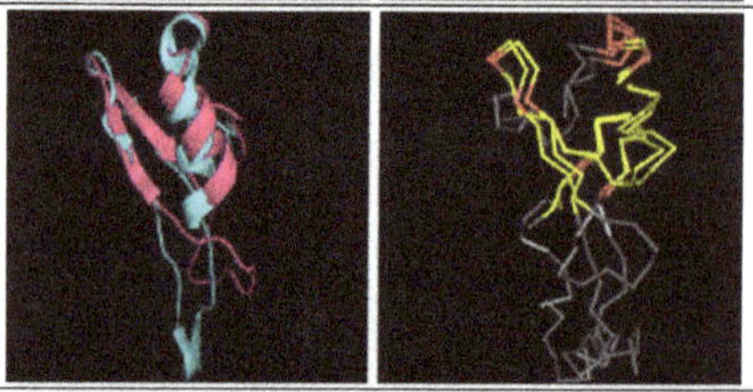	60, 62
Example 4 - 1U85_Chain A: tryptophan-containing CCHH zinc finger-Kruppel like factor 3-DNA binding protein. *Mus musculus* - 1UBD_Chain C: Human YY1 zinc finger domain which binds to DNA. *Homo sapiens.*	RMSD: 1.0 Identity: 15.8%	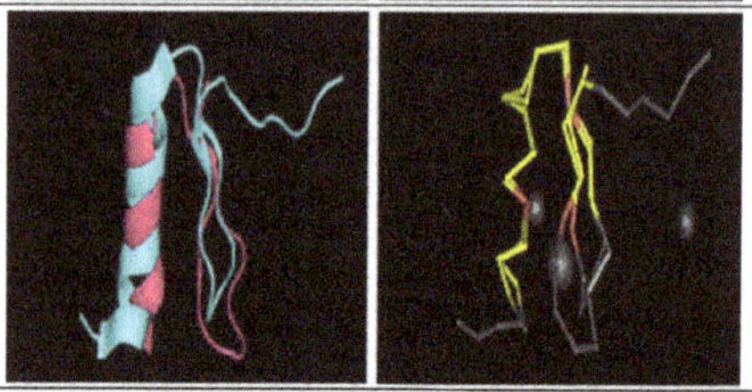	Cram et al., deposited in PDB, not published [63]
Example 5 - 2VY4_Chain A: A splicing protein which Binds the 5 Splice Site of U12-Type intron. *Homo sapiens.* - 3MHH _Chain C: Transcriptional coactivator complex;has multiple roles on role in deubiquitination of histone H2B. *Saccharomyces cerevisiae.*	RMSD: 1.2 Identity: 23.1%	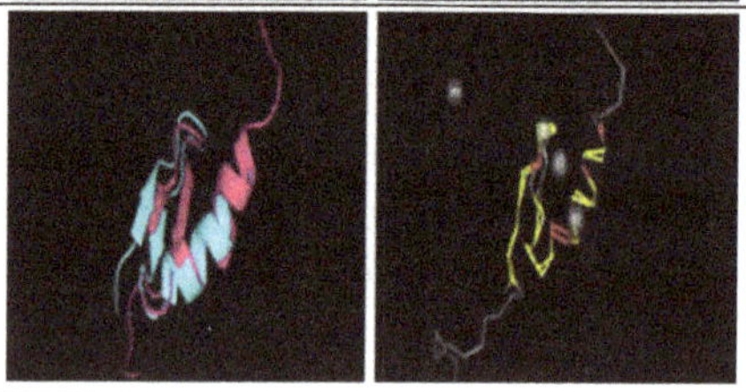	64, 75

Figure 9 Comparison of structure and sequence similarity of sample zinc finger motif proteins according to PDB number. First column: PDB number and a brief description of the protein. Second column: RMSD and amino acid sequence identity as defined by VAST. Third column: Left is the alignment of the two proteins taken by PyMOL. In the structure representation, the first protein is in *pink*, and the second, in *cyan*. Right is the alignment of the two proteins taken by Cn3D. In the sequence representation, *red* indicates the same amino acid, whereas *yellow* indicates differing amino acids. Fourth column: references.

The hydroxylase domain from methane monooxygenase (PDB: 1MHZ) [70] contains a Zn finger which does not bind to DNA. Though, it is structurally very similar (RMSD: 1.5) and their sequence is very different (3%) from the human Zn finger 2 which binds to DNA (PDB: 3ODC) [71]. This is a good example to point out that structures are built up from extensively used raw materials (domains) like the Zn finger even if they are not going to be used as the majority of the other proteins in which these domains are found.

In the third example, and as a follow up from the previous one, the two proteins are monooxygenases (PDB: 1MHZ, 2INC) [70,72] which belong to different species and have Zn finger domains whose structures overlap.

YY1 (PDB: 1UBD) [73] is a protein with four Zn fingers and is structurally similar to kruppel-like factor 3 (PDB: 1U85), which contains a Zn finger with tryptophan as shown in the fourth example.

Finally, the Zn finger in U11/U12 (PDB: 2VY4) [74], which is a RNA-binding protein, has a good structural overlap with SAGA protein (PDB: 3MHH) [75], which is a DNA-binding protein, in spite of the low sequence similarity. In addition, the role of SAGA is to deubiquitinate H2B histone, so the affinity for DNA helps to dock to the nucleosome. This example was selected because these two different proteins bind to two different types of nucleic acids, have different functions, have low sequence identity, but exhibit a good overall structural similarity.

Concluding remarks

In this review, we have selected four protein motifs, which are present in several DNA-binding proteins and in oxygen-carrying and -transporting proteins. Using several comparisons, we show that these motifs exhibit an astonishing degree of structural conservation even though their primary sequence is not similar and even when they are involved in different functions. The examples underscore the importance of structure selection in evolution and a strategy of economy that nature is implementing. Much is to be learned when similar structures have evolved despite unrelated function. It will be interesting to determine how such similar structures have evolved and what could be the possible ancestors be. Eventually, when all structures have been solved, evolution of protein structure will provide valuable information on protein function in general.

Competing interests
The authors declare that they have no competing interests.

Authors' contributions
PAT conceived the idea, analyzed the data, and co-wrote the paper. KS, CEH, JC, and BS performed the search and analysis. KS co-wrote the paper. All authors read and approved the final manuscript.

Author details
[1]Department of Biology, University of Dayton, Dayton, OH 45469-2320, USA. [2]Department of Chemical Engineering, University of Dayton, Dayton, OH 45469-2320, USA. [3]Department of Chemistry, University of Dayton, Dayton, OH 45469-2320, USA.

References

1. Lander ES, Linton LM, Birren B, Nusbaum C, Zody MC, Baldwin J, Devon K, Dewar K, Doyle M, FitzHugh W, Funke R, Gage D, Harris K, Heaford A, Howland J, Kann L, Lehoczky J, LeVine R, McEwan P, McKernan K, Meldrim J, Mesirov JP, Miranda C, Morris W, Naylor J, Raymond C, Rosetti M, Santos R, Sheridan A, Sougnez C, *et al*: Initial sequencing and analysis of the human genome. *Nature* 2001, **409**:860–921.
2. Venter JC, Adams MD, Myers EW, Li PW, Mural RJ, Sutton GG, Smith HO, Yandell M, Evans CA, Holt RA, Gocayne JD, Amanatides P, Ballew RM, Huson DH, Wortman JR, Zhang Q, Kodira CD, Zheng XH, Chen L, Skupski M, Subramanian G, Thomas PD, Zhang J, Gabor Miklos GL, Nelson C, Broder S, Clark AG, Nadeau J, McKusick VA, *et al*: The sequence of the human genome. *Science* 2001, **291**:1304–1351.
3. Illergard K, Ardell DH, Elofsson A: Structure is three to ten times more conserved than sequence–a study of structural response in protein cores. *Proteins* 2009, **77**:499–508.
4. Gibrat JF, Madej T, Bryant SH: Surprising similarities in structure comparison. *Curr Opin Struct Biol* 1996, **6**:377–385.
5. Madej T, Gibrat JF, Bryant SH: Threading a database of protein cores. *Proteins* 1995, **23**:356–369.
6. Ohkawa H, Ostell J: Bryant S. *MMDB: an ASN.1 specification for macromolecular structure. Proc Int Conf Intell Syst Mol Biol* 1995, **3**:259–267.
7. Hogue CW: Cn3D: a new generation of three-dimensional molecular structure viewer. *Trends Biochem Sci* 1997, **22**:314–316.
8. Hogue CW, Ohkawa H, Bryant SH: A dynamic look at structures: WWW-Entrez and the Molecular Modeling Database. *Trends Biochem Sci* 1996, **21**:226–229.
9. Schuler GD, Epstein JA, Ohkawa H, Kans JA: Entrez: molecular biology database and retrieval system. *Methods Enzymol* 1996, **266**:141–162.
10. Schrodinger LLC: *The PyMOL Molecular Graphics System.*: Version; 2006:99.
11. Berman HM, Westbrook J, Feng Z, Gilliland G, Bhat TN, Weissig H, Shindyalov IN, Bourne PE: The Protein Data Bank. *Nucleic Acids Res* 2000, **28**:235–242.
12. Murzin AG, Brenner SE, Hubbard T, Chothia C: SCOP: a structural classification of proteins database for the investigation of sequences and structures. *J Mol Biol* 1995, **247**:536–540.
13. Park SY, Yokoyama T, Shibayama N, Shiro Y, Tame JR: 1.25 A resolution crystal structures of human haemoglobin in the oxy, deoxy and carbonmonoxy forms. *J Mol Biol* 2006, **360**:690–701.
14. Kendrew JC, Bodo G, Dintzis HM, Parrish RG, Wyckoff H, Phillips DC: A three-dimensional model of the myoglobin molecule obtained by x-ray analysis. *Nature* 1958, **181**:662–666.
15. Kendrew JC, Dickerson RE, Strandberg BE, Hart RG, Davies DR, Phillips DC, Shore VC: Structure of myoglobin: a three-dimensional Fourier synthesis at 2 A. resolution. *Nature* 1960, **185**:422–427.
16. Perutz MF, Muirhead H, Cox JM, Goaman LC: Three-dimensional Fourier synthesis of horse oxyhaemoglobin at 2.8 A resolution: the atomic model. *Nature* 1968, **219**:131–139.
17. Perutz MF, Rossmann MG, Cullis AF, Muirhead H, Will G, North AC: Structure of haemoglobin: a three-dimensional Fourier synthesis at 5.5-A. resolution, obtained by X-ray analysis. *Nature* 1960, **185**:416–422.
18. Trent JT 3rd: Kundu S, Hoy JA, Hargrove MS: Crystallographic analysis of synechocystis cyanoglobin reveals the structural changes accompanying ligand binding in a hexacoordinate hemoglobin. *J Mol Biol* 2004, **341**:1097–1108.
19. Ficner R, Lobeck K, Schmidt G: Huber R. *Isolation, crystallization, crystal structure analysis and refinement of B-phycoerythrin from the red alga Porphyridium sordidum at 2.2 A resolution. J Mol Biol* 1992, **228**:935–950.
20. Schirmer T, Bode W, Huber R, Sidler W, Zuber H: X-ray crystallographic structure of the light-harvesting biliprotein C-phycocyanin from the thermophilic cyanobacterium *Mastigocladus laminosus* and its resemblance to globin structures. *J Mol Biol* 1985, **184**:257–277.
21. Rozwarski DA, Gronenborn AM, Clore GM, Bazan JF, Bohm A, Wlodawer A, Hatada M, Karplus PA: Structural comparisons among the short-chain helical cytokines. *Structure* 1994, **2**:159–173.
22. Williams AF, Barclay AN: The immunoglobulin superfamily–domains for cell surface recognition. *Annu Rev Immunol* 1988, **6**:381–405.
23. Bork P, Doolittle RF: Proposed acquisition of an animal protein domain by bacteria. *Proc Natl Acad Sci USA* 1992, **89**:8990–8994.
24. Aronson HE, Royer WE Jr: Hendrickson WA: Quantification of tertiary structural conservation despite primary sequence drift in the globin fold. *Protein Sci* 1994, **3**:1706–1711.
25. Pesce A, Dewilde S, Kiger L, Milani M, Ascenzi P, Marden MC, Van Hauwaert ML, Vanfleteren J, Moens L, Bolognesi M: Very high resolution structure of a trematode hemoglobin displaying a TyrB10-TyrE7 heme distal residue pair and high oxygen affinity. *J Mol Biol* 2001, **309**:1153–1164.
26. Royer WE Jr: Sharma H, Strand K, Knapp JE, Bhyravbhatla B: Lumbricus erythrocruorin at 3.5 A resolution: architecture of a megadalton respiratory complex. *Structure* 2006, **14**:1167–1177.
27. Pesce A, Thijs L, Nardini M, Desmet F, Sisinni L, Gourlay L, Bolli A, Coletta M, Van Doorslaer S, Wan X, Alam M, Ascenzi P, Moens L, Bolognesi M, Dewilde S: HisE11 and HisF8 provide bis-histidyl heme hexa-coordination in the globin domain of *Geobacter sulfurreducens* globin-coupled sensor. *J Mol Biol* 2009, **386**:246–260.
28. Smagghe BJ, Kundu S, Hoy JA, Halder P, Weiland TR, Savage A, Venugopal A, Goodman M, Premer S, Hargrove MS: Role of phenylalanine B10 in plant nonsymbiotic hemoglobins. *Biochemistry* 2006, **45**:9735–9745.
29. Numoto N, Nakagawa T, Kita A, Sasayama Y, Fukumori Y, Miki K: Structural basis for the heterotropic and homotropic interactions of invertebrate giant hemoglobin. *Biochemistry* 2008, **47**:11231–11238.
30. Ouellet Y, Milani M, Couture M, Bolognesi M, Guertin M: Ligand interactions in the distal heme pocket of *Mycobacterium tuberculosis* truncated hemoglobin N: roles of TyrB10 and GlnE11 residues. *Biochemistry* 2006, **45**:8770–8781.

31. Liu JY, Jiang T, Zhang JP, Liang DC: Crystal structure of allophycocyanin from red algae *Porphyra yezoensis* at 2.2-A resolution. *J Biol Chem* 1999, **274**:16945–16952.

32. Murray JW, Delumeau O, Lewis RJ: Structure of a nonheme globin in environmental stress signaling. *Proc Natl Acad Sci USA* 2005, **102**:17320–17325.

33. Nardini M, Pesce A, Thijs L, Saito JA, Dewilde S, Alam M, Ascenzi P, Coletta M, Ciaccio C, Moens L, Bolognesi M: Archaeal protoglobin structure indicates new ligand diffusion paths and modulation of haem-reactivity. *EMBO Rep* 2008, **9**:157–163.

34. Pesce A, Dewilde S, Nardini M, Moens L, Ascenzi P, Hankeln T, Burmester T, Bolognesi M: Human brain neuroglobin structure reveals a distinct mode of controlling oxygen affinity. *Structure* 2003, **11**:1087–1095.

35. Pereira SL, Reeve JN: Histones and nucleosomes in Archaea and Eukarya: a comparative analysis. *Extremophiles* 1998, **2**:141–148.

36. Gangloff YG, Romier C, Thuault S, Werten S, Davidson I: The histone fold is a key structural motif of transcription factor TFIID. *Trends Biochem Sci* 2001, **26**:250–257.

37. Wong JT, New DC, Wong JC, Hung VK: Histone-like proteins of the dinoflagellate *Crypthecodinium cohnii* have homologies to bacterial DNA-binding proteins. *Eukaryot Cell* 2003, **2**:646–650.

38. Decanniere K, Babu AM, Sandman K, Reeve JN, Heinemann U: Crystal structures of recombinant histones HMfA and HMfB from the hyperthermophilic archaeon *Methanothermus fervidus*. *J Mol Biol* 2000, **303**:35–47.

39. Sandman K, Reeve JN: Archaeal histones and the origin of the histone fold. *Curr Opin Microbiol* 2006, **9**:520–525.

40. Tachiwana H, Kagawa W, Osakabe A, Kawaguchi K, Shiga T, Hayashi-Takanaka Y, Kimura H, Kurumizaka H: Structural basis of instability of the nucleosome containing a testis-specific histone variant human H3T. *Proc Natl Acad Sci USA* 2010, **107**:10454–10459.

41. Werten S, Mitschler A, Romier C, Gangloff YG, Thuault S, Davidson I, Moras D: Crystal structure of a subcomplex of human transcription factor TFIID formed by TATA binding protein-associated factors hTAF4 (hTAF(II)135) and hTAF12 (hTAF(II)20). *J Biol Chem* 2002, **277**:45502–45509.

42. Ramakrishnan V: The histone fold: evolutionary questions. *Proc Natl Acad Sci USA* 1995, **92**:11328–11330.

43. Alva V, Ammelburg M, Soding J, Lupas AN: On the origin of the histone fold. *BMC Struct Biol* 2007, **7**:17.

44. Grove A: Functional evolution of bacterial histone-like HU proteins. *Curr Issues Mol Biol* 2010, **13**:1–12.

45. Oberto J, Drlica K, Rouviere-Yaniv J: Histones, HMG, HU, IHF: Meme combat. *Biochimie* 1994, **76**:901–908.

46. Ramstein J, Hervouet N, Coste F, Zelwer C, Oberto J, Castaing B: Evidence of a thermal unfolding dimeric intermediate for the *Escherichia coli* histone-like HU proteins: thermodynamics and structure. *J Mol Biol* 2003, **331**:101–121.

47. Greco C, Fantucci P, De Gioia L: In silico functional characterization of a double histone fold domain from the *Heliothis zea* virus 1. *BMC Bioinformatics* 2005, **6**(Suppl 4):S15.

48. Sondermann H, Soisson SM, Bar-Sagi D, Kuriyan J: Tandem histone folds in the structure of the N-terminal segment of the ras activator Son of Sevenless. *Structure* 2003, **11**:1583–1593.

49. Zhou Z, Feng H, Hansen DF, Kato H, Luk E, Freedberg DI, Kay LE, Wu C, Bai Y: NMR structure of chaperone Chz1 complexed with histones H2A.Z-H2B. *Nat Struct Mol Biol* 2008, **15**:868–869.

50. Greco C, Sacco E, Vanoni M, De Gioia L: Identification and in silico analysis of a new group of double-histone fold-containing proteins. *J Mol Model* 2005, **12**:76–84.

51. Birck C, Poch O, Romier C, Ruff M, Mengus G, Lavigne AC, Davidson I, Moras D: Human TAF(II)28 and TAF(II)18 interact through a histone fold encoded by atypical evolutionary conserved motifs also found in the SPT3 family. *Cell* 1998, **94**:239–249.

52. Hu H, Liu Y, Wang M, Fang J, Huang H, Yang N, Li Y, Wang J, Yao X, Shi Y, Li G, Xu RM: Structure of a CENP-A-histone H4 heterodimer in complex with chaperone HJURP. *Genes Dev* 2011, **25**:901–906.

53. Aravind L, Anantharaman V, Balaji S, Babu MM, Iyer LM: The many faces of the helix-turn-helix domain: transcription regulation and beyond. *FEMS Microbiol Rev* 2005, **29**:231–262.

54. Pabo CO, Sauer RT: Protein-DNA recognition. *Annu Rev Biochem* 1984, **53**:293–321.

55. Zhang AP, Pigli YZ, Rice PA: Structure of the LexA-DNA complex and implications for SOS box measurement. *Nature* 2010, **466**:883–886.

56. Meinhart A, Blobel J, Cramer P: An extended winged helix domain in general transcription factor E/IIE alpha. *J Biol Chem* 2003, **278**:48267–48274.

57. Parkinson G, Gunasekera A, Vojtechovsky J, Zhang X, Kunkel TA, Berman H, Ebright RH: Aromatic hydrogen bond in sequence-specific protein DNA recognition. *Nat Struct Biol* 1996, **3**:837–841.

58. Rupert PB, Mollah AK, Mossing MC, Matthews BW: The structural basis for enhanced stability and reduced DNA binding seen in engineered second-generation Cro monomers and dimers. *J Mol Biol* 2000, **296**:1079–1090.

59. Stella S, Cascio D, Johnson RC: The shape of the DNA minor groove directs binding by the DNA-bending protein Fis. *Genes Dev* 2010, **24**:814–826.

60. Eicken C, Pennella MA, Chen X, Koshlap KM, VanZile ML, Sacchettini JC, Giedroc DP: A metal-ligand-mediated intersubunit allosteric switch in related SmtB/ArsR zinc sensor proteins. *J Mol Biol* 2003, **333**:683–695.

61. Wisedchaisri G, Wu M, Rice AE, Roberts DM, Sherman DR, Hol WG: Structures of *Mycobacterium tuberculosis* DosR and DosR-DNA complex involved in gene activation during adaptation to hypoxic latency. *J Mol Biol* 2005, **354**:630–641.

62. Hong M, Fuangthong M, Helmann JD, Brennan RG: Structure of an OhrR-ohrA operator complex reveals the DNA binding mechanism of the MarR family. *Mol Cell* 2005, **20**:131–141.

63. Liu T, DeRose EF, Mullen GP: Determination of the structure of the DNA binding domain of gamma delta resolvase in solution. *Protein Sci* 1994, **3**:1286–1295.

64. Dangi B, Gronenborn AM, Rosner JL, Martin RG: Versatility of the carboxy-terminal domain of the alpha subunit of RNA polymerase in transcriptional activation: use of the DNA contact site as a protein contact site for MarA. *Mol Microbiol* 2004, **54**:45–59.

65. Krishna SS, Majumdar I, Grishin NV: Structural classification of zinc fingers: survey and summary. *Nucleic Acids Res* 2003, **31**:532–550.

66. Klug A: The discovery of zinc fingers and their applications in gene regulation and genome manipulation. *Annu Rev Biochem* 2010, **79**:213–231.

67. Wolfe SA, Nekludova L, Pabo CO: DNA recognition by Cys2His2 zinc finger proteins. *Annu Rev Biophys Biomol Struct* 2000, **29**:183–212.

68. Lee BM, Xu J, Clarkson BK, Martinez-Yamout MA, Dyson HJ, Case DA, Gottesfeld JM, Wright PE: Induced fit and "lock and key" recognition of 5 S RNA by zinc fingers of transcription factor IIIA. *J Mol Biol* 2006, **357**:275–291.

69. Omichinski JG, Pedone PV, Felsenfeld G, Gronenborn AM, Clore GM: The solution structure of a specific GAGA factor-DNA complex reveals a modular binding mode. *Nat Struct Biol* 1997, **4**:122–132.

70. Elango N, Radhakrishnan R, Froland WA, Wallar BJ, Earhart CA, Lipscomb JD, Ohlendorf DH: Crystal structure of the hydroxylase component of methane monooxygenase from *Methylosinus trichosporium* OB3b. *Protein Sci* 1997, **6**:556–568.

71. Langelier MF, Planck JL, Roy S, Pascal JM: Crystal structures of poly(ADP-ribose) polymerase-1 (PARP-1) zinc fingers bound to DNA: structural and functional insights into DNA-dependent PARP-1 activity. *J Biol Chem* 2011, **286**:10690–10701.

72. McCormick MS, Sazinsky MH, Condon KL, Lippard SJ: X-ray crystal structures of manganese(II)-reconstituted and native toluene/o-xylene monooxygenase hydroxylase reveal rotamer shifts in conserved residues and an enhanced view of the protein interior. *J Am Chem Soc* 2006, **128**:15108–15110.

73. Houbaviy HB, Usheva A, Shenk T, Burley SK: Cocrystal structure of YY1 bound to the adeno-associated virus P5 initiator. *Proc Natl Acad Sci USA* 1996, **93**:13577–13582.

74. Tidow H, Andreeva A, Rutherford TJ, Fersht AR: Solution structure of the U11-48 K CHHC zinc-finger domain that specifically binds the 5' splice site of U12-type introns. *Structure* 2009, **17**:294–302.

75. Samara NL, Datta AB, Berndsen CE, Zhang X, Yao T, Cohen RE, Wolberger C: Structural insights into the assembly and function of the SAGA deubiquitinating module. *Science* 2010, **328**:1025–1029.

Permissions

All chapters in this book were first published in JNI, by BioMed Central; hereby published with permission under the Creative Commons Attribution License or equivalent. Every chapter published in this book has been scrutinized by our experts. Their significance has been extensively debated. The topics covered herein carry significant findings which will fuel the growth of the discipline. They may even be implemented as practical applications or may be referred to as a beginning point for another development.

The contributors of this book come from diverse backgrounds, making this book a truly international effort. This book will bring forth new frontiers with its revolutionizing research information and detailed analysis of the nascent developments around the world.

We would like to thank all the contributing authors for lending their expertise to make the book truly unique. They have played a crucial role in the development of this book. Without their invaluable contributions this book wouldn't have been possible. They have made vital efforts to compile up to date information on the varied aspects of this subject to make this book a valuable addition to the collection of many professionals and students.

This book was conceptualized with the vision of imparting up-to-date information and advanced data in this field. To ensure the same, a matchless editorial board was set up. Every individual on the board went through rigorous rounds of assessment to prove their worth. After which they invested a large part of their time researching and compiling the most relevant data for our readers.

The editorial board has been involved in producing this book since its inception. They have spent rigorous hours researching and exploring the diverse topics which have resulted in the successful publishing of this book. They have passed on their knowledge of decades through this book. To expedite this challenging task, the publisher supported the team at every step. A small team of assistant editors was also appointed to further simplify the editing procedure and attain best results for the readers.

Apart from the editorial board, the designing team has also invested a significant amount of their time in understanding the subject and creating the most relevant covers. They scrutinized every image to scout for the most suitable representation of the subject and create an appropriate cover for the book.

The publishing team has been an ardent support to the editorial, designing and production team. Their endless efforts to recruit the best for this project, has resulted in the accomplishment of this book. They are a veteran in the field of academics and their pool of knowledge is as vast as their experience in printing. Their expertise and guidance has proved useful at every step. Their uncompromising quality standards have made this book an exceptional effort. Their encouragement from time to time has been an inspiration for everyone.

The publisher and the editorial board hope that this book will prove to be a valuable piece of knowledge for researchers, students, practitioners and scholars across the globe.

List of Contributors

Petras J Kundrotas
Center for Bioinformatics, The University of Kansas, 2030 Becker Dr., Lawrence, KS 66047, USA

Ilya A Vakser
Center for Bioinformatics, The University of Kansas, 2030 Becker Dr., Lawrence, KS 66047, USA
Department of Molecular Biosciences, The University of Kansas, 2030 Becker Dr., Lawrence, KS 66047, USA

Zhengwei Zhu
Center for Bioinformatics, The University of Kansas, 2030 Becker Dr., Lawrence, KS 66047, USA
Department of Genetics, Room 716B, Abramson Research Center, University of Pennsylvania, 3615 Civic Center Blvd., Philadelphia, PA 19104, USA

Mehdi Pirooznia and Fernando S Goes
Department of Psychiatry and Behavioral Sciences, Johns Hopkins University, Baltimore, MD 21205, USA

Melissa Kramer and Jennifer Parla
Stanley Institute for Cognitive Genomics, Cold Spring Harbor Laboratory, Woodbury, NY 11797, USA

James B Potash
Department of Psychiatry, Carver College of Medicine, University of Iowa School of Medicine, Iowa City, IA 52242, USA

W Richard McCombie
Stanley Institute for Cognitive Genomics, Cold Spring Harbor Laboratory, Woodbury, NY 11797, USA
Watson School of Biological Science, Cold Spring Harbor Laboratory, Cold Spring Harbor, NY 11724, USA

Peter P Zandi
Department of Psychiatry and Behavioral Sciences, Johns Hopkins University, Baltimore, MD 21205, USA
Department of Mental Health, Johns Hopkins Bloomberg School of Public Health, Baltimore, MD 21205, USA

Konstantinos Sousounis, Rital Bhavsar, Jessica Beebe and Panagiotis A Tsonis
Department of Biology and Center for Tissue Regeneration and Engineering at Dayton, University of Dayton, 300 College Park, Dayton, OH 45469, USA

Mario Looso, Marcus Krüger and Thomas Braun
Department of Cardiac Development and Remodeling, Max-Planck-Institute for Heart and Lung Research, Ludwigstrasse 43, 61231 Bad Nauheim, Germany

Lidiia Zhytnik
Clinic of Traumatology and Orthopaedics, University of Tartu, Puusepa 8, 51014 Tartu, Estonia

Katre Maasalu and Aare Märtson
Clinic of Traumatology and Orthopaedics, University of Tartu, Puusepa 8, 51014 Tartu, Estonia

Tiit Nikopensius and Margit Nõukas
Estonian Genome Centre, University of Tartu, Riia 23b, Tartu 51010, Estonia
Institute of Molecular and Cell Biology, University of Tartu, Riia 23b, Tartu 51010, Estonia

Mart Kals
Estonian Genome Centre, University of Tartu, Riia 23b, Tartu 51010, Estonia

Ele Prans
Institute of Molecular and Cell Biology, University of Tartu, Riia 23b, Tartu 51010, Estonia

Andres Metspalu
Estonian Genome Centre, University of Tartu, Riia 23b, Tartu 51010, Estonia
Institute of Molecular and Cell Biology, University of Tartu, Riia 23b, Tartu 51010, Estonia
Estonian Biocentre, Riia 23b, 51010 Tartu, Estonia

Sulev Kõks
Department of Pathophysiology, University of Tartu, Ravila 19, Tartu 50411, Estonia

Luisa A Wakeling, Laura J Ions, Suzanne M Escolme, Tianhong Su, Madhurima Dey, Emily V Hampton and Dianne Ford
Institute for Cell and Molecular Biosciences, Human Nutrition Research Centre, Newcastle University Medical School, Newcastle upon Tyne NE2 4HH, UK

Simon J Cockell
Faculty of Medical Sciences, Newcastle University Medical School, Newcastle upon Tyne NE2 4HH, UK

Jill A McKay
Institute of Health and Society, Human Nutrition Research Centre, Newcastle University Medical School, Newcastle upon Tyne NE2 4HH, UK

Gail Jenkins and Linda J Wainwright
Unilever R&D, Colworth Discover, Colworth Science Park, Sharnbrook, Bedfordshire MK44 1LQ, UK

Noreen Dugan, Danielle Haley and Victoria L. Vetter
Department of Pediatrics, Perelman School of Medicine at the University of Pennsylvania, Philadelphia, PA, USA
Division of Cardiology, The Children's Hospital of Philadelphia, Philadelphia, PA, USA

Mindy H. Li, Jenica L. Abrudan, Matthew C. Dulik, Joshua Brunton, Vijayakumar Jayaraman, Matthew A. Deardorff, Alisha Wilkens, Sarah E. Noon, Maria I. Scarano and Ian D. Krantz
Department of Pediatrics, Perelman School of Medicine at the University of Pennsylvania, Philadelphia, PA, USA
Division of Human Genetics, The Children's Hospital of Philadelphia, Abramson Research Center, Room 1012G, 3615 Civic Center Blvd, Philadelphia, PA 19104, USA

Ariella Sasson, Mahdi Sarmady and Jeffrey Pennington
Department of Biomedical and Health Informatics, The Children's Hospital of Philadelphia, Philadelphia, PA, USA

Sawona Biswas
Department of Pathology & Laboratory Medicine, Perelman School of Medicine at the University of Pennsylvania, Philadelphia, PA, USA

Ramakrishnan Rajagopalan, Elizabeth T. DeChene, Avni B. Santani, Laura K. Conlin and Nancy B. Spinner
Department of Pathology & Laboratory Medicine, Perelman School of Medicine at the University of Pennsylvania, Philadelphia, PA, USA
Division of Genomic Diagnostics, The Children's Hospital of Philadelphia, Philadelphia, PA, USA

Peter S. White
Department of Pediatrics, Perelman School of Medicine at the University of Pennsylvania, Philadelphia, PA, USA
Department of Biomedical and Health Informatics, The Children's Hospital of Philadelphia, Philadelphia, PA, USA
Division of Oncology, The Children's Hospital of Philadelphia, Philadelphia, PA, USA
Department of Pediatrics, Cincinnati Children's Hospital and Medical Center, and Department of Biomedical Informatics, College of Medicine, University of Cincinnati, Cincinnati, OH, USA

Marco Trerotola, Valeria Relli and Pasquale Simeone
Unit of Cancer Pathology, CeSI, Foundation University 'G. d'Annunzio', Chieti, Italy

Saverio Alberti
Unit of Cancer Pathology, CeSI, Foundation University 'G. d'Annunzio', Chieti, Italy
Department of Neuroscience, Imaging and Clinical Sciences, Unit of Physiology and Physiopathology, 'G. d'Annunzio' University, Chieti, Italy

Anis Karimpour-Fard and Lawrence E. Hunter
Department of Pharmacology, University of Colorado School of Medicine, Aurora, CO 80045, USA

L. Elaine Epperson
Integrated Center for Genes, Environment, and Health, National Jewish Health, Denver, CO 80206, USA

S. Fokstuen, E. Hammar, E. Ranza, M. Albarca-Aguilera, M. E. Poleggi, F. Couchepin, C. Brockmann, C. Gehrig, A. Vannier, T. Araud, S. Gimelli, E. Stathaki, A. Paoloni-Giacobino, A. Bottani, F. Sloan-Béna, L. D'Amato Sizonenko, M. Mostafavi, T. Nouspikel and J. L. Blouin
Service of Genetic Medicine, University Hospitals of Geneva, Geneva, Switzerland

M. Guipponi, K. Varvagiannis, F. A. Santoni and P. Makrythanasis
Service of Genetic Medicine, University Hospitals of Geneva, Geneva, Switzerland
Department of Genetic Medicine and Development, University of Geneva, 1 rue Michel-Servet, 1211 Geneva, Switzerland

S. E. Antonarakis
Service of Genetic Medicine, University Hospitals of Geneva, Geneva, Switzerland
Department of Genetic Medicine and Development, University of Geneva, 1 rue Michel-Servet, 1211 Geneva, Switzerland
iGE3, Institute of Genetics and Genomics of Geneva, Geneva, Switzerland

J. Bevillard and H. Hamamy
Department of Genetic Medicine and Development, University of Geneva, 1 rue Michel-Servet, 1211 Geneva, Switzerland

A. Mauron, S. A. Hurst and C. Moret
Institute for Ethics, History, and the Humanities, Geneva University Medical School, Geneva, Switzerland

Federico A. Santoni, Muhammad Ansar and Hanan Hamamy
Department of Genetic Medicine and Development, University of Geneva, 1 Rue Michel-Servet, 1211 Geneva, Switzerland

Periklis Makrythanasis
Department of Genetic Medicine and Development, University of Geneva, 1 Rue Michel-Servet, 1211 Geneva, Switzerland
Service of Genetic Medicine, University Hospitals of Geneva, Geneva, Switzerland

Stylianos E. Antonarakis
Department of Genetic Medicine and Development, University of Geneva, 1 Rue Michel-Servet, 1211 Geneva, Switzerland
Service of Genetic Medicine, University Hospitals of Geneva, Geneva, Switzerland
iGE3, Institute of Genetics and Genomics of Geneva, Geneva, Switzerland

Michel Guipponi
Service of Genetic Medicine, University Hospitals of Geneva, Geneva, Switzerland

Maha Zaki and Mahmoud Y. Issa
Department of Clinical Genetics, National Research Centre, Cairo, Egypt

Yung-Chang Chen
Kidney Research Center, Department of Nephrology, Chang Gung Memorial Hospital, Linkou Medical Center, Taoyuan, Taiwan

Pei-Chun Fan
Kidney Research Center, Department of Nephrology, Chang Gung Memorial Hospital, Linkou Medical Center, Taoyuan, Taiwan
Graduate Institute of Clinical Medical Sciences, Chang Gung University, Taoyuan, Taiwan

Yu-Sun Chang and Chia-Chun Chen
Molecular Medicine Research Center, Chang Gung University, Taoyuan, Taiwan

Pao-Hsien Chu
Division of Cardiology, Department of Internal Medicine, Chang Gung Memorial Hospital, College of Medicine, Chang Gung University, Taipei, Taiwan
Healthcare Center, Chang Gung Memorial Hospital, College of Medicine, Chang Gung University, Taipei, Taiwan
Heart Failure Center, Chang Gung Memorial Hospital, College of Medicine, Chang Gung University, Taipei, Taiwan
Department of Cardiology, Chang Gung Memorial Hospital, College of Medicine, Chang Gung University, 199 Tung Hwa North Road, Taipei 105, Taiwan

Shui-Ying Tsang, Tanveer Ahmad, Flora W. K. Mat and Hong Xue
Division of Life Science, Applied Genomics Centre and Centre for Statistical Science, Hong Kong University of Science and Technology, Clear Water Bay, Hong Kong, China

Cunyou Zhao
Division of Life Science, Applied Genomics Centre and Centre for Statistical Science, Hong Kong University of Science and Technology, Clear Water Bay, Hong Kong, China
Department of Medical Genetics, School of Basic Medical Science, Southern Medical University, Guangzhou, Guangdong 510515, China

Shifu Xiao
Department of Geriatric Psychiatry, Shanghai Mental Health Center, Shanghai Jiaotong University School of Medicine, Shanghai 200030, China

Kun Xia
The State Key Laboratory of Medical Genetics, Central South University, Changsha, Hunan 410078, China

Shrey Gandhi
GN Ramachandran Knowledge Center for Genome Informatics, CSIR Institute of Genomics and Integrative Biology (CSIR-IGIB), Mathura Road, Delhi 110 025, India

Saakshi Jalali and Vinod Scaria
GN Ramachandran Knowledge Center for Genome Informatics, CSIR Institute of Genomics and Integrative Biology (CSIR-IGIB), Mathura Road, Delhi 110 025, India
Academy of Scientific and Innovative Research (AcSIR), CSIR-IGIB South Campus, Mathura Road, Delhi 110025, India

Elina A. M. Hirvonen, Esa Pitkänen and Outi Kilpivaara
Genome-Scale Biology Research Program, Research Programs Unit and Department of Medical and Clinical Genetics, Medicum, University of Helsinki, Helsinki, Finland

Kari Hemminki
Division of Molecular Genetic Epidemiology, German Cancer Research Center (DKFZ), Heidelberg, Germany

Lauri A. Aaltonen
Genome-Scale Biology Research Program, Research Programs Unit and Department of Medical and Clinical Genetics, Medicum, University of Helsinki, Helsinki, Finland
Department of Biosciences and Nutrition, Karolinska Institutet, SE-17177 Stockholm, Sweden

Itsuki Taniguchi, Chihiro Iwaya and Hiroki Shibata
Division of Genomics, Medical Institute of Bioregulation, Kyushu University, 3-1-1 Maidashi, Higashi-ku, Fukuoka 812-8582, Japan

Keizo Ohnaka
Department of Geriatric Medicine, Graduate School of Medical Sciences, Kyushu University, 3-1-1 Maidashi, Higashi-ku, Fukuoka 812-8582, Japan

Ken Yamamoto
Department of Medical Biochemistry, Kurume University School of Medicine, 67 Asahi-machi, Kurume, Fukuoka 830-0011, Japan

Alexander M. Vaiserman and Alexander K. Koliada
Laboratory of Epigenetics, Institute of Gerontology, Vyshgorodskaya st. 67, Kiev 04114, Ukraine

Ahmad H. Mufti, Anas Dannoun, Hassan Kordi and Mohammed T. Tayeb
Department of Medical Genetics, Medicine College, Umm Al-Qura University, Mecca 21955, Saudi Arabia

Nasser A. Elhawary
Department of Medical Genetics, Medicine College, Umm Al-Qura University, Mecca 21955, Saudi Arabia
Department of Molecular Genetics, Faculty of Medicine, Ain Shams University, Cairo 11566, Egypt

Essam H. Jiffri and Osama H. Jiffri
Department of Medical Laboratory Technology, Faculty of Applied Medical Sciences, King Abdul-Aziz University, Jeddah, Saudi Arabia

Samira Jambi
Department of Pediatrics, Al Hada Military Hospital, Al Hada, Saudi Arabia

Asim Khogeer
Department of Plan and Research, General Directorate of Health Affairs, Mecca Region, Ministry of Health, Mecca, Saudi Arabia

Abdelrahman N. Elhawary
Department of Pediatrics, Faculty of Medicine, Cairo University, Giza, Egypt

Hui He, Zhiping Hu, Han Xiao, Fangfang Zhou and Binbin Yang
Department of Neurology, 2nd Xiangya Hospital, Central South University, No 139, Renmin Road, Changsha, Hunan Province, China

Mathew W Wright
HUGO Gene Nomenclature Committee (HGNC), EMBL-EBI, Wellcome Trust Genome Campus, Hinxton, Cambridge CB10 1SD, UK

Edmund Maser
Institute of Toxicology and Pharmacology for Natural Scientists, University Medical School, Schleswig-Holstein, Campus Kiel, Brunswiker Str. 10, Kiel D-24105, Germany

Marie-Eve Lalonde, Manon Ouimet, Mathieu Larivière and Ekaterini A Kritikou
Division of Hematology-Oncology, Research Center, Sainte-Justine Hospital, 3175 Chemin de la Cote-Sainte-Catherine, Montreal H3T 1C5, Canada

Daniel Sinnett
Division of Hematology-Oncology, Research Center, Sainte-Justine Hospital, 3175 Chemin de la Cote-Sainte-Catherine, Montreal H3T 1C5, Canada

Department of Pediatrics, Faculty of Medicine, University of Montreal, Montreal QC H3T 1C5, Canada

Konstantinos Sousounis, Carl E Haney and Panagiotis A Tsonis
Department of Biology, University of Dayton, Dayton, OH 45469-2320, USA

Jin Cao
Department of Chemical Engineering, University of Dayton, Dayton, OH 45469-2320, USA

Bharath Sunchu
Department of Chemistry, University of Dayton, Dayton, OH 45469-2320, USA

Index